Black Islanders

Douglas Oliver

Emeritus Professor Oliver has spent a lifetime in the study of the Pacific peoples.

Born in Ruston, Louisiana, on 10 February 1913, he trained at Harvard, USA, and the University of Vienna, Austria. In 1950 he became Professor of Anthropology at Harvard University and he held the Chair of Pacific Islands Anthropology at the University of Hawaii until he retired in 1978.

It was in 1938 as a Research Associate in the Peabody Museum, Harvard, that he first visited Bougainville where he worked among the Siwai of southern Bougainville. Thus began his friendship with the peoples of Bougainville.

Professor Oliver is married to an Australian anthropologist and has three children and four grandchildren. He lives in Hawaii.

By the same author

The Pacific Islands
Planning Micronesia's Future
Studies in the Anthropology of Bougainville
Somatic Variability and Human Ecology on Bougainville
A Solomon Island Society: The Siuai of Bougainville
Invitation to Anthropology
Ancient Tahitian Society (three volumes)
Bougainville: A Personal History
Two Tahitian Villages
Aspects of Westernization on Bougainville
*Oceania: The Native Cultures of Australia and the Pacific
    Islands* (two volumes)
Native Cultures of the Pacific Islands
Return to Tahiti: Bligh's Second Breadfruit Voyage
The Pacific Islands (revised edition)

Black Islanders

A Personal Perspective of Bougainville
1937–1991

Douglas Oliver

UNIVERSITY OF HAWAII PRESS
HONOLULU

Published in North America by
University of Hawaii Press
2840 Kolowalu Street
Honolulu, Hawaii 96822

Published in Australia by
Hyland House Publishing Pty Limited
10 Hyland Street
South Yarra
Melbourne
Victoria 3141

ISBN 0-8248-1434-7

Library of Congress Catalog Card Number 91-057969

Cover designed by Rob Cowpe
Typeset by Butler Graphics Pty Ltd, Hawthorn 3122
Printed by Australian Print Group, Maryborough, Victoria

Contents

List of Illustrations

Pig feeding at hamlet house (1938)
Women at work in taro garden (1938)

Nasioi potters
'Shell money' near Panguna
Ceremonial feast

Project (but not yet mine) conference in hut at Kobuan, 1966
Catholics and Methodists engage in a Bible Knowledge contest (1939)

Village elder, 1988
Ceremonial mask
Village gardening, 1988 *between pages*
Sunday outing, 1988 76–7

Geologist Ken Phillips with Dateo of Pakia village, October 1964 *facing page*
Shares in Bougainville Copper Limited were available to local residents and 100
employees

Fijian Prime Minister visits Panguna, 1974 101
Arawa township, 1988

Panguna, 1988 148

Resettled village, early 1970s 149
Shovel and haul truck

Leo Hannett 196
Father John Momis
Dr Alexis Sarei
Bishop Gregory Singkai

Francis Ona 197
Sam Kauona
Pylon toppled by explosive charge, 15 April 1989

Damage by militants, Kobuan, 10 July 1989 244
An emotional Mrs Cecilia Kenevi embraces Sir Michael Somare on his return from
signing the 'Endeavour Accord', 5 August 1990

Don Carruthers 245
Peace ceremony Arawa, 27 October 1989

Acknowledgements

Adequate and proportionate acknowledgement of assistance is impossible when the help received extended over half a century and included everything from official permits to cold beer, from hours of an informant's time to gifts of (usually undercooked) pork.

It has been my privilege to talk with (and to be instructed, and occasionally corrected by) many of the individuals named in these pages, and with scores of others not named. To all of those still alive, please accept my thanks.

In learning about Bougainville I have also utilized the published words of several perceptive journalists and academics not actually met. To them also I proffer thanks.

More specifically, I acknowledge with gratitude the indispensable assistance provided by three individuals in the writing of this particular book: to James Byth (now of Melbourne), who helped immeasurably to reduce my ignorance of the mining industry; to Gillian Andrew (now of Sydney), who sent me Australian and Papua New Guinea news reports about the Bougainville Crisis, and who shared with me her own extensive knowledge of Papua New Guinea personalities and politics; and to my wife, Margaret McArthur, who applied her Australian-bred candour and her own wide New Guinea anthropological experience to help correct my narrow focus on Bougainville.

Douglas Oliver

To
Frank Fletcher Espie

Note on Currency

Unless otherwise specified, pounds and dollars in this book refer to Australian currency. The Australian pound was worth the same as, or a little less than, the pound sterling. There were 12 pennies in one shilling, and 20 shillings in one pound. When Australia changed to decimal currency in 1966, $2 was equal to £1.

Australian currency was used in Papua New Guinea until April 1975, when that country introduced its own, the kina and toea (1 kina = 100 toea). At that time K1 was equal to $1; thereafter the exchange rate varied. The kina was usually higher, mainly because of devaluations of the Australia dollar.

Introduction

T HE WRITER OF THIS 'HISTORY' IS AN ANTHROPOLOGIST, NOT an academically qualified historian. (Nor am I an expert in the complexities of the mining industry—as the reader will doubtless judge!) Some day, it is to be hoped, a properly qualified historian, with access to all the relevant documents, will write an authoritative account of the events that constitute and led up to the 'Bougainville Crisis', but that may be years away. Meanwhile, as an anthropologist I hold the view that something more than a chronicle of events is required for describing a situation as complex as the Crisis—especially in the case of Bougainville, where the cultural background is so unlike those of the kinds of societies historians usually write about. Which obligates this anthropologist to explain why he is qualified to write about this particular crisis.

My first view of Bougainville was in early February 1938 from the deck of a small schooner that had left Rabaul two days before. The sight of the island's 800-metre Emperor Range was spectacular even through seasick eyes—due to engine breakdown we had wallowed for a day and a night in the Solomon Sea. A few hours later we entered the mercifully calm waters of Buka Passage, docked at Sohano to off-load some goods, and proceeded along Bougainville's northern and eastern coasts to Kieta. There my wife and I were met by one of the district's four administrative officials, by its one qualified physician, and by the manager of the district's one Burns Philp store—in other words, by three-fifths of Kieta's adult male whites. (We exchanged grins with the scores of Bougainvillians who were also there observing that infrequent event, but were not 'introduced' to them.) In the absence of anything resembling a hotel we were welcomed, warmly and sincerely, as house guests of the medical officer and his wife. In those days few tourists visited New Guinea, and anthropologists were so rare as to be considered 'interesting' and possibly even 'useful' house guests—a situation that has since changed!

We spent a few days in Kieta, where we met the other resident and nearby whites, including the most congenial (American) bishop of the Marist mission. Then we charted a small launch, which eventually unloaded our supplies and us (including Manao our versatile Muschu Islander 'boy'—servant, friend and mentor), onto the beach at Kangu, the official 'port' of Buin. There we were greeted only by sandflies, since the sub-district officer who was stationed there was away. Luckily, we managed to move ourselves and our supplies into the nearby rest house—built, if that is the

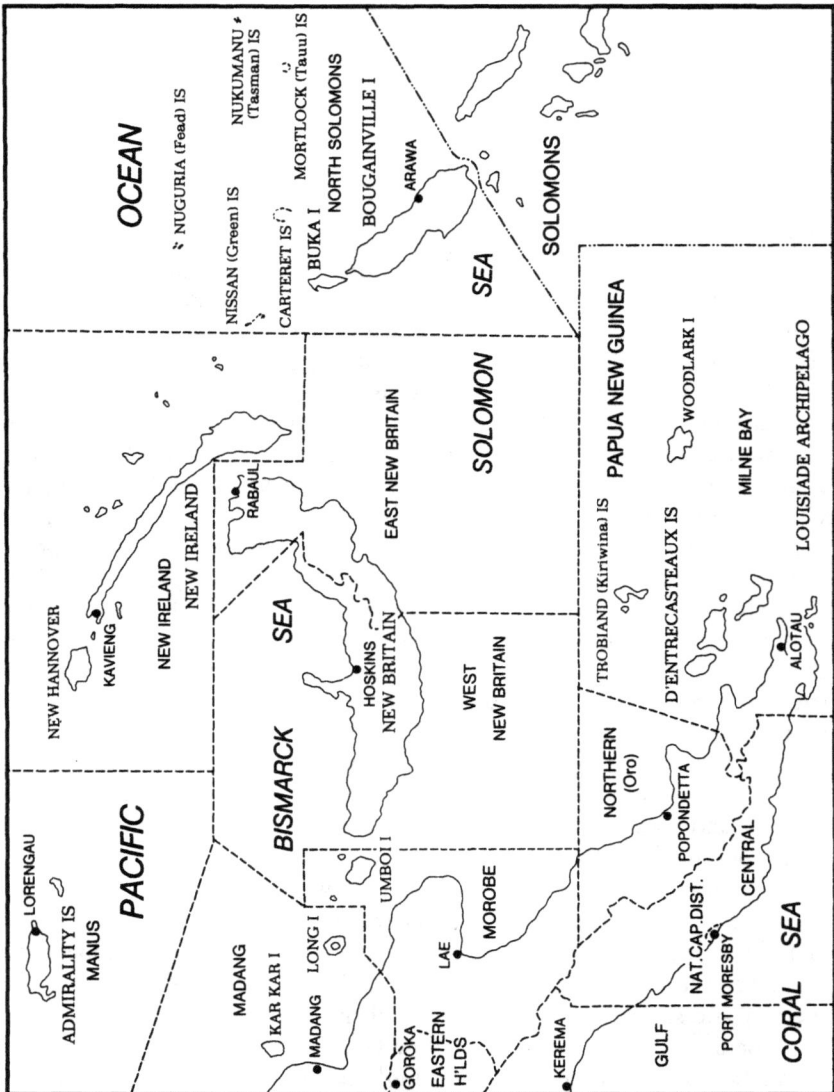

Figure 1   Eastern Papua New Guinea

word, to accommodate hypothetical visiting whites—just before the afternoon rain set in. Because all of Buin's native settlements were located, for sound reasons of health and comfort, several miles inland, our only company was swarms of mosquitoes. That night would have been filled with somber misgivings had I not been at least partly prepared for the work ahead.

When Harvard University's Peabody Museum sent me to New Guinea in 1937 to undertake research, it was intended that I do so in one of the main island's recently 'discovered' and still largely 'pristine' societies. To that end I was invited to accompany the geologist, G. A. V. Stanley (of Australia's Oil Search Ltd), who was at the time surveying the mountainous region southwest of Wewak. It was a marvellous experience for this fledgling—a perfect introduction to travelling and working in New Guinea's remote interior, under guidance of a veteran explorer. In the course of that expedition we passed through many Stone Age villages just waiting to be 'anthropologized'—unfortunately, however, not by this anthropologist. One of the several facts of life I had come to accept during that learning period was that my thin budget ($US70 a month) would not begin to support the cost of a long residence in that remote area, including the services of the Niuginians (New Guineans) needed to carry in supplies and to maintain the kind of security that might be called for among those feuding and to me still unpredictable people.

Consequently I returned to Rabaul to rejoin my wife and to select another research site that we could afford. Because so little anthropological research had been done in the Territory up to then the choice was very wide; in the end I accepted the advice of another New Guinea veteran and chose the Siwai people of southern Bougainville. This veteran was E. W. P. Chinnery, who was head of the Territory's Department of Native Administration, so knew whereof he spoke. According to him the Siwai were 'very interesting and friendly', and the cost of working among them relatively low. Both observations turned out to be accurate. An added advantage was that during the two months spent in Rabaul waiting for boat transport to Bougainville—no daily inter-island flights during that era!—I was able to make a beginning in learning the Siwais' (unrecorded) language by means of daily sessions with a Pidgin-speaking Siwai on Chinnery's staff.

Returning to our rain-sodden introduction to southern Bougainville, the lessons learned from Stanley and Chinnery enabled me to secure enough porters to take us and our supplies, by stages, inland and to the eastern border of Siwai. There we were met by a large crowd, who not only welcomed us warmly—news of our approach having long preceded us—but vied with one another in invitations to stay. (This attitude is not likely to greet anthropologists in 1991.)

We finally settled in a village in northeastern Siwai and made that our home for the next year and a half, except for a brief visit to 'civilization' for rest and medical care, and excursions, by foot or boat, to other parts of the island. (Our observations are recorded in several of the publications listed in this book's bibliography.)

My second trip to Bougainville took place in early 1944, when I visited the American military beachhead at Torokina as head of the South Pacific branch of the United States Board of Economic Warfare, whose mission was to relieve the military of certain civilian jobs, including producing fresh vegetables for the troops and securing trade goods for natives in areas occupied by our forces. On that visit I renewed acquaintance with the few Siwais who had found safe haven within the beachhead but could not revisit Siwai itself, which was still occupied by the Japanese.

In July 1965 I made another brief visit to Bougainville in order to obtain official permission for Harvard University scientists to carry out multi-disciplinary research there. Several of my students took part in this research, whose objectives and general findings are described in an article by Albert Damon listed in the bibliography.

From 1968 to 1978 I was employed by Bougainville Copper Ltd as a part-time consultant, to advise the company on how to conduct their operations so as to shield Bougainvillians as much as possible from the harms that inevitably accompany such mining. During that period I visited the island once or twice every year. In addition I organized and super-vised a research programme designed to provide the province's leaders with information relevant to their planning for the populace's social and economic well-being. The programme was sponsored by the University of Hawaii and included scholars from that institution, from the Australian National University, from New Zealand's Canterbury University, and from the University of Papua New Guinea. It was financed in large part by Bougainville Copper Ltd, which, however, exercised no control over it. During its five-year course the programme involved more than a score of researchers—including geographers, anthropologists, economists, political scientists, and a psychologist—and yielded dozens of published reports. A summary of its findings is contained in Oliver 1981.

I have not visited Bougainville since 1978, but have kept fairly up to date on trends and events there by means of newspapers and journals, and through communications with several Bougainvillians, including two of the province's former premiers.

Much of this book was written in Honolulu, at a distance in space and time from the events it attempts to record.

I have sought to bring out a factor which has been noticeably neglected by much contemporary reporting: the extent to which past events and the facts of geography and demography continue to influence current events, for good or bad.

My task has been immeasurably assisted by friends from Bougainville and Buka, from Papua New Guinea and from Australia, who have shared with me their hopes and fears and confidential details of their involve-ment in recent events.

Anthropologists share a common duty with historians and journalists: to understand how human beings in different societies see the world and their part in it, and to translate that vision into our own understanding with sympathy and precision so that others may begin to comprehend. It is therefore no accident that many of the observations quoted in this book are those of other anthropologists. One conclusion to be drawn from recent events on Bougainville and Buka is that few people in politics, government administration and business comprehend the cultural and individual diversities that obtain in these South Pacific islands.

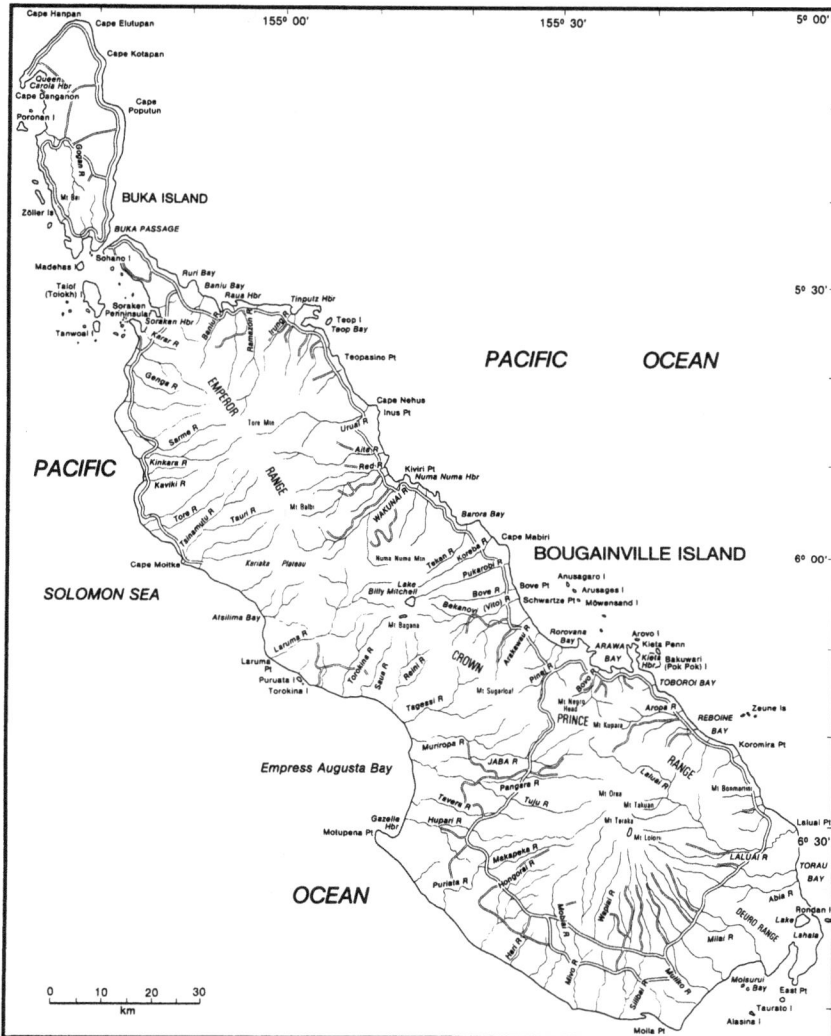

Figure 2   Bougainville and Buka—general

The barriers to comprehension begin with language, and I point out in Chaper 7 the past and present difficulties of communication. It remained as true in 1991 as in 1971 that few expatriates, whether foreign diplomats or businessmen or aid workers, had more than elementary competence in Melanesian Pidgin. Almost none could converse in a local language — there being about fifteen on Bougainville–Buka. A second barrier is deeper and to many 'educated' white people very difficult to cross: the concept that the internal reality of family and clan and regional loyalties and identity, which determines for all of us the question 'Who am I?', may differ between cultures. It is clear from events that individual Bougainvillians themselves have no uniform answer to this question, nor necessarily those which have been attributed to them.

Yet this is much more than simply cross-cultural difficulty, as I discuss later in describing the gift process. Perhaps the greatest barrier to comprehension, in PNG as well as in Australia, is an incapacity to understand the distinctive differences of climate, land, demography and social pressures within Bougainville itself. My earlier chapters attempt to explore the long-standing nature of these differences and their present consequences.

Truth is the first casualty of war, and journalists attempting to report events on Bougainville have had understandable problems in separating truth from propaganda and positive disinformation by both sides. It will be thirty years at least before documents tell the official version of events, and (as in World War II) many of the realities will be recorded only in oral memory rather than on paper.

In PNG communities, oral witness is often more persuasive than the printed word. By far the most effective day-to-day reportage of events, in my opinion, has come from Sean Dorney of the Australian Broadcasting Corporation. The spread of ABC radio, and the availability of Australian television transmissions to people in Port Moresby, have made his precise and thoughtful reporting authoritative. At times it has been the only source of immediate news for those directly involved. He has recently published a book, *Papua New Guinea*. Within PNG, I have been enlightened by the dogged persistence of Wally Hiambohn and Angwie Hriehwazi. Amongst Australia-based reporting I have found Rowan Callick's articles in the *Australian Financial Review*, and Mary-Louise O'Callaghan's in the *Sydney Morning Herald*, very helpful.

With some honourable exceptions, most Australian and other foreign reportage has been superficial sensationalist 'colour'. Of these, the most egregious was the rewrite desk of *The Economist*, 7 July 1990, which attributed Oxford accents to the people of Bougainville!

By far the most balanced historical commentary over the period since exploration has been that of Professor James Griffin. Unlike many, he has meticulously distinguished between historical facts and his preferred view of their significance, and has engaged in dialogue with both expatriates and people of all shades of opinion in PNG.

In view of my 52-year association with Bougainville, the reader will not be surprised that I have a personal, somewhat sentimental, interest in its current Crisis, which provides an informative case study of radical social change—in this case brought about by an uncommon combination of factors.

Most visible of these factors has been the presence of the copper mine, which was established in what up to then had been a sleepy, economically backward and culturally non-Westernized human setting. And not only did the mine bring about radical changes in the everyday lives of the Bougainvillians but, through its injection of huge new revenues into the coffers of the colony's central administration, it undoubtedly hastened the whole Australian decolonization process, thereby presenting Bougainvillians (and other Niuginians) with unprecedented and sometimes unsettling political choices. However, large industrial projects have been established worldwide in other economically backward and politically fragmented dependencies, and although many of these have also been accompanied by political turmoil, none has resulted in the particular kind of crisis that now grips Bougainville.

A second factor involved in the Bougainville Crisis is the rapid rate of natural increase of the district's (now province's) native populations in relation to their limited amount of agriculturally useful land. However, that situation also prevails on several other South Pacific islands.

A third factor is the sharp rise in expectations that has recently taken place among Bougainvillians: the widespread increase in people's desires for, indeed claims for, imported material goods. These demands are not just for more cloth and steel tools and tobacco and tinned beef and such like (which characterized their wants a generation or two ago) but for automobiles, refrigerators, VCRs and the like (which many Bougainvillians developed a wish for in the late 1980s). Contributing to that rise has been an increase in their Western-type schooling (including the tertiary education of many of them) and people's sight of the more affluent living standards enjoyed by those of their compatriots on copper company payrolls. In addition, many Bougainvillians continue to hold deep-seated indigenous beliefs in *cargo*, the millenarian-flavoured anticipation of supernaturally supplied material riches. Here again, factors similar to these operate in several other parts of Melanesia, without having resulted in a Bougainville-type crisis.

Fourthly, while a few other Melanesian societies have had colonial histories roughly similar to that of Bougainville, without culminating in a similar crisis, the colonial history of Bougainville did include some fairly distinctive features. One which is especially noteworthy is the dominating influence, past and continuing, of Bougainville's Catholic Church.

Finally, among the combination of factors contributing to the Bougainville Crisis one stands out as unique. Reference here is to the skin colour of the Bougainvillians, which is darker by far than that of any other of

Papua New Guinea's many peoples. Throughout most of their colonial history this factor served at most to distinguish Bougainvillians from other Niuginians, either favourably or unfavourably, but in recent years it has served also to unify them and to detach them politically. While the native peoples of Bougainville–Buka have been geographically separated from other Niuginians for many thousands of years, their hundreds of autonomous communities did not begin to unify — to become 'Bougainvillian' — until about two decades ago. Moreover, there are grounds for concluding that they are still not as fully unified as some of their present leaders — military, political and clerical — would have the rest of the world believe.

1

The First 28,000 Years

THE FIRST HUMANS TO SET FOOT ON BOUGAINVILLE–BUKA, some 28,000 years ago, came from the northwest — either directly, from southeastern New Ireland or, more probably, by stages from there via the Feni and Nissan Islands. The present open-sea distances between New Ireland and Buka, via Feni and Nissan, are no wider than 72 kilometres. Their coasts may have been even closer during the Pleistocene period, when the sea level throughout this area had been lowered appreciably as a result of the impounding of much of the earth's waters in vast continental ice sheets. But even 72-kilometre stretches of ocean were well within the seafaring range of those pioneers: their canoes were certainly seaworthy enough, and the inter-island distances were within visibility range (Lewis 1972).

There is no mystery about how the ancestors of those pioneers came to be in New Ireland at that early date. Archaeological evidence from both New Guinea and Australia (which were periodically joined together during low-sea-level phases of the Pleistocene) shows that humans had begun to cross over from insular Indonesia as early as 50,000 years ago, and that some of them had spread eastward into New Britain and New Ireland by about 30,000 years ago. It is also quite likely that some of the early descendants of the first Bougainvillians pressed further southeastward, at least as far as the island of San Cristobal.

There is no means of knowing why those pioneers made their ways to Bougainville, or beyond: escape from victorious enemies? deliberate search for richer food supplies? need for safe landings during stormy fishing expeditions? or perhaps, in a few cases, curiosity about unfamiliar shores?

More certain is how they subsisted. From archaeological evidence it can be inferred that their diet consisted of forest vegetables, fish and shellfish, birds, lizards, fruit bats, and rats. Included among the vegetables they gathered and ate were two species of taro, *Colocasia* and *Alocasia*, some

1

of which may have been 'tended' (i.e., semi-domesticated) long before the indigenes began to cultivate them in gardens.

For some 25,000 years after initial settlement, the Bougainvillians appear to have had little contact with their northwestern homelands, except, for example, for their import of the *galip* nut (*Canarium indicus*), which they proceeded to plant and use as a favoured food supplement. Even opossums (*Phalanger orientalis*), which were to become highly favoured hunting prey, did not reach Bougainville–Buka until about 3200 years ago. Thus, the first Bougainvillians were to remain almost entirely isolated from their northwestern homelands for nearly twenty-five millennia. Although they were isolated, they were evidently not unified, and most certainly not homogeneous, either physically or culturally. During those many millennia, the pioneers' descendants proliferated and dispersed, mostly in the larger island, where indigenous terrestrial food resources were richer and more diverse. In time those little bands of food gatherers and hunters (and in some places, fishermen), dispersed so widely and remained so scattered that they evolved into many, sharply different, societies, each with its own language. (A society as herein defined is a social unit composed of people who reside adjacently, speak the same language, or languages, and who share, in large measure and more or less distinctively, a common set of cultural principles, values, and practices. In some parts of Melanesia a single community constituted a whole society as well, but in most cases a society contained two or more communities.) Some of those earlier languages may in the course of time have died out, but in the year 1939 there were nine of them:

Northern stock	*Southern stock*
Rotokas family	Nasioi family
1. Rotokas proper	5. Nasioi
2. Eivo	6. Simiku
3. Kunua	7. Nagovisi
4. Keriaka	Terei family
	8. Buin
	9. Siwai

This classification is based mainly on the degree of similarity between the languages' vocabularies. In addition there are some significant differences between the northern and southern stocks with respect to grammar. For example, the languages of the southern stock classify their numerals into forty or more categories, according to the nature of the objects they count; the northern languages lack such a classification but share a complicated kind of verbal system that differs markedly from the one found in those of the south. Linguists have not yet calculated how long the two stocks have been separated, but clearly it must be reckoned in thousands of years. During this time there developed several other marked differences between the cultures of the northern and southern societies (including cannibalism and male initiation rites in the north but not in the south). On the other hand, both northerners and southerners retained their common practice of affiliating individuals into matrilineal clans — supra-familial social units made up of persons related by maternal, rather than paternal, kin ties. This is evidently a cultural heritage of their common ancestry from New Ireland and New Britain, where such matriliny also prevails.

Another trait shared by the present-day descendants of both northerners and southerners is their skin colour, which is *very* black. Indeed, it is darker than that of any population of present-day Pacific islanders, including the present-day indigenes of New Ireland, the larger homeland of the first Bougainvillians. The presence of Bougainville as a 'black spot' in an island world of brownskins (later called redskins) raises a question that cannot now be answered. Were the genes producing that darker pigmentation carried by the first Bougainvillians when they arrived? Or did they evolve, by natural or by 'social' selection, during the millennia in which the descendants of those pioneers remained isolated, reproductively, from neighbouring islanders? Nothing now known about Bougainville's physical environment can support an argument for the natural selection of its peoples' distinctively black pigmentation; therefore a case might be made for social selection, namely, an aesthetic (and hence reproductive) preference for black skin. This preference has, by the way, surfaced recently with added political meaning.

While alike in their distinctive skin colour (and in the Melanesia-wide frizziness of their hair), the descendants of Bougainville's pioneer settlers eventually became differentiated into two major types with respect to some other bodily traits: a taller and broader northern type, and a shorter, slenderer southern one. This distinction corresponds to the language differences noted earlier.

Bougainville's long-lasting isolation was not ended until about three to four thousand years ago. Then, people having different physiques, speaking entirely different kinds of languages, and bearing many cultural innovations, surged from the west into the Pacific and on into or through New Britain, New Ireland, the Solomons and the New Hebrides. (From the Solomons some of the descendants of these newcomers moved on into the Gilbert Islands, and thence on to the Marshalls and Carolines. From the New Hebrides others moved into Fiji and Tonga and Samoa, where they evolved into the people now known as Polynesians.) Meanwhile, beginning about 3200 years ago, some bands of those newcomers settled on Buka and on Bougainville's northern and southwestern coasts. Much later, the descendants of some of those who had settled on the islands immediately south of Bougainville, resettled along Bougainville's eastern coast; the most recent of these movements founded the present-day community of Roruana, only about a century ago.

The descendants of the newcomers who settled on Buka and on the fringes of northern Bougainville eventually superseded or mixed with

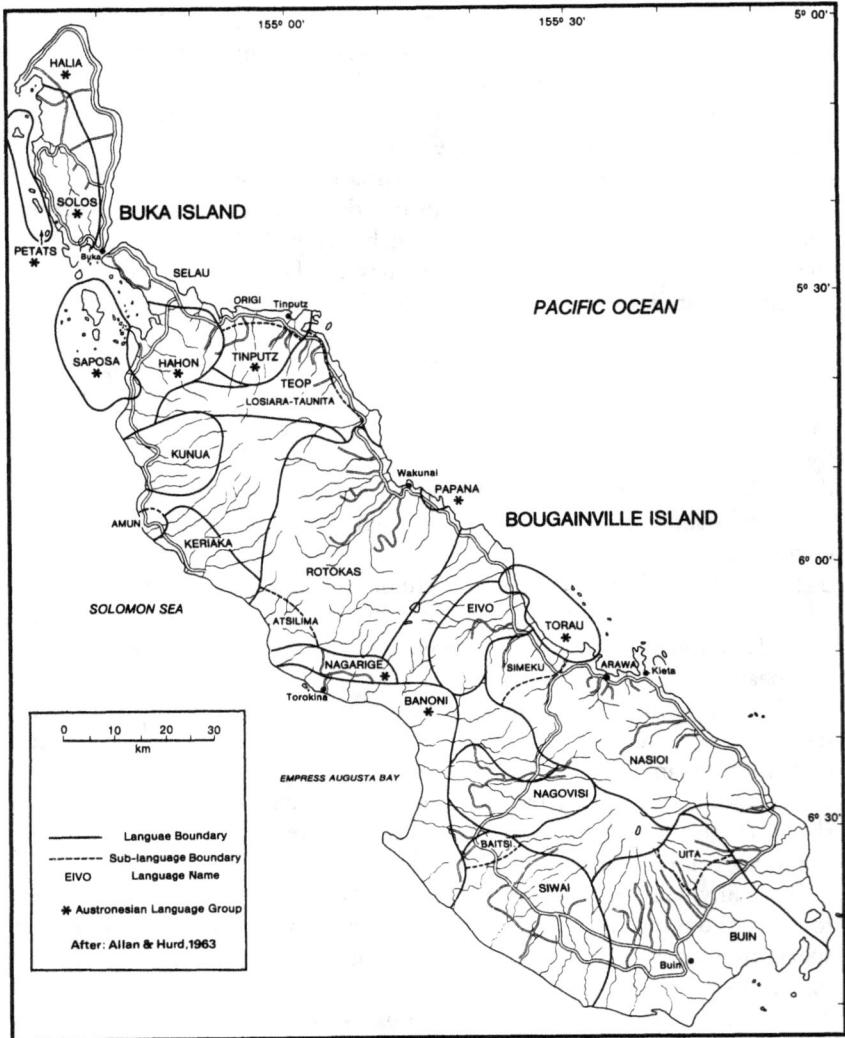

Figure 3   Language groups

whatever firstcomers still remained there, as revealed by the entirely different kinds of languages spoken there today. These newer languages are all interrelated and they are as different from the earlier ones as is, say, English from Arabic. They are divided into two groups (see Figure 3): one consisting of Tinputz, Teop, and Hahon; the other of Petats, Halia, Solos, and Saposa. All of these newcomer languages are members of a vast family of languages labelled Austronesian, which originated in South China and/or Formosa. Austronesian languages proliferated and spread throughout Southeast Asia (with one branch in far-off Madagascar) and all over the Pacific. They are found in the islands of Micronesia and Polynesia, and of Melanesia, except for most of New Guinea and pockets of earlier, non-Austronesian languages elsewhere, including those of Bougainville. (In this connection, some linguists believe all or most of Melanesia's non-Austronesian languages to be members of a single, 'genetically' interrelated group which they label *Papuan*, but expert opinion is not unanimous on this point.)

As Figure 3 shows, Austronesian languages are spoken also on Bougainville's central coasts, both east and west. Banoni and closely related Nagarige-Amun, spoken on the west coast, are direct and fairly recent offshoots of the island's northern Austronesian languages; on the east coast Torau (also called Roruana) and Papapana are spoken by people whose ancestors migrated there from the Shortland Islands only a few generations ago. When the author was on Bougainville in 1938–9 the present site of the town of Arawa was occupied by a small community speaking an Austronesian language also derived from the Shortlands at a time somewhat earlier than the arrival of the speakers of Torau. Now, fifty years later, that language, called Uruava (also Arawa), has become virtually extinct. Its former speakers have died out and their offspring have adopted the more prevalent language of their Nasioi-speaking neighbours, a transformation doubtless furthered by marriages between immigrants and earlier residents. Some of Bougainville's languages, both Austronesian and pre-Austronesian, are somewhat mixed, in that they contain certain words and even grammatical features borrowed from neighbouring languages.

Like the Uruava, several other bands of Austronesian-speaking immigrants may have lost both their language and their physical (i.e. genetic) distinctiveness after settling on Bougainville, but those who did not do so (including the Banoni, the Roruana, and the present-day residents of Buka and northernmost Bougainville) remain somewhat lighter in skin colour and generally taller in stature than their non-Austronesian neighbours.

Accompanying the new languages and genes that the Austronesian speakers brought to Buka and Bougainville were several other innovations; these included pottery, obsidian tools, domesticated pigs and chickens, and probably domesticated dogs. It is likely that they also introduced new crops and new techniques of gardening, although the idea of producing food plants — rather than merely collecting or tending wild ones — may have spread to these islands before then. Moreover, while the languages and the genes of the newcomers remained mostly on Buka and in the coastal areas of Bougainville, the cultural innovations brought in by them diffused throughout the larger island. Thus by the time Europeans 'discovered' Bougainville, all of its inhabitants were growing most of their vegetable food while continuing to collect a few wild-growing ones, such as the starchy pith of the sago palm. While they continued to fish, and to hunt such wild animals as opossums, flying foxes, birds and bats, they also raised pigs and chickens for their occasional feasts, and kept dogs as pets and for assistance in hunting the pigs that had escaped domestication and gone wild. Doubtless there were always regional differences in food-getting: fishing figured larger in the lives of coast dwellers than of inlanders, and gardening required more effort among mountaineers than among plains-dwellers. But rather than attempt to reconstruct the changes that had taken place in Bougainvillians' cultures from the early days of settlement, let us focus on what they had become just prior to European 'discovery' and colonization.

On evidence that will be given in later chapters, the number of persons living on Bougainville–Buka just prior to their 'discovery' by Europeans was about 45,000. This number had been reached several centuries earlier and had remained, thereafter, about the same. Most of those 45,000 resided in small and widely dispersed hamlets; it was only in a few places (for example, on beaches adjacent to good fishing, on tiny offshore islands) that larger settlements, nucleated villages, were to be found. Except for the island's southeast tip—a heavily forested but otherwise habitable area—the uninhabited, blank, spots on Figure 4 correspond with terrain wholly unsuitable for gardening (e.g. very high mountains or extensive swamps).

Within both hamlets and villages, the basic social unit (for sleeping; for getting, processing, and consuming food; for raising children, etc.) was the family-household. This consisted in some cases of three genera-tions of family members, or of a man and his two, or three, wives and their offspring; but in most cases it consisted of a monogamous couple and their own offspring. In addition, every Bougainvillian became at birth a member of his or her mother's clan, a kind of social unit which had many shapes and many functions (with regard to property rights, choice of spouse, religious practice, etc.), as will be described later on.

What then, many readers will wish to know, about the Bougainvillians' *tribes*—about their membership and rights and duties in more-or-less autonomous governmental units? This complex question will also be addressed in future chapters of this book, but a few prefatory points will be useful here. First of all, those tribes were in most cases very small, consisting of no more than a single village, or a few neighbouring hamlets; in other words, a tribe's 'citizenry' ranged in number from about twenty persons to seldom more than 300. Secondly, the normal relationship between neighbouring tribes was characterized by some degree of hostility, ranging from constant wariness to active warfare. And thirdly, 'chieftainship'—i.e. the kind of leadership characteristic of Bougainville tribes—varied from place to place, with respect to *who* the leaders were and *what* they did.

Such were the human conditions on Bougainville and Buka when whites first landed there: in doing so they precipitated, within a few decades, greater changes than had occurred during the previous 28,000 years.

We turn now to the physical features and climate of the area. How Bougainvillians (including the natives of Buka) see their home islands will

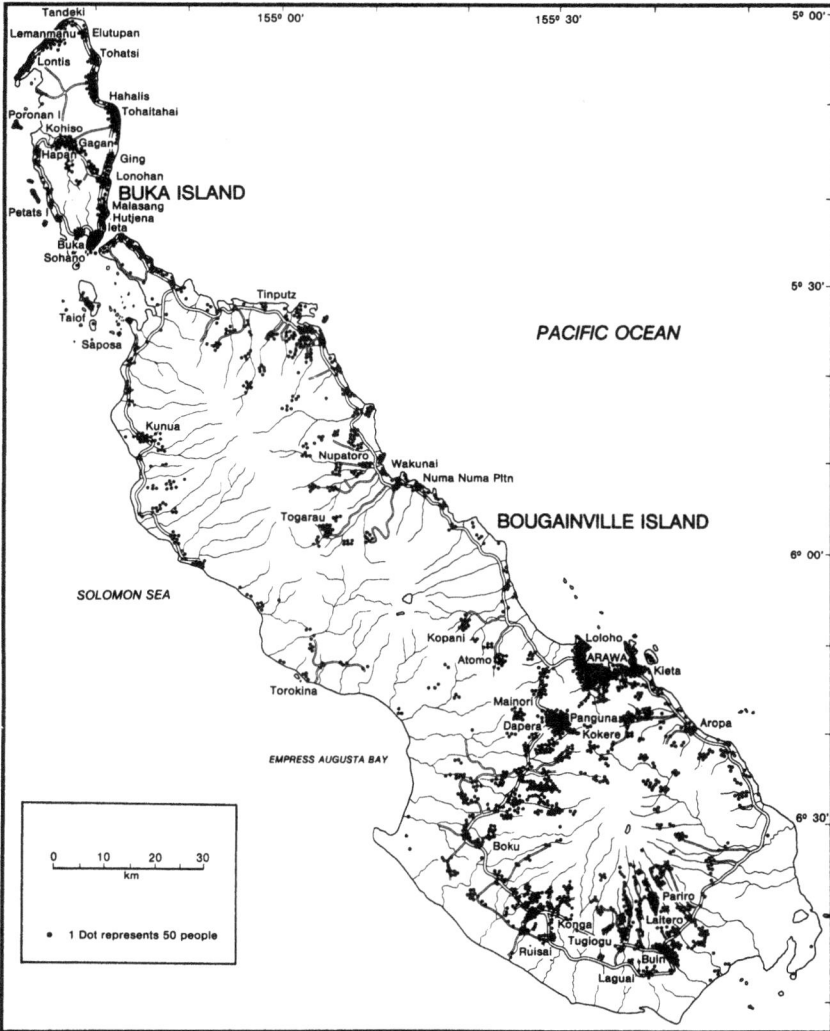

Figure 4   Population (latest census)

be considered later on; the concern of the present section is how the physical features of these two islands are viewed by expatriates.

Every adult visitor to Bougainville–Buka these days will know that they contain large quantities of valuable ores, but a first view of the visible landscape is likely to leave two impressions. One is that the topography is monotonously uniform, a jumble of hills and mountains and some flat coastal plains. The other superficial impression is that the soil is everywhere fertile, as indicated by the thick mantle of vegetation that covers all but the two active volcanic peaks.

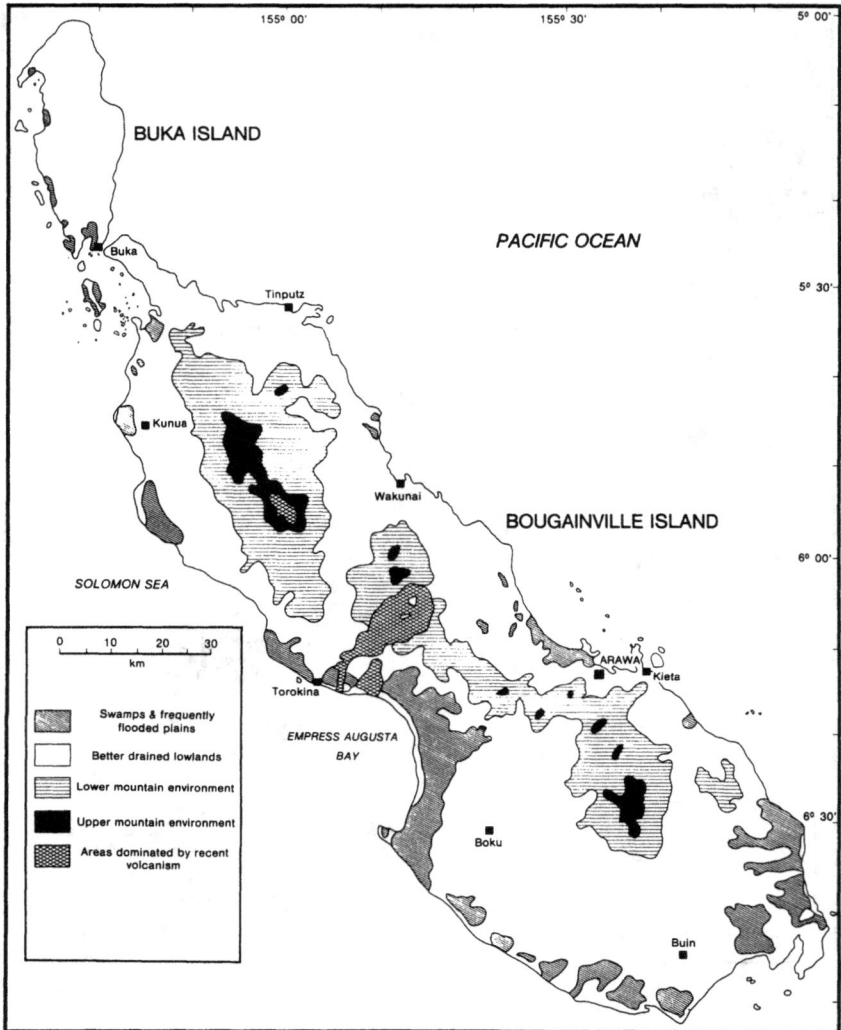

Figure 5   Major environments

After a while the initial impression formed of the topography will be
confirmed, but even the most unperceptive visitor will learn that the mantle
of vegetation is extraordinarily varied, and that the underlying soils also
vary greatly in the kinds and amounts of vegetation they can support.
While the islands' rich ores last, and are profitably mined, their native
residents will probably continue to share, directly or indirectly, in that
source of wealth. But when the rich ores are all gone and the two islands'
residents have to depend again on what they can grow in their soils, their
standards of living will inevitably return to levels simpler than they were
when mining began.

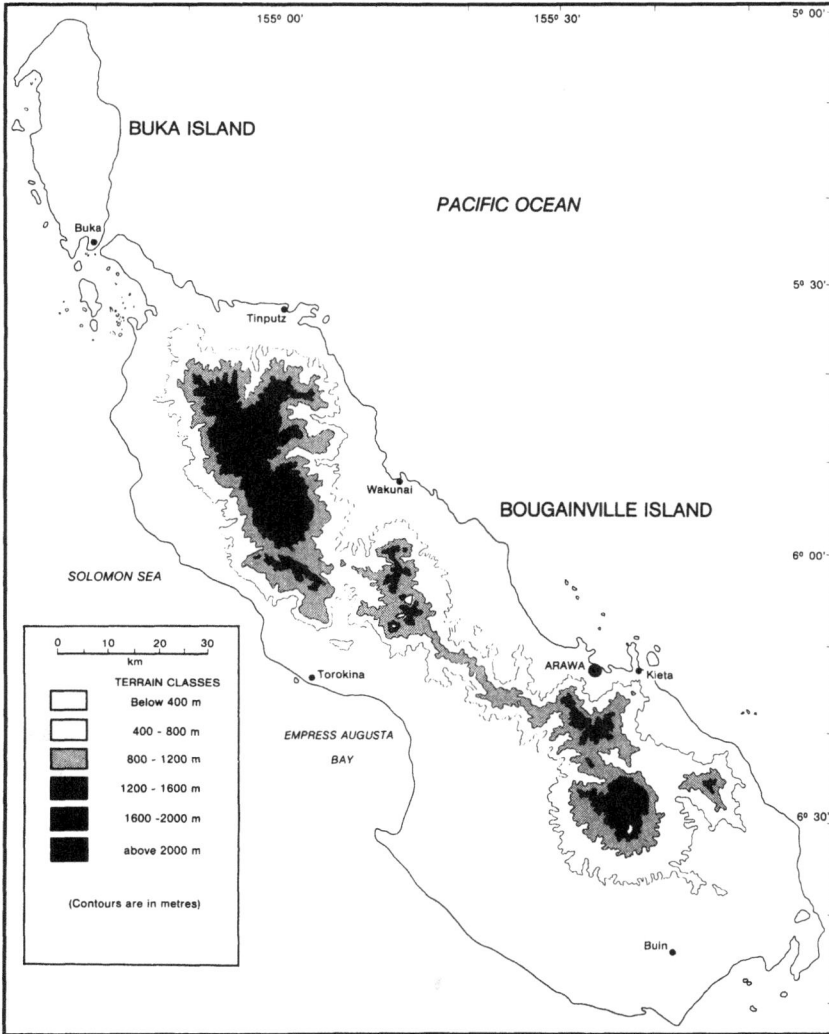

Figure 6   Altitudes

Bougainville and Buka Islands form a single land mass separated from one another by a shallow strait 800 metres wide. Together they are about 240 kilometres long, and about 64 kilometres across at their widest point. They are located along a northwest–southeast axis, and are, geologically, part of the Solomon Islands chain.

Their total land area is approximately 9000 square kilometres, minus some 13 square kilometres of lakes and some other expanses of freshwater swamp. About half of the land area is hilly or mountainous, with peaks rising to 1500 to 2400 metres, including several active, dormant, or inactive volcanoes, along with remnants of a geologically ancient plateau of uplifted

coral limestone (Figures 5, 6). The distribution of the islands' major types
of landscapes is shown in Figure 5. This coarse-grained classification of
natural environment is detailed enough for some purposes, but it is
inadequate for anyone seeking deeper understanding of the islands'
geographic history and economic prospects. For such purposes scientists
have devised a much finer-grained classification composed of 'land units'
and 'land systems'. According to this scheme a land unit is one characterized by 'a particular association of topography, soil, and vegetation' such
as, for example, 'a beach with an average slope of about 10 degrees composed
of white sand and supporting mixed herbaceous vegetation'; or 'a drainage
depression of low gradient composed of submerged peats (up to three feet
deep) and supporting tall forest trees chiefly of the *Terminalia brassi* species'.

Using 'units' in this sense, a land 'system' is an area characterized by
a recurring pattern of specific land units (Figure 7). For example, the
so-called Jaba land system consists of several of the particular kind of beach
units just described, along with other distinctively defined types of coastal
units labelled 'outermost beach ridge', 'beach ridge', 'swale', and 'tidal creek'.
The so-called 'Siwai' land system (which is not limited to the cultural unit
of that name), is defined as a class of broad, high-lying alluvial plains that
consist of four types of land units: plain, terrace, flood plain and river
bed, each of which is characterized by a particular kind of relief, soil and
vegetation.

Needless to say, land systems vary widely in the kinds of human activity
they can support. For example, of the two just delineated, the Siwai system
can and does support fairly intensive growth of indigenous food plants
as well as certain kinds of cash crops; the coastal Jaba system appears to
be suitable only for coconuts and plants with similar growing requirements. There are many other kinds of land systems on these islands that
can support no food or cash-crop plants at all—and indeed no other
conceivable form of human activity except perhaps swatting mosquitoes
or admiring distant views.

Some general conditions can be drawn from a map of the islands' forty
distinctive land systems. First, the environmental diversity helps partly
to explain the cultural diversity that obtains among the islands' several
types of subsistence technologies: between coast-dwellers and plains-dwellers
and mountaineers, between coast-dwellers of the north and the east, etc.
While no human society has its way of life determined in all details by
its physical environment, none is wholly independent of environmental
influences. And for societies with less-developed technologies, including
those of the indigenes of Bougainville–Buka, such influences tend to be
more decisive.

Second, and more relevant to present-day concerns, a land-systems map
reveals, in a way that no amount of guesswork and wishful thinking can
deny, how very limited are these islands' surface land resources in terms
of economically feasible agriculture. This needs to be asserted here at the

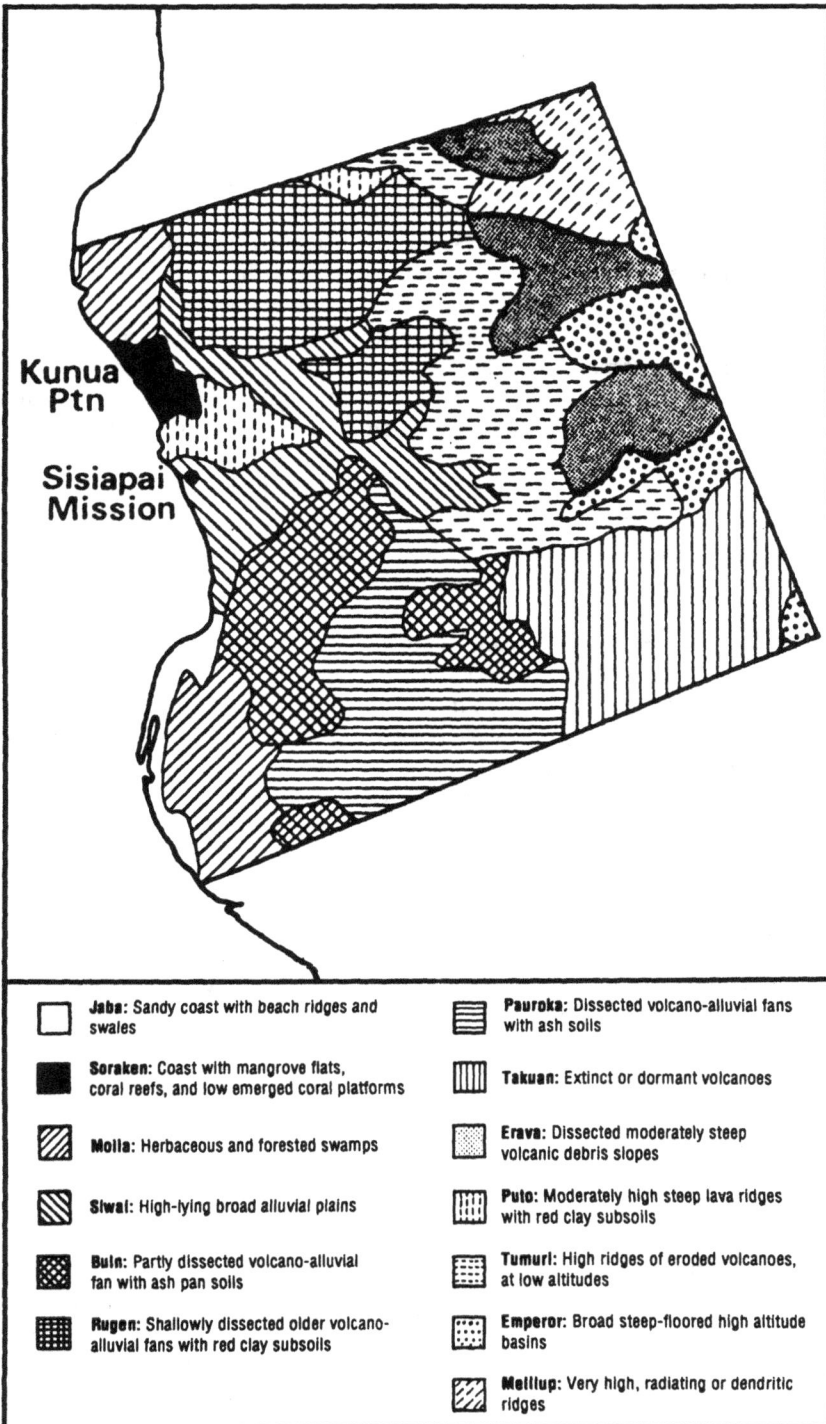

Figure 7 Land systems of Kunua census division

Kunua Ptn

Sisiapai Mission

Jaba: Sandy coast with beach ridges and swales

Soraken: Coast with mangrove flats, coral reefs, and low emerged coral platforms

Molla: Herbaceous and forested swamps

Siwai: High-lying broad alluvial plains

Buin: Partly dissected volcano-alluvial fan with ash pan soils

Rugen: Shallowly dissected older volcano-alluvial fans with red clay subsoils

Pauroka: Dissected volcano-alluvial fans with ash soils

Takuan: Extinct or dormant volcanoes

Erava: Dissected moderately steep volcanic debris slopes

Puto: Moderately high steep lava ridges with red clay subsoils

Tumuri: High ridges of eroded volcanoes, at low altitudes

Emperor: Broad steep-floored high altitude basins

Meilup: Very high, radiating or dendritic ridges

outset, as a caution against the widespread and erroneous impression that in the seemingly verdant soils of Bougainville and Buka, 'anything can be made to grow'.

The land 'systems', as just defined, owe their similarities and diversities to several factors, including the islands' geology and climate, and the land-altering activities of its indigenous residents — which, it will be recalled, have been taking place for 28,000 years.

The geological history of these islands has been marked by four land-forming processes: volcanism, coral-limestone growth, tectonic movements, and weathering. At least three periods of major volcanism can be distinguished in the remote past — one prior to the Miocene epoch and two during the Pleistocene. As the fiery cone of Mount Bagana attests, volcanism continues to take place and to alter nearby landscapes. The growth of coral limestone is also a continuing process along the islands' shores; throughout two large areas — northern Bougainville and Buka, and the Keriaka Plateau — raised limestone constitutes the entire bedrock. Heavy rainfall and year-round tropical temperatures have served to mould all these formations, as well as to build alluvial plains, to cut deep stream beds, and to create economically useless swamps. This brings us to the topic of climate.

The climate of the two islands is of the wet-tropical or tropical-rainfall type, and it is remarkably equable the year round. The mean annual temperature at sea level is about 26.7 degrees Centigrade; the monthly sea-level mean temperatures vary only a degree or so above or below that mark, and the average diurnal range at sea level is only about 10.6 degrees. Temperatures are lower at higher elevations (according to records from comparable places, mean temperature undergoes a drop of 1.35 degrees with every 300 metres), but here also they change within quite narrow monthly and diurnal ranges, and nowhere reach conditions of frost.

The alternating wind systems that affect these islands consist of a variable set from the northwest, which occurs between December and April, and a stronger, more continual set from the southeast, which prevails from May to December. These changes in wind have little discernible effect on (sea-level) temperatures but they exert some influence on patterns of rainfall, particularly in the north.

Average rainfall at sea level is higher in the south (about 3353 millimetres per annum) than in the north (about 2667 millimetres per annum), and regional topographic factors serve to extend these differences somewhat. (Rainfall also tends to increase with elevation, but there are too few records available to indicate how much.) The northwesterly winds (December–April) distribute about the same amount of rainfall over all parts of both islands. During the southeast season (May–December), however, the moisture-laden winds deposit more of their water on the southern slopes of Bougainville's mountains, thereby accounting for the higher average annual rainfall in the south. As a result of this circumstance, Buka and north Bougainville undergo a dry season during this part of the year. But it is only relatively dry — or rather, relatively less wet; the longest recorded period without rain anywhere in these islands is only sixteen days, and the mean duration of rainless days for both islands is three days or less for all months of the year.

The equability of the climate is also indicated by figures for relative humidity. On the basis of observations at sea-level stations, the mean monthly recordings range between only 75 and 86 per cent, and diurnal variations for any one station are even smaller.

We began this discussion of climate by considering it as one of a number of factors that affect the landscapes of the islands. In addition, of course, climate exercises more direct influences on the lives of the human inhabitants. For example, the unremittingly high temperature and relative humidity undoubtedly affect the health and activities of expatriate visitors from temperate climes. It is reasonable to assume — although difficult to prove — that these climatic factors have some deleterious effects on the indigenes as well, not only in the encouragement they offer to some kinds of diseases but also in their influence upon levels of energy.

The ways in which human activities have altered and diversified the islands' landscapes will occupy much of the remainder of this book, but it is essential once again to caution against too simple an assessment of those results.

To the unsuspecting eye viewing these islands from the air or the sea or even from the ground, humans appear to have made very little impression upon the soils and the profuse vegetation, except for the areas directly affected by mining activities (including the new urban centres and sprawls), and the pockets of expatriate-developed plantations. Even some general awareness of the economic uselessness of many of the land systems cannot entirely erase the visual impression that the indigenous residents have barely begun to exploit the economic potential of their land. To correct this impression, one has only to compare two other maps concerning the islands' agriculture. Figure 8 shows land-use capability and Figure 9 shows actual land use, including areas under actual cultivation and those in which cultivation has taken place during the recent past.

The most striking conclusion to be drawn from such a comparison is that in terms of existing types of cultivation and their present ratio of mix, the agricultural potential has already been pushed almost to its limit. Further discussion of this sobering conclusion will be offered later in this book; it is stated here mainly to curb the optimism with which many persons regard these seemingly thinly populated and luxuriantly vegetated lands.

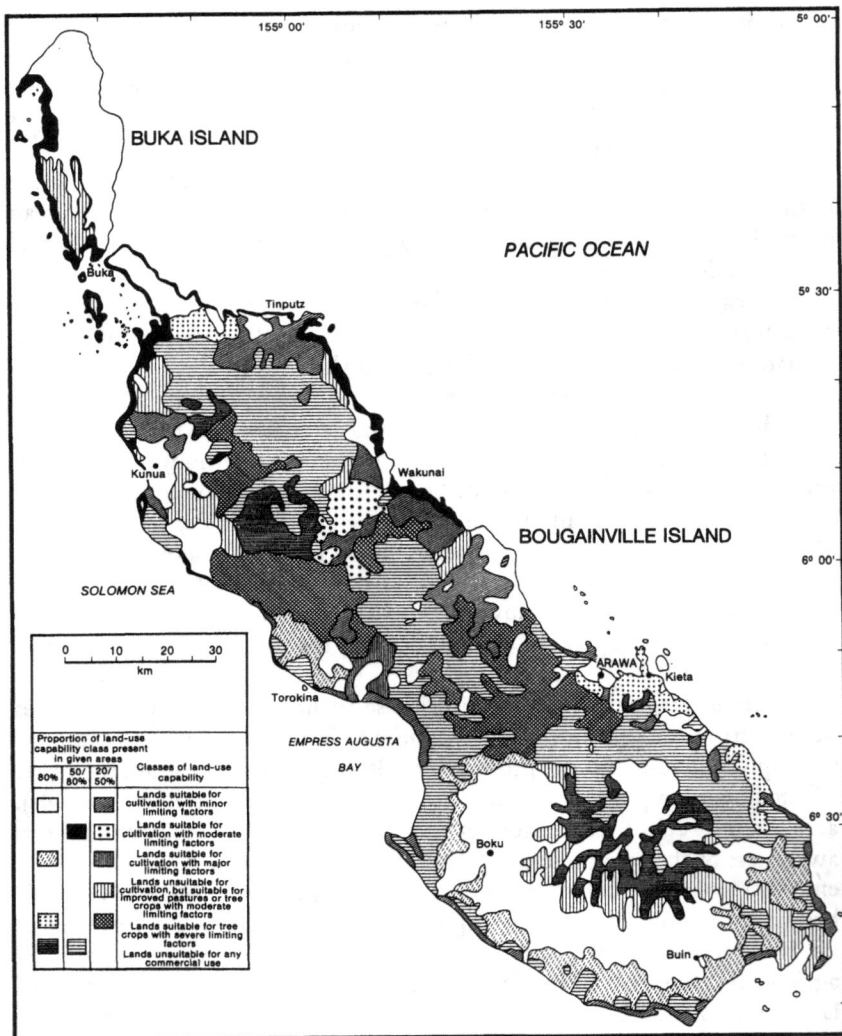

Figure 8   Land use capability

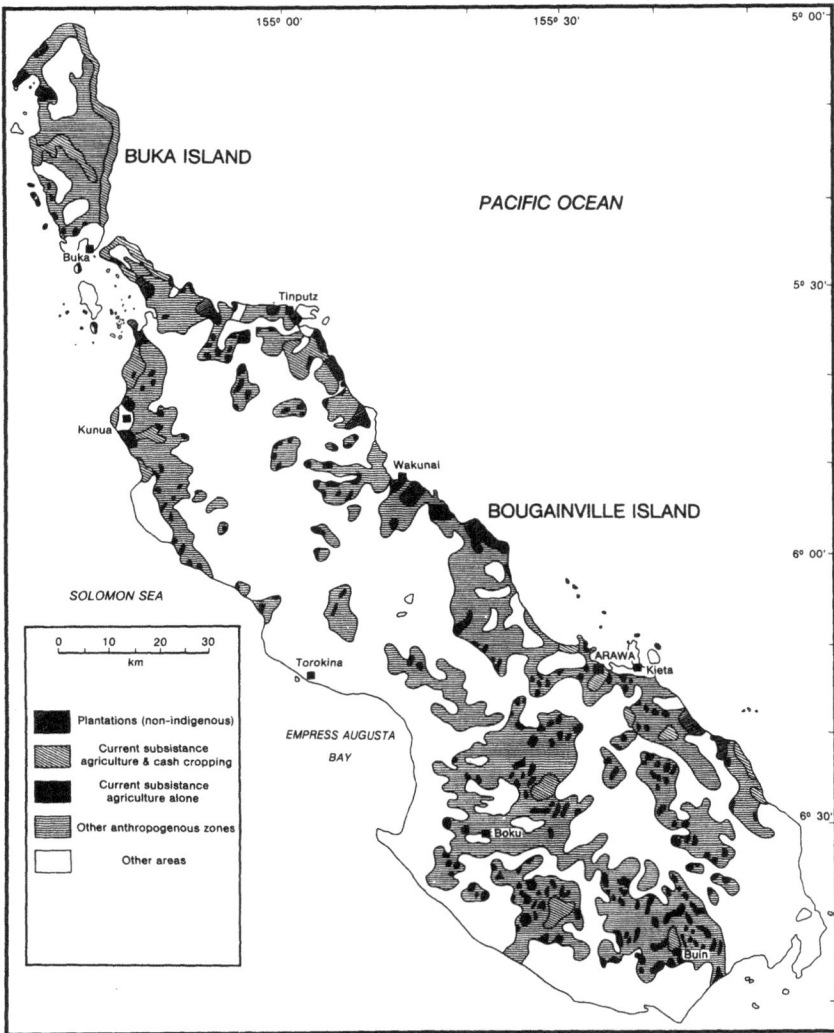

Figure 9 Actual land use

2

First Contacts with Europeans

THE FIRST EUROPEANS KNOWN TO HAVE SIGHTED EITHER Bougainville or Buka were those aboard the British ship *Swallow*, commanded by Philip Carteret. The *Swallow* passed within sight of Buka Island on 25 August 1767, but did not approach its shores.

Bougainville Island itself was first sighted on 4 July 1768 when the French ships *La Boudeuse* and *L'Etoile* sailed along the eastern coasts of both islands and anchored briefly off Buka. Here is the account of their encounter with the indigenes, written by the expedition's commander, Louis de Bougainville:

> After leaving the passage (west of Choiseul), we discovered to the westward a long hilly coast, the tops of whose mountains were covered with clouds. . . . The 3d in the morning we saw nothing but the new coast, which is of surprising height, and which lies N.W. by W. Its north part then appeared terminated by a point which insensibly grows lower, and forms a remarkable cape. I gave it the name of Cape *l'Averdi*. On the 3d at noon it bore about twelve leagues W, 1/2 N, and as we observed the sun's meridian altitude, we were enabled to determine the latitude of this cape with precision. The clouds, which lay on the heights of the land dispersed at sun-setting, and shewed us mountains of a prodigious height. On the 4th, when the first rays of the sun appeared, we got sight of some lands to the westward of Cape l'Averdi. It was a new coast [Buka], less elevated than the former, lying N.N.W. Between the S.S.E. point of this land and Cape l'Averdi, there remains a great gap, forming either a passage or a considerable gulf [Buka Passage]. At a great distance we saw some hillocks on it. Behind this new coast we perceived a much higher one, lying in the same direction. We stood as near as possible to come near the low lands. At noon we were about five leagues distant from it, and set its N.N.W. point

bearing S.W. by W. In the afternoon three *periaguas* [canoes], in each of
which were five or six negroes, came from the shore to view our ships.
They stopped within musket shot, and continued at that distance near
an hour, when our repeated invitations at last determined them to come
nearer. Some trifles which were thrown to them, fastened on pieces of
planks, inspired them with some confidence. They came along-side of
the ships, shewing cocoa-nuts, and crying *bouca, bouca, onelle*! They
repeated these words incessantly, and we afterwards pronounced them as
they did, which seemed to give them some pleasure. They did not long
keep along-side of the vessel. They made signs that they were going to
fetch us cocoa-nuts. We applauded their resolution; but they were
hardly gone twenty yards [18 metres], when one of these perfidious
fellows let fly an arrow, which happily hit nobody. After that, they fled
as fast as they could row; our superior strength set us above punishing
them.

   These negroes are quite naked; they have curled short hair, and very
long ears, which are bored through. Several had dyed their wool red,
and had white spots on different parts of the body. It seems they chew
betel as their teeth are red. . . . This isle, which we named *Bouka*, seems
to be extremely well peopled, if we may judge so by the great number
of huts upon it, and by the appearance of cultivation which it has. A
fine plain, about the middle of the coast, all over planted with cocoanut
trees, and other trees, offered a most agreeable prospect, and made me
very desirous of finding an anchorage on it; but the contrary wind, and
a rapid current, which carried to the N.W. visibly brought us further
from it.

During the quarter-century after Bougainville's brief visit, other European
vessels sailed within sight of these islands. The first recorded shore visit took
place in 1792, when d'Entrecasteaux's vessels lay off the west coast of Buka
for a few hours and carried on a lively trade with the indigenes who came
to meet them in their canoes. According to one journal of this voyage, the
islanders were more eager to obtain red cloth than iron. They are described
as astute in bargaining, as well as cheerful and friendly:

> M. de Saint-Aignan played them a fairly lively air on the violin, and the
> sound of this instrument, new to them, appeared to please them greatly;
> they laughed and jumped on the benches of their canoes. They offered in
> exchange for this violin not only the bow which we had already asked of
> them, but also some clubs they had not yet showed us. [Rossel, *Voyage*,
> vol. 1, p. 110, quoted in Dunmore 1965, p. 302]

During this period other European vessels may have made contact with
the indigenes of Buka. When *Sarah*, an English whaler, lay off northern Buka
in 1812, the inhabitants traded with the visitors with some degree of

familiarity and with apparent appreciation of the utility of the glass bottles and iron they received in exchange for their coconuts and weapons. Thereafter, until the end of the century, Bougainville and Buka were visited by whites for four different purposes: by whalers in search of provisions and fresh crews; by traders in search mainly of coconuts and copra; by labour recruiters; and by explorers, English and German.

Between 1820 and 1860 British, French and American vessels hunted sperm whales in the waters of the northern Solomon Islands, and through them Bougainvillians acquired quantities of weapons, metal tools, cloth and tobacco. During this period some Bougainvillians accompanied the vessels as crew members, sometimes as far as Australia. As a by-product of these contacts, foreign diseases and a liking for liquor were also introduced.

Bougainvillians had been trading with other islanders long before Europeans appeared on the scene. Those in the north traded with Nissan and Kilinailau, and those in the south with Shortland, Mono and Fauro.

More is known about the southern trade, in which the Bougainvillians exchanged pigs, vegetables, pottery and decorated weapons for shell money, fish and lime (for betelnut chewing). When European traders appeared on the scene, the Bougainvillians began to trade smoke-dried copra in return for steel axe- and adze-blades, machetes and calico. Sometimes a venturesome European trader would cast anchor off the southern Bougainville coast and barter direct with local islanders, but in the beginning most of this trade was carried out through Bougainville Strait islanders acting as middlemen. Occasionally the latter also acquired live Bougainvillians, by kidnapping or trade, to serve as menials or concubines or for religious sacrifice—and probably for sale to Europeans as labourers. During the 1880s the Strait Islands were under the suzerainty of Gorai, a famous Shortland Island chief, who professed great respect and liking for Europeans. Gorai's influence, though not his actual rule, extended up Bougainville's eastern coast as far as Cape L'Averdy. On one occasion he sent a fleet of his war canoes to the village of Numa Numa, 160 kilometres north of Shortland, and killed a score of its people to avenge the killing of a white trader with whom he was friendly.

A detailed description of a foreign trading visit is provided by the German museum collector, Carl Ribbe, who accompanied a white trader, based on Shortland, on some of his voyages to Bougainville in 1894–5. I reproduce here my translation of an abridged account of one of these visits—this one to the Buin coast just east of Kangu—because of the picture it gives of the manner in which commercial relations between Europeans and indigenes were conducted during that era:

> Mr. Tindal and I left Faisi [the main port on Shortland Island] in a small two-masted cutter. . . . Around four in the afternoon we drew near to the Bougainville coast . . . From where we were the land looked flat in every direction except for two or three 100-metre-high hills directly on

the coast [i.e., Kangu Hill]. The whole southern Bougainville plain was canopied by tall trees . . . and channelled by numerous full-flooded streams. Far to the north-east the horizon was dominated by several high and steep-sided mountains, comprising the Crown Prince Range. . . . The narrow strip of hill country between mountains and plain are, like the latter, covered with a high stand of forest but the mountains themselves appeared to be only partly forested. . . .

On first seeing these mountains I thought to myself what a rich field of research they must offer to the naturalist. Unfortunately, it would be virtually impossible to get to them, because the country approaching them is said to be densely populated by inhospitable and warlike tribes, whose opposition would prove even more difficult to overcome because of their ignorance of the power of firearms. Cases have been reported in which such inlanders as these have ridiculed their fire-arm bearing opponents, asking what possible effect the latter's noise-making bamboo sticks could have against their own formidable spears and bows and arrows.

It has been my experience that when accompanying a small-sized expedition into the interior of islands in this part of the world, one has less to fear from the hostility of the natives who have already experienced some contact with whites than from those who have never before seen them. Typically, when a small party of explorers is opposed by indigenes with no prior experiences with firearms it is apt to be wiped out in the first assault. Since the thick undergrowth conceals the attackers struck down by bullets, their unwounded fellows remain unaware of the deadliness of firearms and so press their attack fearlessly and relentlessly . . .

From where we lay at anchor off shore no houses or canoes were to be seen along the beach, the native villages being located some five to six kilometres inland. . . .

We remained aboard our cutter that night, then, shortly after sunrise, four of our Shortland Island servants took off for the villages of Suriei and Takerei to inform the villagers that we had come to trade for copra. Each of our messengers was of course armed with rifle and revolver, for we could not rule out the possibility that they might meet up with hostile mountaineers and be obliged to fight. . . .

Shortly after noon we were hailed from the shore, where we saw many indigenes alongside several piles of copra. The cutter's boat was rowed to the beach to bring back some villagers and their copra, and the trading then began — during which, I should add, we kept our firearms constantly at the ready. . . .

The bartering indigenes were permitted to board the cutter from one side only. While one of us whites occupied himself with the trading, the other kept a close watch to guard against attack. This kind of trading is no great pleasure — indeed, it is long drawn out and boring. However, it is essential not to give up or lose patience; otherwise the blacks would not bother to return to trade another time.

The owners of the long strings of dried coconut chunks usually delegate negotiations to one or two of their number, who invariably, in the beginning, demand exorbitant prices. Then, before any transaction is concluded, each of the villagers present is asked whether he agrees with the terms offered. They appear to have no conception of the monetary value of the various trade items offered to them. It often happens that they will first demand a ridiculously high trade price for their copra, and then in the end be satisfied with a very modest return. Thus, one can obtain 100 coconuts for 65 pfennigs worth of calico, or for 10 coconuts they will accept either a clay pipe worth 2½ pfennigs or 2 sticks of tobacco worth 5. A short machete costing 40 pfennigs will purchase 50 coconuts, while a long one costing 1 mark will purchase 100. A box of matches worth 4 pfennigs will obtain 10 coconuts, a Jew's-harp worth 15, 30 coconuts, and an axe worth 1 mark will obtain 100. From these few examples one can see . . . that the indigenes have no idea of the relative values of the trade goods they obtain with their coconuts. This state of affairs, which often results in a disadvantage to the white traders (when indigenes demand too highly priced goods for their coconuts) is the fault of the traders themselves. [How much *more* often, one may inquire, did this 'state of affairs' result in disadvantage to the indigenes?]

The trader has to exercise special care to protect his own interest when the indigenes demand calico for their copra, since the customary method for measuring cloth can work to the latter's advantage. The unit of measure used here is the 'fathom', the span between fingertips of a person's outstretched arms. The length of this span can of course be varied according to the extent that one stretches the arms and the way one holds the cloth. And it is not surprising that the indigenes insist upon having the measurement done by the man with the longest arms. (One can well understand what diplomacy the trader is called upon to exercise in winning agreement to use a shorter-armed man as a measure.) Distance from outstretched fingertip to nipple also serves as a unit of measure, as does the distance between the outstretched tips of thumb and index finger — these measurements being used when the indigenes exchange their coconuts, etc. for strings of shell money (*mauwai, perasali*).

Thus, the whole commerce is a form of barter — which incidentally, is highly profitable to the white traders in this part of the Solomons. The copra, which the trader sells to the schooners that ply these waters, at seven and eight pounds sterling a ton (i.e., 5000 coconuts), he is able to buy from the indigenes at three pounds, thereby realizing a profit of four to five pounds per ton.

Among Bougainvillians the trade goods in most demand are hatchets, axes (with metre or half-metre long handles), pocket knives, large blue and red beads for necklaces, small red, blue and white beads used for making ornaments, porcelain bracelets, tobacco and pipes, thin, patterned or red or white calico, plane blades for wood-working, mirrors and Jew's-harps.

The indigenes also extend this form of trade among themselves. Some goods obtained from the white traders end up in distant mountain villages, being bartered from tribe to tribe and at increasingly high prices, so that a distant inlander will pay 300 to 400 coconuts for a hatchet that was obtained at the beach for only 100.

Ribbe went on to say that it was the copra trade that had made the southern Bougainville coast a safer place for outsiders to visit.

After about 1870 Bougainvillians were recruited in large numbers for plantations in Queensland, Fiji, Samoa, and New Britain. Those of Buka were in especially heavy demand because of their reputation for trustworthiness and industry. For example, the German trader-planter, Richard Parkinson, found his Buka labourers to be invaluable protection from his hostile indigenous neighbours in the Blanche Bay area of New Britain: 'I always licked them fearfully with my Bouka boys of which I have 150.' (quoted in Sack 1966, p. 120) Some of the Bougainvillians (including Bukas) went voluntarily with the European recruiters, evidently eager for the European goods to be earned, or to escape from dangerous situations at home. But others went under duress, as in the case of those kidnapped by the Melbourne vessel *Carl* for work on Fiji.

The *Carl* was owned by an Irish physician, Dr James Murray, who embarked upon his South Seas adventures in 1871 after a series of scandalous scrapes in Australia. After 'recruiting' — that is, kidnapping — nearly eighty indigenes from various islands in the New Hebrides and Solomon Islands and imprisoning them in the vessel's holds, he sailed his ship to northern Bougainville and Buka. Here is Edward Docker's recent reconstruction of what ensued:

Even King Ghorai of the Shortlands with his mighty war fleet never dared attack Buka, and with visiting ships the big, very black Bukamen paddled out in their twenty-man canoes prepared either to trade or fight as the mood suggested. They never had such a shock in their lives. Large lumps of pigiron or cannon slung in ropes crashed down on the canoes; then immediately, as they struggled in the water, with many of them badly gashed and bruised, the boats were among them, hauling them in like tuna. The score was forty the first day; forty-five the next. The earlier captives were now stowed right forward and aft, with the eighty-five Bukamen under the main hatch. Not one was either handcuffed nor leg-ironed.

That evening there was much recrimination on deck among Murray's party about these methods of recruitment — with no attention being paid to what was happening below. Here the Bukamen had broken up their bunks and were using them as implements to force open the hatch. Before long the clamour from the hold drowned out all sounds of the dispute on deck and settled the argument among the white men, at least for the time being. The best-corroborated version of the events of that evening are supplied by a seaman, Devescove. He later testified:

'I was awakened about ten by the boy Fallon coming to my bunk, and asking me for God's sake to come on deck, as the ship was on fire, and they would be all dead men. I went on deck, and to the main hatch, where I found the passengers and others assembled, called out to the natives to keep quiet. I saw no signs of fire, and went below to the cabin for a minute. While away I heard sounds of firing, and returned on deck, and saw William Scott, Dr Murray, Captain Armstrong, and others firing down into the hold. The natives were fighting amongst themselves, and trying to break open the hatchways, Mount and Morris were firing with revolvers.

'After the natives had been fighting a bit they would stop for a few minutes, and then the firing would cease, and be resumed when the row began again. I went to the cabin after the first row was quieted. I saw Morris there loading a rifle, and Dr Murray loading a revolver. There was firing off and on during the night. I fired myself, once or twice, before I saw Morris and Murray in the cabin. At one o'clock in the morning the mate raised a cry that the natives had charge of the deck, and Dr Murray called out, "Shoot them, shoot them, shoot every one of them."

'When daylight broke, everything was quiet. The shooting continued, off and on, until about three o'clock, or half-past three, when we knocked off altogether. The firing was resumed at intervals of five, ten and fifteen minutes, and sometimes half an hour elapsed between the rows. At four o'clock everything was quiet, and I went into the galley and served out some coffee to the men and passengers. After a bit Dr Murray came aft. Lewis, the second mate, said, "What would people say to my killing twelve niggers before breakfast?" Dr Murray replied, "My word, that's the proper way to pop them off."

'Everything was then quiet, and breakfast was got ready. After breakfast the ladder was put down the hold by the passengers and crew, and the natives were told to come on deck. Some of the wounded natives came up; they were wounded in the back, arms, and legs. Those who had a narrow wound were put on one side, and those more dangerously wounded on the other. All the wounded natives who could come up, came up. Two of the good natives were sent down by Dr Murray with ropes, which they fixed round those who were dangerously wounded, so that they could be hauled up. The wounded were separated as I have described by Dr Murray's directions. The passengers were looking on all the time, and Mount and Morris told the natives to do their work.

'I heard them tell them to lay the wounded down, and make fast their hands.

'Dr Murray went forward to the starboard side of the ship, and said, "Well, boys, what do you think of doing with these men?" Mount asked, "What do you think of doing?" "Well," said Murray, "I think that the best we can do is to go the leeward of the island and land them there." A man said, "How far are we from land?" Dr Murray answered, "I don't know, but not very far." Mount said, "You have been gaffer all this time, what are you going to do?" Dr Murray then took four or five of the friendly natives and went aft, and told them to pick up a man and throw him overboard. There was a boy with six fingers and six toes, who was wounded in the wrist, and he was the first thrown overboard. When Dr Murray told the friendly natives to pick up the boy, the other natives screamed "No, no, no!" He was lifted onto the rail, and Dr Murray pushed him overboard. He was the first who was thrown overboard. At this, all the Bougainville men who could do so, jumped overboard.'

In the end the total of natives killed outright or tossed badly wounded into the sea amounted to seventy. Another fifteen or so of the unwounded may have swum safely ashore, which now left on board the seventy-six so-called 'friendly' natives. One result of the abortive mutiny was that the Malaitamen had completely abandoned their former over-hasty ideas of escape.

Some of the Europeans who took part in these outrages were eventually arrested and sentenced to death or terms in prison, but the agile and ingratiating Dr Murray turned Queen's evidence and escaped punishment altogether.

The most detailed account of labour recruiting on Bougainville–Buka is that of Douglas Rannie, who accompanied a recruiting expedition on board an Australian vessel as government agent. The vessel stopped twice off Bougainville–Buka in search of recruits for the Queensland sugar fields; first at an unspecified point off Bougainville's northwest coast, and then off Buka. The different receptions accorded the vessel at these two points serve to show how different the inhabitants of the two islands had by then become in terms of their experience and sophistication in dealing with whites:

On the morning of the 25th of June we lowered our boats about eight o'clock and made towards the shore. This being the weather side, a very heavy surf was breaking on the beach; so heavy, indeed, that for some time we thought we should have to give up all idea of getting into communication with the natives, whom we saw in large numbers lined up on the sand.

There appeared to be two tribes assembled. They did not seem to be upon amicable terms, as they held aloof from one another. They were all heavily armed with very long bows and sheaves of arrows. Besides these

weapons some carried spears, and each man had suspended from his
shoulder a tomahawk, club, or heavy wooden sword. The tribes were
distinguished by the colour of their head-dress. This was composed of a
hat exactly resembling an egg-shaped lamp-globe and of similar size.
These hats were made of basket-work, and beautifully covered and sewn
with the skin of the pandanus leaf. They reminded me more than
anything else of the baskets used in billiard-rooms for pool and pyramid
balls. The opening was not much wider, although it might have
admitted a cricket ball; into this the natives stowed their long, woolly
hair. The large amount of hair they managed to stuff in caused the hats
to stick up jauntily on the side of the head. The hats worn by one tribe
were all white, while those worn by the other were stained a bright red.

Pulling along the coast we came to a smooth part, and were able to
approach nearer the islanders. After a lot of persuasion we induced them
to approach nearer to each other as well as to us.

Both tribes wished to enter into communication with us, and both
had stuff for barter. As neither would entirely trust the other, they each
left a strong armed party immediately behind them in the scrub as
guards.

The mate, with his boat stern first, cautiously approached what
seemed to be the most moderate break in the surf, and I directed his
attention to the heavy break which occurred with every third or fourth
wave outside the ordinary surf. As a man came out neck-deep in the
water, holding a young sucking pig over his head, the mate ventured
too much. A huge wave broke over the bows of his boat, filled her and
swept her right up on the beach. The boat's crew leaped out before she
grounded, having first secured their rifles. Many of the islanders ran for
their weapons, but others professed to offer assistance. In the meantime
we were outside the influence of the surf, and covered the other boat's
crew with our rifles. The natives ashore seemed to be of two minds.
Some, I thought, desired to assist our men, while others were inclined
to go for loot. But the fact that our men still retained their arms ashore,
and we were almost out of range of their arrows, and had them well
covered, decided them to help us in our difficulty and trust to our
generosity for remuneration. A number of them turned to with a will;
and after the mate had given them all the print and calicoes, besides
beads, pipes, and tobacco, which he had in his trade box (the axes,
tomahawks, long knives, and butcher's knives were in the bottom of the
boat), they re-launched the boat. But alas! before the boat's crew could
get her under way with their oars, a great rolling sea caused her to
broach-to and capsize, and surge in towards the beach, bottom up, with
the crew underneath. One by one they struggled out. The mate was
dragged out with a horrible gash on the back of his head and neck, from
which the blood flowed freely. Hastily we unbent the painter and the
sheet of our lug-sail, and backing the boat in as far as we deemed safe

through the surf, we threw the boat's crew the rope. They made it fast
to the mate, and we were able to draw him through the surf to us.
Pulling out to a safe distance beyond the breakers, we rendered what
first aid we could to the wounded man.

A terrible scene ensued ashore. The natives of both tribes rushed
down to the boat, dragged her up on to the beach, and fought savagely
for the axes, tomahawks, and knives that were lying in from two to
three feet [60 to 90 centimetres] of water. Two natives would be
struggling for an axe. One would manage to free his arm, with the axe
aloft; and the next instant it would be brought crash, down through the
skull of the other unfortunate one. Several could be seen fighting and
slashing each other with the long knives and butcher knives, as they
rolled over and over each other in the water. Those ashore along the
fringe of scrub took up the fight, and a general battle ensued. The
arrows were flying in the air like showers of hail.

Presently a large body of men charged out from the scrub, on those
nearest the boat (they had manoeuvred round through the back of the
scrub from the tribe of the white hats), and making a wild dash among
the bowmen of the red hats, mowed them down with tomahawks and
hardwood swords before the red hats had time to unsling their weapons.

The red hats then took to flight, but were followed by the white hats
with showers of arrows until the bush gave them shelter. There must
have been upwards of a thousand engaged in the fray, and the casualties
were very numerous.

Seeing that we could not do much more until our second boat was
patched up, we made for the north end of Bougainville and came to
anchor at Buka Island.

We were visited at Buka Island by large numbers of islanders in many
canoes. The canoes carried from ten to sixty men in each. As many of
them were as high in the sides as our own little vessel, we made a rule
that canoes were to be allowed on one side only, and that the starboard.
The port side was to be kept clear, as well as the main deck on the port
side; so the ship was roped off fore and aft amidships. We had also to be
constantly on the watch and always armed; for, on the slightest show of
carelessness on our part, or of being off guard, we should all have been
massacred for the sake of loot.

One of our boatmen told me that on a previous visit he had been
shown on a clear day the hull and masts of a vessel lying on the bottom
in deep water. She had been taken and looted by the natives and then
sunk. We secured the services of an islander here as an interpreter. He
was the only one able to speak English. He told us that his name was
'Maggy', and that he had worked for a Mr Farrell in Samoa.

Maggy piloted us to quite a number of villages, but found no one
anxious to emigrate to Queensland. The villages were kept as clean and
ship-shape as any in the Shortlands, and the natives displayed as much

taste in the manner in which their plots of flowers and flowering plants about their houses were attended to. At the Shortlands I noticed that the dead were buried in the ground and large cairns of stones were piled over the graves; these again were filled in with soil, and the interstices planted with bright and fragrant flowers. But here the dead were disposed of in quite a different manner. We had an opportunity while on a visit to one of the villages of seeing their strange funeral ceremony. The corpse was carried down to and over the reef by a few of the deceased's comrades, followed by a crowd of women wailing and performing strange antics. At the edge of the reef the remains were placed in a canoe, paddled out some hundred yards or so, and with a few heavy stones attached were sunk to the bottom. Although free from any particular amount of general sickness, and physically as fine a race of people as we had so far met, 80 per cent of them seemed to be afflicted with a disagreeable skin disease they called 'buckwah' [?]. This disease breaks out in patches on the body in the form of a number of small dry rings, resembling ringworm. They spread till the whole body becomes covered with a mass of dry, scaly rings, which comes off in flakes and dust. I have cured many of the sufferers with a mixture of sulphur and kerosene, applied with a large paint brush. Clean-skinned natives seem to have a horror of contracting the disease.

We found the islanders very skilful in the manufacture of spears and arrows, and many of their weapons were tastefully inlaid with mother-of-pearl.

Pearl-shell appears to be fairly plentiful in those regions. From the many patches of reefs we sailed over, I believe large quantities of shell could be obtained.

Three days were spent in visiting villages scattered here and there, but all our recruiter's eloquence could not induce any of the natives to engage in the Queensland sugar industry. So the skipper decided to make a move the following day. That evening two large canoes came along from some foraging expedition. Their crews, numbering about forty each, were quite jubilant over some foray, or success. They clicked their paddles on the side of their canoes, keeping time to a wild chant or war song.

The paddles of these canoes had each the design of a dancing demon stained on it in permanent black and red dyes. Crouched despondent in the bottom of one of the two canoes, we noticed, as they came alongside, a wild, powerful-looking man.

After an animated conference with the savages in the canoes, our interpreter Maggy approached the skipper and me, and told us that the savages had a captive in one of their canoes whom they wished to dispose of by selling him to us. I said that the strict meaning of the Act would not allow any such mode of recruiting. Yet as the circumstances of the case seemed very peculiar, I determined that I would go into them very carefully.

Impressing upon Maggy that he must speak the truth, and nothing but the truth, I elicited through him from the canoe savages that the man whom they now wanted us to take was a captive they had made upon their present expedition.

They were taking him home with them, there to be dealt with in a way that even Maggy hesitated to describe. I inferred that he was to be put to death, and eaten. I got Maggy to explain to the captive that if he chose to come of his own free will on board of us he could do so, and that if he chose to leave the ship at any place in the islands no one would prevent him. As he came on board under peculiar circumstances, the same circumstances would allow him to go ashore anywhere he liked where the ship should touch before leaving for Queensland. As we had an interpreter on board who could speak his own language, the whole of the nature of the work expected of South Sea Islanders on Queensland plantations would now be fully explained to him, together with the nature and terms of agreement. But that he would not be called upon to enter into that agreement until some time during the trip when others might be signed on.

All this I was confident Maggy faithfully explained to him, and he came very joyfully aboard. In return, the savages, at their captive's hands, received a bundle of fancy-coloured print, in which were rolled up some glass beads, paint, tobacco, a couple of butcher's knives, and a tomahawk. I made the parcel of trade come from the captive as a ransom paid by himself, and not as the price paid for a slave.

Thus we got our first recruit, and he was entered on the Passenger List as No. 1, Cheeka of Buka, Bougainville. He was about twenty-five years of age, well built and muscular-looking, with a huge head of hair hanging in a mass of ringlets down to his shoulders. Each ringlet was plastered thick with lime and cocoa-nut. We soon set one of the crew to work with the scissors and his locks were consigned to the deep. Cheeka was quite pleased with the change, and was anxious to adopt European habits at once, so great was his delight at escaping from his enemies. And yet, he told me afterwards, he had never seen a white man in his life before.

The effects of these early encounters between Bougainvillians and whites must have varied widely. Some of the former, mainly coast-dwellers, and especially those of Buka and northern Bougainville, became well acquainted with the material goods and customs of whites, and with their characters, both good and bad. Many, however, experienced nothing of the new alien influences except the occasional steel tool that filtered to them through coastal intermediaries.

One of the most detailed accounts of that period was written by H. B. Guppy, the naval surgeon attached to a British exploring expedition to the Solomon Islands in 1882. This writer tarried for several months in the islands of the Bougainville Strait and made several visits to the south coast of

Bougainville itself. Guppy collected much useful information concerning the indigenes and the natural resources of southern and eastern Bougainville, including specimens of ore that led him to make the prophetic statement: 'A sample of stream tin from the southeast part of Bougainville was given to me by the Shortland chief. Copper will not improbably be found in association with the serpentine rocks of these islands.'

Until 1884 Bougainville–Buka continued to remain outside the administrative domain of any European power, although British subjects (including some Australians) were most in evidence there, as visiting traders and labour recruiters. This situation began to change in 1884 when Germany annexed northeast New Guinea (Kaiser-Wilhelmsland) and the Bismarck Archipelago. This action moved Queensland, and eventually Britain, to annex Papua (i.e., southeast New Guinea) — but that is another story. Bougainville and Buka were not officially added to the German colony until 1899, but by an exchange of notes with Britain, in 1886, these islands (along with Shortland, Choiseul, and Isabel) were declared to be within the German sphere of influence. In fact, German influence began to extend to Bougainville and Buka some years earlier in the persons of traders, explorers and recruiters of labourers for plantations on Samoa, the Bismarck Archipelago and elsewhere. The best-known of those early Germans was the Richard Parkinson referred to earlier. Parkinson had moved to New Britain from Samoa in 1882. (His wife was sister to the much-married Emma Forsyth — 'Queen Emma' — who had gone from Samoa to New Britain earlier and had established extensive trading and plantation enterprises in the Duke of York Islands and on Blanche Bay.) From his New Britain base Parkinson made many trips to Bougainville and Buka, trading, recruiting, and collecting natural history specimens; he recorded his observations in several scientific papers and in his lengthy book: *Dreissig Jahre in der Südsee* (Thirty Years in the South Seas). In summarizing his findings, Parkinson reported that by the turn of the century the coastal inhabitants had become fairly familiar with Europeans, through trading with them or serving on their plantations on New Britain and elsewhere, but that the interior of the larger island remained 'virtually closed-off'.

Ignorance about Bougainville's inland areas during that era can be attributed partly to its physical inaccessibility, partly to their inhabitants' ways of life, and partly to the behaviour of the white visitors themselves.

During the nineteenth century, and probably for centuries and millennia before, the native people were separated into numerous minute tribes whose interrelations, if any, were typically hostile, with the exception of occasional instances of intertribal trade. Moreover, this normal state of hostility was more often than not intensified in specific cases where one tribe was made up of coast-dwellers and the other of inlanders. This antagonism was for a time reinforced by the appearance on the scene of white traders and labour recruiters. Some coastal tribes used their newly acquired steel tools as weapons against their traditional enemies. For some inlanders the only way to acquire

Traditional mountain stream fishing.

Preparing betel nut chewing mixture.

Women in taro garden (1939).

Sounding a slit gong.

Siwai flutes for music and signals.

Mount Bagana (1747 metres) is an active volcano.

Differing physical characteristics:

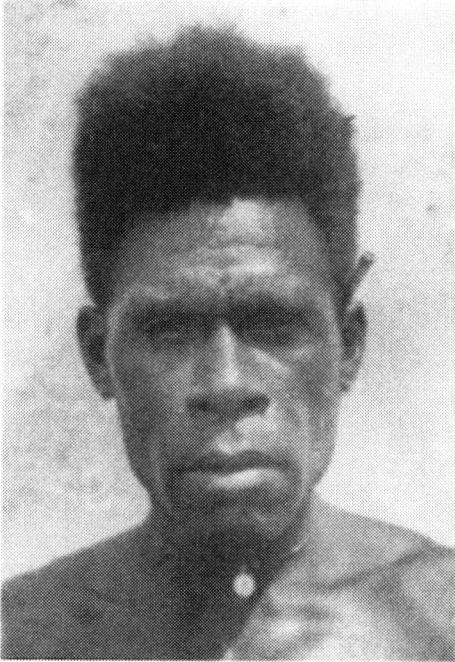

Eastern coastal man.

Intermediate coastal-mountain woman.

Southern mountain man.

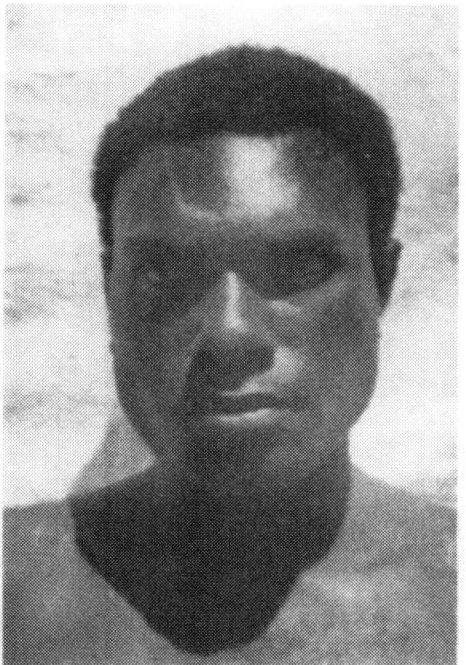

North-central mountain man.

'Blackbird taming' (1872).

'The Blackbirds tamed' (1872).
(Both by courtesy National Library of Australia from Illustrated Monthly Herald
28 December 1872)

the eagerly sought European trade goods was by raids against coastal settlements. Also it is likely that many of the inlanders who ended up in the hands of labour recruiters arrived there through kidnapping by coastal middlemen.

In addition, much of the initial hostility shown to whites was the direct result of the latters' behaviour. For every Parkinson visiting these shores — for every white who viewed the indigenes with intellectual curiosity and treated them with some degree of fairness and humanity — there were many others who considered them to be subhuman and handled them fraudulently and brutally. Before some measure of colonial authority was established, the only constraint exercised by most traders and recruiters was their wish again to trade and recruit there some day. Here is Parkinson's description of the labour recruiters' part in this contact:

> The recruiters concentrated their efforts on the filling of their ships. From place to place they went, searching the coast up and down with their boats, and, whether or not, came into conflict with the natives who could not make themselves understood, and who knew from experience and hearsay the methods of recruiting labourers which they regarded as pure kidnapping.

No wonder, then, that the Bougainvillians of that era earned reputations for hostility against whites, *all* whites. As Parkinson recorded:

> Murders of white men were recorded every year, murders that were brought about by the victims' own fault, or, as was unfortunately the case, done to avenge the misdeeds of other recruiters. Every white person was regarded as an enemy, recruiter, trader or missionary; the crime of another has often caused the death of a perfectly harmless and peaceful man.

In 1902 the Catholic Society of Mary extended its missionary endeavours in the Solomon Islands by setting up a station on Bougainville's eastern coast, near Kieta. Then, in 1905 the German colonial administration at Rabaul established a post at Kieta, and at about the same time a few European planters and traders began to settle along the eastern and northern coasts of Bougainville and along Buka's western and southern coasts. Between 1899 (when these islands were officially annexed) and 1905, German political control — such as it was — was administered by means of occasional visits of officials from Rabaul.

3

The German Era

GERMAN MERCHANTS AND SHIPPING FIRMS BEGAN TO MOVE INTO the Pacific in the 1850s, intent upon building up a trade empire to equal or surpass Britain's. Their first South Seas base was established at Apia, Samoa, in 1856. Within a few years they had extended their trading activities, including shore-based stores, to the Marshalls, the Gilberts (Kiribati), the Ellices (Tuvalu), Tonga and Fiji. About 1870 their agents became the first traders to brave the frontier hardships of New Britain, thereby becoming the forerunners of German sovereignty there.

For a number of years after that the Germans' commercial operations were carried on without government backing; even the unification of Germany did not immediately change that, Bismarck having been initially opposed to colonialism. In time, however, German merchants and patriots had their way, and the government adopted a policy in favour of empire and world-girdling naval power, in deliberate competition with Britain.

The first product of this policy change in the South Seas was in the form of even stronger support for German firms in New Guinea. In 1884 this led to annexation of northeast New Guinea (Kaiser–Wilhelmsland) and the Bismarck Archipelago — which, as noted earlier, prompted Britain's annexation of Papua. For a number of years thereafter the German government left it to the *Neuguinea-Kompagnie*, the new colony's largest plantation and trading firm, to govern it. This arrangement worked more or less successfully (i.e., profitably) for a few years, mainly because of profits from tobacco-growing around New Guinea's Astrolabe Bay. However, the company's losses elsewhere in the colony, plus the cost of trying to govern, forced it to turn administration over to the government in 1899, the year in which the colony's boundaries were extended to include Bougainville and Buka. After that, occasional visits were made to the latter by German officials but a permanent government station was not established there (at Kieta) until 1905.

Official German policy valued the colony principally as a source of raw materials and as a strategic outpost in Germany's expanding commercial and political empire. Insofar as the area's indigenous peoples figured in these objectives, they were looked upon mainly as private producers of raw materials, as labourers in European enterprises, as consumers of European manufactures, and as accessories in the civilization and 'development' of the colony. In extenuation of this official policy it should be added that it was not indifferent to the indigenes' 'welfare'; it merely reflected the widespread European view of that era, that the best thing one could do for 'primitives' everywhere was to inculcate in them *waitman's* habits of work, thrift, civic orderliness, sexual morality, hygiene, religion, etc. Some whites in the colony dealt with the indigenes as if they were less than human and hence to be exploited like domesticated animals, but official policy was more positive and humane, especially under the aegis of Dr Albert Hahl, whose tenure of governorship lasted from 1896 to 1914.

From 1899 to 1914 the principal commercial activity on Bougainville and Buka was growing coconuts for export. By 1914 nearly 30,000 hectares of land on Bougainville–Buka had been alienated by whites, principally for coconut plantations. This represented only 3.3 per cent of the islands' total land area, but some 10 per cent of all the areas suitable for growing coconuts, and a much larger proportion of such land favourably situated for commercially feasible production.

The manual labour used in copra production was supplied mainly by the two islands' own indigenes. (Some efforts were made to employ Asians for such work, but this source proved in time to be too expensive and unreliable.) In addition, many of these islands' indigenes were employed to work on plantations elsewhere in the Pacific, especially in Samoa and, for a while, in the British Solomon Islands. Some plantations were able to obtain some or all of their labour from nearby villages, either on a contract or non-contract basis, but all 'overseas' labour, that is all individuals having to be transported by boat from home area to place of employment, had to be employed under contract and was subject to Administration supervision.

The usual term of contract was three years and the legal minimum wage was five marks a month plus keep. (At that time a mark was roughly equal to a shilling.) Most employers were officially licensed to punish their labourers physically, usually by flogging, for breaches of discipline; and when runaways were caught they were forcibly returned to work, if necessary by armed police. On the other hand the Administration attempted to see to it that such labourers were fed, housed and doctored well enough to keep them active and reasonably healthy; and their employers were required to repatriate them at the end of their contracts.

The Administration attempted to ensure that any individual entering into a labour contract did so voluntarily. However, the methods whereby unsophisticated indigenes were usually recruited—by inducements that never materialized, or in terms that they did not comprehend—rendered this measure meaningless. For many indigenes the first inkling of what a contract meant occurred only when they found themselves forcibly detained at work in places without native women and far from home.

Contemporary apologists for this indenture system asserted that forced disciplined labour of this kind was a civilizing influence, the best way to transform barbarous and inherently lazy natives into useful and presumably happier members of the wider colonial society. It was even held that such work was essential to save them from the mental and physical stagnation that allegedly resulted when the stimulus of intertribal warfare was denied them. (As we shall see, the white stereotype that 'natives' are 'lazy', like those concerning their sexual morals and religious beliefs, is both ancient and durable.) As for other effects of working and living on a white-owned plantation, it is questionable how much civilization rubbed off onto men who were herded together in barracks and work gangs, and consigned to such tasks as bush clearing and coconut splitting.

The long-term absence of a labourer also affected his home community. Most indigenous communities of Bougainville–Buka were so small and closely knit that the absence of several of their productive male members left them economically and socially upset; households were left without male food producers, and families without husbands and fathers. In some instances the proportion of absent males was so large and birth rates fell so sharply that the authorities attempted to limit recruiting, though as much from concern for future labour supply as for the welfare of the communities themselves.

Some plantations were able to draw much of their labour from neighbouring settlements, with or without contract. From the indigenes' point of view, this arrangement was far better; the labourers were able to earn some cash income without giving up their familiar satisfactions, and their communities were able to maintain a more normal way of life.

Although a large proportion of the colony's copra exports was produced by the indigenes in their own small groves, the German authorities were concerned mainly with the white-owned plantations. Very little was done to transform the indigenes into successful independent producers, or to stimulate other forms of indigenous economic enterprise. Instead the German authorities sought to civilize their charges by organizing them into an administrative hierarchy, and by requiring them to pay a head tax and to work without compensation on public projects.

As soon as any indigenous community was brought under 'control', i.e., as soon as it was made reasonably safe for whites to visit it to trade or recruit labour, one of its residents was appointed to the office of *luluai* (the word for chief in one of the languages of New Britain). The German official in charge usually made an effort to appoint the community's established leader, or at least one of its more respected elders, but often the office was given to the most ingratiating man. (In many instances a community's real leader chose not to occupy this office and put forward a nonentity instead.) The duties given to a *luluai* were varied: collection of the annual head tax, supervision of local public works projects, provision of quarters for white travellers and government recruiters, arbitration of minor local disputes, etc. Instead of a salary, such officials received 10 per cent of the tax money collected by them, and they were given badges of office in the form of a hat and a silver-headed stick.

To assist the *luluai* there was also appointed an interpreter (*tultul*), and a medical orderly (*doctorboy*). The former was a man with some fluency in Pidgin, usually an ex-plantation labourer. The latter's job was to dispense bandages and simple medicines and to enforce elementary sanitation measures.

Under German administration all physically fit indigenous males past childhood were required to contribute unpaid work on public projects for up to four weeks a year. Such projects included road-building and maintenance, and work on government plantations and stations. In addition, forced labour of this kind was used as a means of punishing fractious individuals and, more generally, as a device for inculcating indigenes with the civilized value of sustained physical work on behalf of 'public welfare'. The public welfare in question was, of course, chiefly that of the colonial authorities and businesses.

The system of taxation introduced by the German authorities was regarded by them more as a civilizing device than a source of revenue. When an area was first brought under administrative control it became subject to taxation, but on a graduated scale. In highly colonized areas, like the area around Rabaul, where the indigenes had many opportunities to earn money, the tax rate was ten marks a year. Elsewhere it was seven, or five, marks a year, according to the area's stage of commercial development. Where taxation applied, every physically fit male over about twelve years of age was assesed, except for those currently working under indenture or those who had worked for a white for at least ten months during the current tax year. For those taxable individuals unable to pay, the alternative was work on a public project for about fourteen extra days a year.

Whether or not this taxation helped to civilize the indigenes, it undoubtedly contributed to the economic progress of the colony, both by encouraging work on white-owned enterprises and in the production by indigenes of cash-earning crops. It is doubtful, however, that it served to educate those taxed in the virtuous necessity of taxation as a duty of responsible citizenship.

As for other measures of education, the German authorities left formal schooling, such as it was, almost entirely in mission hands.

What effects did these colony-wide policies and practices have on Bougainville–Buka?

By 1914, large numbers of Bougainvillians were employed on plantations, both locally and in the Bismarck Archipelago. They were generally

known as 'Buka boys', and easily identified as such by their darker skin
colour; they had become favourably known for their industriousness and
what seemed to be eagerness to learn new skills. But exactly how many
were so engaged and what proportion of them worked away from home
is not recorded. What is certain, however, is that these two islands were
only partially under the Administration's control. Buka, being smaller,
less mountainous, and nearer to Rabaul, was better explored and subjugated,
as was the northern end of Bougainville and the coastal areas immediately
north and south of Kieta. The rest of the larger island was uncontrolled,
including the populous and relatively accessible southern part of the island —
the Greater Buin Plain — where the usual state of intertribal warfare was
further complicated by the illegal recruiting of labourers for plantations
in the British Solomon Islands. At one point it was proposed to set up
a government station on the Buin coast to assist 'in pacifying the tribes,
who even at the present time have pitched battles, and render accessible
to [legal] recruiting this virile stamp of men. Occasional [official] tours
and punitive expeditions cannot create quietude in these regions.' (From
the official *Report on New Guinea*, quoted in Rowley 1958, p. 5)

The first whites to establish permanent residence on Bougainville–Buka
were Marist missionaries, who founded a station at Kieta in 1902. Three
years previously the mission had set up headquarters on Shortland Island,
and prior to the move to Kieta had begun to win Bougainvillian converts
in the persons of the many young people who became workers and trainees
at the Shortland station. (Then and previously, it will be recalled, there
was frequent contact between Shortland Islanders and the Buin-speaking
peoples of southeast Bougainville. Many of the latter lived on Shortland
in a stage of benign semi-bondage or of concubinage.) The German authorities encouraged the Marists to extend their influence on Bougainville–Buka
itself — including the acquisition of land — but more with a view to economic
development than to evangelization.

By 1906 relations between whites and coast-dwelling Bougainvillians
had reached a state of peaceable interaction. Here are some extracts from
an account by Parkinson, who, it will be recalled, was the German planter
based in Rabaul who made frequent visits to these islands trading and
recruiting labourers for New Britain plantations. This account, which I
translate freely, is in the form of a travelogue describing the coasts of the
larger island. Only those parts dealing with the inhabited strips of the
coasts are included here. It may be safely concluded that the remaining
coastal areas contained no indigenous settlements of any size, except those
along the southwest coast which Parkinson did not include in his circuit.
Travelling north from Bougainville's southeast point, Cape Friendship,
he first mentions indigenes in his description of the mouth of the Luluai
River, where he records the presence of:

> small canoes drawn up along the banks, which indicate the proximity
> of indigenes — a conclusion that is borne out by the sight of some

native gardens along both banks of the river . . .

At the mouth of the Luluai there are usually many indigenes to be seen, and although they are invariably armed with their bows and barbed arrows they are not as dangerous as they seem. They come to this spot to fish, their actual settlements being north of here in the Kaianu district. . . .

North of the Luluai the steep and deeply fissured foothills of the Crown Prince Range reach almost to the coast, and in Kaianu they terminate abruptly at the coast itself. The palm-shaded houses of the villages in this area are built on the hillside slopes, and far inland the sight of forest clearings and rising columns of smoke indicates the presence of native gardens. . . .

North of Kaianu is the district of Koromira whose coastal area is well populated. According to reports, inland Koromira is also well peopled, and by indigenes who are described by the coast dwellers as being so warlike that the latter must keep themselves continuously armed. (This is the characteristic way that coast dwellers describe their inland neighbours in this part of the world.) . . .

Proceeding north along the coast we come to the village of Toboroi whose inhabitants are a peaceful folk who came originally from Shortland Island. During the period when the great Shortland chief, Gorai, was extending his rule over much of south Bougainville, the Toboroi people constituted his northern-most outpost. . . .

Next comes the Toboroi Roman Catholic Mission which was established in 1902 (the first permanent European settlement on the island), and after that Kieta, where a police station has recently been set up by the German administration.

Further along, in Arawa Bay, one is treated to a sight which is becoming increasingly rare in the South Seas. From time to time the mountaineers who live inland from Arawa come down to the coast, either to fish or to view with wonder the sight of a European vessel. They come in throngs, of both sexes and all ages, mainly for mutual protection but perhaps also because it would be unsafe to leave anyone behind if the men alone were to come (since no village is able to trust its neighbours). They arrive completely naked, their black bodies painted red or white, and carrying their bows and arrows and spears. These wild-looking bands rush at the visitors with loud cries, but they turn out to be quite harmless. Their shouts and gesticulations are simply their way of expressing excitement and astonishment at the unfamiliar sight of whites. Everything we possess excites their amazement and wonder, whether it be a piece of coloured calico, a bright-hued bead, a mirror, a knife, an axe, a fishhook, or whatever. They readily exchange their finely wrought weapons for cheap trinkets, behaving like children who have just been given a long-wanted toy. In due course even the females overcome their initial shyness and crowd around us to clamour for their share of the

beautiful new things. Speaking of the women, while many of the
young girls are favoured with strong slender bodies and pleasant faces
and ivorywhite teeth, the older ones, with their wrinkled skin and
deeply furrowed faces, look like nothing so much as Blocksberg
witches. . . .

In recent years it has been possible to recruit some of these
mountaineer males to work on plantations in the Bismarck
Archipelago. After they have served out their indentures and returned
home they will probably, through their example, influence a large
number of their fellows to engage in works away from home. . . .

Continuing north along the coast some fourteen kilometres past
Cape Mabiri, one comes to the village of Bagovegove which is located
in what evidently is a very vulnerable position. When I first visited
this village in 1886 it had just been rebuilt after having been
destroyed by hostile mountain-dwellers. In 1889 it was again wiped
out by the latter, to be rebuilt in 1894, and again burnt to the
ground by the same enemies in 1895. However, since its last
reconstruction in 1898, it has survived unscathed, mainly because of
its reinforcement by immigrants from north Bougainville and east
Buka. On my visit to Bagovegove in 1902 I counted eighteen large
wár canoes and over fifty ordinary ones, which bore witness to a
large population and was confirmed by the sight of swarms of men,
women and children around the houses built along the beach. In 1889
I also spotted a small village, Sapiu, about one kilometre south of
Bagovegove, but this was subsequently destroyed by the mountain
people and evidently not rebuilt. . . .

Inland from Bagovegove and Sapiu and south of the latter is an
extensive area of swampy terrain, and the inlanders who live on the
higher ground around it are described by the coastal peoples as being
fierce and unrelenting cannibals, ever eager to capture victims either
by open attack or ambush. Now whether it is the inlanders who are
the real aggressors, or the coastal dwellers themselves, I am not able
to prove. I am however inclined to the belief that it is the latter who
are the original aggressors, in their eagerness for bloodletting and
booty, and that the actions of the inlanders are nothing other than
reprisals. . . .

The inhabitants of this strip of coast, known as the Numanuma
area, have on occasion been hostile to whites as well as to their inland
neighbours. In the 1870s, for example, the small trading steamer
Ripple was attacked here by the local people; its captain, a Mr
Ferguson, was murdered, along with several of his crew. Although
the handful of surviving crew members were badly wounded, they
managed to pull up anchor and escape. Revenge was not long in
coming. It happened that Captain Ferguson enjoyed the friendship of
the Shortland chief, Gorai, and the latter sent a fleet of warriors who

wiped out Numanuma and its inhabitants during a month long campaign. Since that time the indigenes hereabouts have remained more or less peaceful [toward Whites], but they continue to bear the reputation of being the most untrustworthy people of Bougainville.
. . .

Between Numanuma and Point Nehus [now the site of Inus plantation] are many small inlets and flat stretches of beach which are ideally suited to native settlement. Indeed, before 1888 there were numerous settlements just here, but now the only remains of them consist of their coconut palms. . . .

Just north of Point Nehus the mountains rise abruptly behind the narrow coastal plain and form the site for many native settlements, their well-built houses, laid out in rows, being clearly visible from the coast. Continuing north towards Cape l'Averdy, the coastal plain broadens and the foothills become less steep. In the waters off these shores one almost always sees canoes, engaged either in fishing or in trade expeditions to nearby settlements. . . .

Off Bougainville's northeast cape lies the inhabited island of Teop, but on the mainland opposite Teop one has to go a considerable distance inland before reaching any settlements, on account of the continual state of warfare between Teop and the inlanders. The latter are industrious gardeners; on occasion they bring large quantities of produce, mainly taro, down to the beach to trade. In addition they are also very warlike and are rarely to be seen unarmed. On the various occasions when I have visited them in them in their mountain villages, I have invariably found them to be friendly and hospitable, but such relations do not obtain between people of separate villages. In every settlement I have visited I saw spears and bows and arrows leaning against the huts, to be ready at hand in case of alarm. Should a [white] visitor enter a village unexpectedly he must not consider it to be a sign of hostility to him if he finds himself suddenly confronted with a crowd of men threatening him with their weapons; for, as soon as he is recognized, his hosts' hostility will immediately give way to joyous greetings. Future exploring expeditions to this region need fear no great difficulties from the local indigenes, provided that their leaders exercise tact and maintain calm. However, a high-handed attitude on the part of the visitors, or an unjust action or display of violence, will quickly have the effect of turning friendship into hostility, and will force the expedition to turn back.
. . .

The harbour at Cape l'Averdy could become an excellent base for opening up the nearby countryside, which contains large areas suitable for cultivation and which could be developed without damage to the interests of the indigenes. In fact, in my opinion there are many places on Bougainville where the local indigenes would welcome the establishment of plantations, in which their own labour would be involved. For not only would this kind of development contribute to more peaceful relations among the different tribes in the areas in question, but it would provide markets for the indigenes' own garden produce. . . .

Some four kilometres west of the harbour at Cape l'Averdy lies the small harbour and village of Tinputz. Then, for the next twelve kilometres or so up to Laua Harbour, the coast itself is uninhabited, the nearest settlements being a long way inland. Within this uninhabited stretch of coastland are many thousands of hectares of excellent agricultural land, the best in all northern Bougainville. Here also are several fine harbours, numerous year-round streams, good soil, regular rainfall — and no indigenous settlements to be disturbed by the establishment of plantations. Moreover, the area further inland and the nearby districts contain a sizeable population already accustomed to sending young men away to work. . . . West of Bantu Bay the coastline becomes high and steep, but here and there are to be seen shallow little bays bordered by sandy beaches which provide sites for a number of indigenous settlements. Here along this coast one frequently meets fleets of twenty and thirty canoes, each one containing twenty to thirty people, there being a lively trade between here and Buka. In addition to the beach villages found along this stretch of coast there are a number of others located along the top of the seaside cliffs. In fact, for a number of years this area has been a major source of plantation labour; the local people are less timid of whites than their fellow islanders elsewhere, and it is possible to communicate with them in Pidgin English. . . .

Continuing westward through Buka Passage and south of the island of Sohano we enter a very large bay, bordered on the east by the Sailo Peninsula and protected on the west by Taiof and a number of other smaller islands. Here we are visited by indigenes whom we have met before, on the other side of the peninsula. This time, however, they are travelling not in their large war- or sea-going craft, but in small outrigger canoes, or even on roughly constructed rafts. They have crossed the narrow peninsula to fish in these relatively calm waters, and evidently find the canoes and rafts better suited to this activity than their larger craft. The whole of the peninsula is given over to cultivation, mostly of taro and banana. . . .

Southward along Bougainville's west coast one passes the foothills of the mighty Emperor Range, and in some of the coastal valleys of the foothills are to be seen small garden clearings. The mountains hereabouts are said to be well populated, but the inhabitants are reputed to be hostile to all outsiders. . . .

Farther along one enters broad Empress Augusta Bay, which acquired an evil reputation during the era of uncontrolled labour recruiting for the plantations of Australia and Fiji. Time after time recruiting vessels were attacked by the local indigenes and all their people killed. Since that era, however, the situation here has greatly changed. The coastal villages, now largely depopulated through emigration, are threatened by the inland mountaineers. Scarcely a quarter of the inhabitants of this once thickly populated coastal area now remain, and some whole villages have entirely disappeared. And what used to be a dangerously unfriendly populace has become far less so; in my frequent visits to the surviving villages I invariably meet with a hospitable reception. These villages are regularly visited by traders from Shortland Island, and for the past few years no whites have been attacked.

Turning now to the island of Buka, it is quite densely populated, and for this reason alone does not provide much opportunity for the establishment of European plantations. The indigenes of Buka belong to the same race as those of Bougainville, and have for many years been offering their services as labourers. . . .

Except for labour, however, neither Bougainville nor Buka has much to offer in the way of local products; and such produce as there is is usually acquired as a sideline by recruiting vessels. With respect to local goods, southern Bougainville is more productive than the rest of these islands, but commerce in that area is largely in the hands of English traders based on Shortland Island, and is thus of little value to us Germans.

By 1914 additional Marist mission stations had been established at Patapatuai (Buin), Koromira, Torokina, and Burunotui (Buka); and the headquarters of the bishop had been transferred from Shortland to Kieta. Up to that point the Marists, mostly of French and German nationality, were the only missionaries at work on Bougainville–Buka, but their exclusive hold on the field was soon threatened from the Solomon Islands where Methodists were reaching out towards the north.

Yet another kind of *waitman* presence on Bougainville–Buka during the German era which must be mentioned was the handful of journalists, scientists, traders and recruiters whose visits were brief but whose influence may have been considerable.

This was the situation in 1914, when World War I convulsed Europe, and its effects were felt even in distant colonies. Shortly after the outbreak of war, Australian authorities rounded up a motley band of volunteers and shipped them to Rabaul. Their commander accepted the surrender of the handful of German residents, and announced to the bewildered crowd of onlooking indigenes: 'No more 'um Kaiser; God save 'um King'.

This declaration revealed just how unprepared they and their political leaders were to govern this huge, alien, and seemingly intractable addition to empire. From then until May 1921 the colony was administered by Australia under a military regime, but throughout this period most of the rules and practices established by the Germans were continued. All German civilians taking an oath of neutrality were permitted to return to their properties and regular pursuits. Such arrangements were not only in accord with the principles of international law of the time; they were also necessitated by the Australians' small numbers, their lack of tropical colonial experience, and their views on colonialism, which seem to have been almost identical with those of their predecessors. After a while whites were deprived of the personal right to punish their indigenous employees corporally, but in most other respects the German-established laws and practices regarding relations between whites and indigenes were maintained. During this period of military occupation the colony came to be viewed as a protective bastion for Australia, but its natural resources and native peoples continued to be treated mainly as colonial economic assets.

The Australian force at Rabaul was so small that three months passed before it was able to extend the occupation to Bougainville–Buka. On 9 December 1914 two companies of infantry and a machine-gun section landed at Kieta without opposition and accepted the surrender of the German district officer there. The German officials were imprisoned and returned to Rabaul; other German residents — missionaries, planters and merchants — were permitted to return to their regular pursuits after taking the oath of neutrality.

A small military garrison was stationed at Kieta, but its main efforts throughout the occupation period were aimed at punishing fractious indigenes. Little was accomplished in the way of extending the areas under administrative control. Plantations and trading stores continued to operate as before, including even the German-owned ones, whose managers were allowed to remain. By war's end, of the 30,000 hectares which had been alienated on Bougainville–Buka, including 1650 owned by the Marist mission, only one-third had been brought under cultivation.

The only other major change that occurred during this period was brought about by missionaries. In 1911 the Methodist mission reached out from its New Georgia headquarters and established a station on Treasury Island, just south of the Shortlands. It would be hard to decide whether the Marists then put more energy into converting islanders from heathenism or saving them from the threat of Protestantism. Some of the Marists were under the impression that the German authorities had granted their mission exclusive rights on Bougainville–Buka; this was not officially so, but until the Australian occupation the Marists had the field to themselves. From their base on Treasury Island the Methodists had made a brief sortie into the Siwai area of southern Bougainville in 1916, and then settled down to stay in 1920. Their entry there had been facilitated by the traditional relationship that existed between the Treasury Islanders and the Siwai; a similar factor had earlier provided the Catholic mission based on Shortland Island with entry into Buin.

4

Mandated Territory

A S WORLD WAR I DREW TO A CLOSE, AUSTRALIA WAS FACED with the question of what to do with German New Guinea, or rather, how to ensure the continuance of Australia's control there. It was officially recognized that the act of military occupation did not legally constitute the establishment of sovereignty, but it was widely assumed, and publicly demanded, that the colony would remain in Australian hands for both military and economic reasons. In Australia a few voices were raised against outright annexation, holding that it would be a betrayal of the Allied commitment to 'no territorial gains'. A few others spoke up for the principle of 'self-determination', implying that the colony's native populace should be consulted in the matter—an unrealistic proposal, to say the least! But when the Prime Minister, William Hughes, left for the peace conference in Paris, even most of the Opposition in the Australian parliament supported his wish to convert the de facto military control into outright sovereignty.

That development, however, was not to take place. In the face of the new anti-colonialist philosophy that prevailed at the peace conference, the best that Hughes could do was to have the former German New Guinea proclaimed a ward of the new League of Nations, under mandate to Australia. As mandatory power, the Australian government agreed to administer the territory according to several general principles, including the undertaking to 'promote to the utmost the material and moral well-being and social progress of the inhabitants' (Article 2 of the mandate agreement).

In order to obtain expert advice on how to carry out its mandate, the Australian government appointed a royal commission to investigate and make recommendations. As it turned out, the recommendations of the commission served to shape events in the territory (including Bougainville–Buka) for the ensuing twenty years.

41

The composition of the commission indicated at the outset what kinds
of recommendations the government wished to have made, since the
members' histories and attitudes concerning colonial matters were widely
known. The chairman was the top colonial official in the Territory of
Papua, the veteran Lieutenant-Governor, Hubert Murray, a champion of
indigenes' welfare. Lined up against him were W. H. Lucas, Island Manager
for Burns Philp (the largest Australian mercantile firm then operating in
the islands), and Atlee Hunt, Secretary of the Department of Home and
Territories.

Papua, it will be recalled, had been under Australian administration since
1884. Because of its discouraging terrain, scattered population, and what
then appeared to be meagre natural resources, Australian governments and
Australian citizens in general showed little interest in developing Papua
economically. Some mining and plantation enterprise became established
there, but under Murray's long governorship the Papua Administration
gave first consideration to the indigenes' welfare. Partly due to this policy
and partly to geography, white businesses did not prosper as well in Papua
as in neighbouring German New Guinea. In any case, Murray wished to
extend this policy to the new Mandated Territory, and in fact to join the
two territories into a united 'Papuasia'. (He also wished to become the
first governor of the new union, but probably more out of a desire to
extend the Papuan-style policy to it, than to advance himself personally.)
As components of this policy he recommended nationalization of the
German-owned properties (title to which had been awarded to Australia
by the peace treaty); the leasing rather than the sale of appropriated lands;
and direct access to European and Asian markets in place of transshipment
through Australian ports. These and similar measures were designed to
favour development of a healthy, educated and affluent indigenous citizenry,
even at the expense, if need be, of Australian enterprise.

For a variety of reasons the views of Lucas and Hunt, the majority,
prevailed. The new Mandated Territory remained separate from Papua,
and plans made for its administration and development were aimed principally at furthering Australian national security and welfare. Australian
attitudes differed, however, as to the best way to proceed. Some large
firms active or interested in the Territory favoured the continuation and
expansion of large-scale businesses there; other persons, including members
of the Opposition Labor Party, spoke up for dense colonization by
independent white settlers. (For political reasons both advocated that
Australian war veterans be given preference for jobs and land in the Territory.) Throughout these debates and the actions that followed them, little
or no attention was devoted to the cause of the indigenes themselves: either
the matter was considered unimportant, or it was assumed that what was
good for Australia was good for its colonial wards as well. As for the
possibility that the indigenes of the new Territory would be treated

humanely, it was asserted that the officials, being Australian, would be incapable of acting otherwise.

Thus the principal and almost only concern of the new civil administration in the Mandated Territory (which assumed control in May 1921) was with Australian economic affairs. To begin with, all German-owned non-mission properties were expropriated and sold outright to Australians; by the terms of the peace treaty it was left to the German government to compensate its citizens for their losses. Most German missionaries were permitted to remain at their posts, but most other Germans were repatriated. Many of the officials in the military administration doffed their uniforms and joined the civil administration. In many other respects, administrative policies and practices and Territory life in general went on as before. New areas were opened up by systematic exploration and patrolling, so that by 1941 all but about 78,000 square kilometres, or roughly one-third of the total land area of the Territory, had been brought under some measure of administrative influence or control. Established missions continued to widen their nets, and new mission bodies entered the field. Economically, coconuts continued for a decade to dominate, with the result that the whole Territory's money economy, including the Administration's revenue and expenditure, was strongly affected by rises and falls in the price of copra.

Later on, following the discovery of new ore beds around Wau, Bulolo and Edie Creek, gold mining became the principal revenue-producing industry of the Territory. Hundreds of whites and thousands of indigenes were employed in the new mining towns, which could be reached from the coast only on foot or by plane. By the end of the 1930s the major ore bodies had been nearly worked out, and since extensive exploration had turned up no other significant deposits of gold or any other minerals (including oil), it appeared that the Territory's future would, as in the past, depend upon agriculture. This meant coconuts; other crops had been experimented with but abandoned, or proceeded with on a very small scale.

By 1941 there were about 4600 whites in the Territory. About 10 per cent of these consisted of Administration personnel (including their dependants); another 15 per cent were associated with Christian missions; and nearly all the remainder were involved with plantations, mines, shipping and merchandising. The 2000 or so Asians in the Territory, mainly Chinese, were engaged largely in commerce. As for the indigenes, about 800,000 had been actually counted or their number estimated by 1941, with some parts of the mainland still largely unexplored. Meanwhile, the role of the indigenes in the Territory's developing economy had changed very little since the German era.

Although criticism had begun to be voiced both in New Guinea and Australia regarding the labour indenture system, it continued much as before. Many critics pointed to the anachronism in the twentieth century of a form of employment that was coercive, requiring government-enforced sanctions to keep labourers at their jobs throughout the term of their contracts. Others criticized the low wages paid to the labourers, and the diet and living conditions found in plantation and mine labour-camps. Still others pointed out the harmful effects produced in the labourers' own communities by their long absences from home. No one, however, was able to propose a workable alternative, given the nature of the Territory's economy and the requirement that the Administration pay for itself.

The Territory's agricultural economy — consisting mainly of copra production by crude technological means — depended upon large and constant amounts of cheap manual labour. Moreover, since most plantations were located far from the areas where most of the prospective labourers lived, and since the latter possessed neither the knowledge nor the means to travel to distant places of work, the cost of recruiting and transport had to be borne entirely by the employers. Under these circumstances it appeared at the time that few plantations could have maintained production and remained solvent without some form of enforceable labour contract. In the Territory as it then was, physical coercion was thought to be the only means of achieving that.

Critics of labourers' diets and living conditions were also rebutted with the item of cost, and with the observation that the living conditions were no more primitive than the ones the labourers were accustomed to at home. Employers were also able to point to the improvement in health experienced by many labourers under plantation regimes. On the other hand, no one could deny nor offer cures for the unhealthy social conditions that obtained in the compounds of labourers cut off from women and from most other interests and activities of their usual ways of life. Homosexuality and gambling were common, as were drunkenness and brawling; but worse than these in long-term effect were the kinds of attitudes such a life fostered in both employers and employees, between whites and indigenes. At its best, the 'master–boy' relationship was tinged with feelings of paternalism and dependency, neither conducive to partnership. And at its worst, it was characterized by contempt, fear, envy and hatred. As for the widespread argument that employment on *waitmans* plantations served to inculcate valuable attitudes towards work and to introduce indigenes to other aspects of civilization as well: the jobs most of them performed would have produced only profound boredom and distaste, and the little bit of civilization they saw, from a distance, was not always edifying, and almost entirely inaccessible. Moreover, except for the blankets, knives, lanterns, etc., and their scant wages (most of which were used eventually to pay head taxes for themselves and their relatives), the things earned or learned on plantations were of little use in their lives at home.

At the height of large-scale gold dredging, many thousands of indigenes were employed at the Wau-Bulolo mines. Although the work there was more diversified and the white communities larger, the lives of the indigenous labourers, and the things earned and learned, were not very different from those of the plantation workers.

As for the effects of the indenture system upon the labourers' home communities, some areas became so denuded of able-bodied males that regulations were imposed to restrict recruitment from them. The absence of many men for long periods of time did not perhaps lead to any long-term population decline as some officials feared, but in some instances it did disrupt family and village life to a demoralizing degree.

However, for all its adverse effects upon the Territory's indigenes, the indenture system cannot be judged solely from that point of view. While it did protect employers from labour desertions, it also protected the labourers from periodic lay-offs and arbitrary dismissal. While it took many labourers far from home, it also ensured their eventual return at the employer's expense. Government taxation did undoubtedly force many otherwise unwilling indigenes to become labourers — the only means most of them had for earning tax money. But perhaps just as many sought employment, voluntarily and eagerly, for reasons of their own: to experience a new kind of life, to earn money for buying enticing new things, and to escape troubles at home.

During the mandate, the Administration imposed a tax on most adult male indigenes residing in 'taxable' areas, i.e. in areas that were firmly controlled and deemed economically capable of supporting the tax. In 1938–9, for example, 42,000 indigenes in the Territory paid the tax, out of an enumerated adult male population of 241,600. Several categories of persons were exempted, however: village officials; those individuals currently under indenture; mission teachers and students; fathers of four or more living children by one wife; and those physically unfit (which generally included all men judged to be over forty years old). In retrospect the tax seems rather high, since it represented about 10 per cent of the average indentured labourer's annual cash wages; but with all its exemptions it was levied on less than 20 per cent of the total enumerated adult male indigenous population of the Territory. Moreover, although it undoubtedly served to impel many indigenes into the ranks of indentured labour, and thus to support white enterprise, its official purposes were more widely developmental and educational: it was viewed as a device for encouraging indigenes to take up cash-crop production and to adopt responsible attitudes towards citizenship. In fact, while the measure may have had a small but positive measure of success in fulfilling the former objective, it failed altogether in relation to the latter.

Throughout the period of the mandate the civil authorities maintained almost unchanged the system of local government established by the Germans. As soon as an area was brought under control a chief (luluai, kukerai), interpreter (tultul), and medical orderly (doctorboy) were appointed in each of its indigenous communities. In some places an effort was made to encourage larger administrative units by placing several communities and their local officials under the supervision of an appointed paramount (number one) chief (nambawan luluai), but usually such measures proved effective only in places where they served to consolidate traditional tribal groupings. In addition, even larger multi-tribal councils were encouraged in more

Europeanized Rabaul and Morobe, but otherwise the Territory's indigenous population remained as administratively atomized and politically voiceless as before.

Rabaul continued to be the Territory's capital throughout the mandate era, even after gold mining and the opening up of vast new areas to control had served to raise the mainland to greater commercial and administrative importance. At the top of the governmental hierarchy was an administrator and three departments — secretariat, public health and native affairs. In practice, because of great distances and infrequent communications, most governing was left to officials in each of the Territory's seven districts (of which Bougainville, Buka, Nissan, Nukumanu, Tauu, Kilinailau, and Nuguria constituted one).

The Christian missions supplied their indigenous members with some medical aid, but most health services, such as they were, were provided by the Administration. Each district had its government hospital, but these were able to serve only a very small proportion of the indigenes requiring medical care. In addition, European medical officers made periodic, usually annual, tours of outlying areas, but were able to attend to only the most obvious, and current afflictions. As for all the other maladies which shortened and made painful the lives of the Territory's indigenes, few preventive or therapeutic measures could be undertaken by an Administration having so few medical or public health resources at its command.

An even drearier picture is presented by the Administration's education programme, if it can be so dignified. There were only six government-operated schools for indigenes (four on New Britain, one on New Ireland, and one on the mainland), comprising in 1940–1 a total of 466 pupils of an estimated total population of 800,000. In fact, the Administration, with its limited resources, seemed more than content to leave schooling in the hands of the missions — an arrangement with which the latter evidently agreed. Some 70,000 pupils were enrolled in mission schools during 1940–1, but the impressiveness of this number must be deflated somewhat, since most of these pupils were in sub-primary village schools, where instruction was rudimentary and casual, to say the least.

Before turning more specifically to Bougainville–Buka, some explanation should be offered as to why so little headway was made by Australia in carrying out its mandate to 'promote to the utmost the material and moral well-being and social progress' of the Territory's indigenes.

Anyone familiar with the Territory during the mandate era could not blame the Administration itself overmuch for the failure to live up to the mandate. Administration forces undoubtedly included many individuals who were 'coon-bashers' through ignorance or calculation; but there were many more who viewed the indigenes with liking and sympathy, and worked faithfully, even heroically, on their behalf. Nor should one blame the white settlers overmuch for the plight of the indigenes. Coon-bashers were probably just as numerous among the ranks of plantation managers, recruiters and traders, but among these were also numerous individuals of fairness and goodwill. Moreover, the latter had their own livelihoods to earn, under conditions wherein low labour costs were essential for economic survival—or so the prevailing theory held.

In retrospect one must hold the policy-makers in Melbourne and Canberra, and ultimately the Australian voters as a whole, responsible for the shortcomings in indigenous welfare and progress. It was they who set the policy of requiring the Territory to pay for itself on a year-to-year basis. Such a policy could have yielded more immediate resources for education, etc., by granting subsidies as investments; but this was not done to any significant degree.

The inevitable consequence of these official policies is indicated by the fact that the Administration's annual expenditure—which derived almost entirely from imports and other forms of local taxation—remained between £450,000 and £500,000. Of this only about one-third (by the most liberal estimate) was spent on matters described as 'essentially native'; in other words, some £150,000 to £170,000 annually among an enumerated population of over 800,000! No wonder that the harassed Administration was pleased to accept the Christian missions' assistance in matters of health, and mission substitution in matters of education.

By the operation of this requirement that the Territory pay for itself, the indigenes were brought to the boundaries of civilization, white style, but were fated not to cross.

In 1941 there were about 175 non-indigenous (white and Chinese) adults residing on Bougainville and Buka. Of these, about half were associated with Christian missions, and the remainder with plantations, retail stores, a small gold-mine at Kupei, and the Administration. The indigenous population of that period is more difficult to characterize and quantify.

According to the census of 1941 there were about 45,000 indigenes residing in native settlements on Bougainville–Buka, and another 3000 residing in European enclaves, mainly plantations. Virtually all of the former were locally born; of the latter, most were locally born, and the remainder had come from other parts of the Territory. Finally, the statistics for that period record that some 850 locally born Bougainvillians were working and residing elsewhere in the Territory.

By themselves these figures provide little sociological information; they need to be supplemented by facts of another kind. Firstly, *not one* of the indigenes occupied a position of authority or social status equal to that of *any* of the whites with whom they associated in administrative, commercial or religious organizations. And secondly, although many had been conditioned to wish for it, not one indigene was able to match any of the whites in the attainment of a Western-style standard of material well-being. In other words, after many decades of contact with whites, including two decades of officially pledged noblesse oblige, the native proprietors of these islands had been obliged (or at least encouraged) to give up many of the

satisfactions of their old ways of life without being able to taste the fruits
of the new. The limbo in which they were stranded contained a poignant
juxtaposition of old and new.

By 1941 every part of Bougainville–Buka had been visited by Administration officials and declared to be 'under control'. Most indigenous settlements of any size had been located and their populations counted or estimated.
High up on the slopes of the northern Emperor Range lived people to whom
the sight of a white was a rare event, but even these had learned to acknowledge the latters' superior power and to want some of their goods.

Kieta remained the administrative headquarters for the whole district, with
sub-district headquarters at Sohano (Buka Passage) and Kangu (Buin coast).
Among all these centres there were half a dozen officials, along with several
'police boys' (indigenous members of the Territorial police force). They issued
rules, judged all but the most serious of civil and criminal cases, punished
malefactors, supervised labour contracts, and carried out periodic tours to
record census changes and collect taxes. In addition to a body of criminal
law, which applied to indigenes and all other residents as well, there were
numerous native regulations concerned with everything from sorcery and
adultery to path maintenance and village hygiene. The system operated in
such a way as to place immense powers, along with overburdening responsibilities, upon a handful of whites of widely varying capabilities and
temperaments. Consider the case of southern Bougainville.

In 1938 the Buin sub-district contained an indigenous population of 16,500
indigenes, scattered over about 3100 square kilometres of difficult terrain
accessible only by footpath or bicycle trail. At that time another 1200 to 1300
residents of the sub-district were working and living elsewhere, mainly under
indenture. Four different languages were spoken in the sub-district, none
of them ever having been systematically recorded in published grammars or
dictionaries. Though alike in some respects, many customs of the sub-
district's populace varied even more widely than their languages; and only
a few years previously the whole region had been riven by local feuding and
warfare. Such was the realm which *one* official, usually of very junior rank,
was required to govern: to count its people, collect taxes from them, introduce them to an alien set of rules, police them, adjudge them, and punish
them. In addition, this official was frequently required by circumstances to
arbitrate conflicts involving indigenous customs about which he knew next
to nothing and was insufficiently trained to investigate. And all of this had
to be done by means of a lingua franca, Pidgin, which was quite incapable
of conveying many nuances of meaning from native vernacular to English
or vice versa. No wonder that many Administration officials regarded their
indigenous wards as contrary and stupid, or that the latter came to view the
kiap (captain) as arbitrary and inscrutable.

One of the more arbitrary and incomprehensible measures undertaken by
kiaps on Bougainville–Buka was to create large nucleated villages where they
had not previously existed. The most characteristic settlement pattern in pre-

European times consisted of small hamlets of one to four households; each was located near its members' groves and gardens, and separated by some hundreds of metres from other such settlements. In some coastal areas somewhat larger and denser settlements were found, but the bulk of the pre-European population of Bougainville–Buka was widely and thinly scattered. In order to facilitate administration, the Australian officials brought together the people of neighbouring hamlets and required them to build 'line-villages', consisting of straight rows of narrowly spaced houses surrounded by a village fence made strong enough, hopefully, to exclude pigs (and thus help keep the villages 'clean'). Whenever possible these line-villages were designed to incorporate hamlets of people sharing close kinship ties, and so many of them were and have since remained. In other instances, however, the officials juxtaposed kin units and individuals who had no interest in being together, with the result that such people occupied their line-villages only during the infrequent visits of the *kiap* and lived the rest of the time in their hamlets.

As in the rest of the controlled parts of the Territory, the Administration maintained on Bougainville–Buka the system of organization introduced by the Germans, wherein each village was placed under the supervision of a *kukerai*, a *tultul*, and a *doctorboy*. All three of these village officials were appointed by the area's Australian Administration officer, or *kiap*, although an effort was usually made by the latter to conform to popular choice in the naming of the *kukerai*. In some places one local resident was so prominent that the selection offered no difficulty. In other places rival factions made the choice more difficult and it had to be resolved by a vote—i.e., an alien and not always satisfactory way of resolving differences. In still other places, where the whole alien institution of *gavman* (government) was viewed with suspicion or outright hostility, the locally recognized leaders caused some henchman to be appointed *kukerai*, and then proceeded to rule as before, unencumbered by the obligations and conspicuousness attached to official appointees.

In most parts of Bougainville–Buka the villages were grouped by the Administration into large units, each under the supervision of an appointed *nambawan luluai*, and his interpreter-executive officer (*bossboy*). These two officials served mainly as links between village officials and the Administration. In some instances they exercised considerable authority, with or without Administration approval; in others their influence reached no further than the boundaries of their own village. All of the officials were exempt from paying head tax, and in addition a *nambawan* received an annual salary of sixty shillings. All of them were presented with peaked caps to wear when carrying out their official duties, hence their label of *hat-men*.

Village officials were responsible for maintaining law and order in their communities; for keeping a record of births, deaths, arrivals and departures; for constructing and maintaining the portion of the Administration road that passed through their areas; for supporting European travellers with porters along assigned stretches of road; for sending the sick and wounded to hospitals; and for carrying out any other duties that might be imposed by their *nambawan*.

'Maintaining law and order' was a broad mandate indeed. Specifically the village *kukerai* was charged to 'arrest natives belonging to their tribes or villages whom they suspect to be guilty of wrongdoing or an offence', and to 'bring them to the nearest court in the district, or before the district court, to be dealt with according to law'.

The courts in question were those presided over by Administration officials. Theoretically the indigenous officials were not permitted to try cases or execute judgements; actually many of them did, with or without the tacit consent of white officials. The actions that constituted offences or wrongdoing were numerous and comprehensive, including, for example, gambling, sorcery and threats of sorcery, use of intoxicating liquor, 'careless use of fire', unsanitary practices, burying bodies too near to dwellings, behaving in a riotous manner, using obscene language, spreading 'false reports tending to give rise to trouble or ill feeling amongst the people or between individuals', and the wearing (by males) of clothes over the upper part of the body. In other words an undigested mixture of Australian laws, public hygiene guidelines, and Victorian pruderies.

Not surprisingly, the opportunities for abuse of authority under such a regime were numerous. The Administration of course regulated against this situation as well, as in the following:

> Any luluai, kukerai, tul-tul, patrol medical tultul (doctorboy) or other native upon whom the Government has conferred authority, who uses such authority for the purpose of blackmail or wrongfully to get any property or benefit for himself or any other person or wrongfully to injure any other person shall be guilty of an offence.
>
> Penalty: Three pounds or six months imprisonment, or both.

In the long run, however, local indigenous sanctions were probably more effective than Administration regulations in checking such abuses, provided that the victims knew enough about the regulations to realize that the *hat-men* were in fact exceeding their authority.

Portering was not required of Bougainvillians very frequently—perhaps three or four times a year in most areas—but it was nevertheless unpopular. Even the relatively high payment received, a twist of tobacco for about two to three hours' work, did not serve to make such work any less distasteful.

Even more unpopular was the construction work required periodically of most men. This consisted mainly of building and maintaining roads, bridges and rest houses (for whites). In the Buin Plains region of southern Bougainville where the writer observed it most closely, this activity required a large amount of the indigenes' time. The age-old trails there which served the purpose of foot travel were considered by the Administration to be too narrow and circuitous. To facilitate patrolling, the Administration obliged the indigenes to construct straight trails some 2.75 metres wide and suitable wherever possible for bicycle travel. Bridges were ordered to be constructed over the smaller streams, and these were required to be roofed over to prolong the life of the spans. At intervals along the road network, rest houses were built to house Administration officials and other white travellers.

The original work for these projects involved weeks of labour, but maintenance would have required only a few hours' work a week of every adult male had it not been for the ambitions of some indigenous officials. In order to curry favour with the Administration, some officials caused their stretches of road to be widened, graded and smoothed—all by hand and machete— far beyond the standards set by the Administration. Bridge timbers were shaped and planed as flat as a floor—first-rate carpentry but hardly necessary for their infrequent use. The climax to such endeavours was reached with the rest houses. A small one-or two-room affair would have served well enough, but in some villages the indigenous officials ordered enormous barn-like structures to be built, up to twenty-seven metres long and suitable for permanent residence of a whole troop. In such instances the competition which characterized indigenous politics was carried into the new activities, and many *kukerai* endeavoured to build bigger and better than their rivals elsewhere.

A head tax of ten shillings a year would appear at first glance to be burdensome for men with so little opportunity to earn Australian currency. Actually the exemptions were so numerous that the burden was spread quite thinly; of the whole adult male population of Bougainville–Buka, only about one in three had to pay. Nevertheless this did not make taxation any less unpopular. Administration officials were at pains to explain the impost in terms of the theory of 'taxation for responsible citizenship', but it is doubtful that this logic was widely understood, much less accepted. Even those few indigenes who comprehended the general theory seemed to feel that it hardly applied to them; after all, they did not elect the Administration officials nor participate in making the rules. Also, many Bougainvillians were heard to express resentment that whites with their great treasures of money should take away money from indigenes who had so little. Finally, although it may not have been officially intended that taxation would force Bougainvillians to work, and thereby serve white commercial enterprise, it often had that effect.

Alongside the governmental structure imposed by the Administration, there continued to exist numerous indigenous ones; some had lines of authority that corresponded to the local administrative hierarchy and others had an entirely separate leadership. In places where the two systems corresponded, where the indigenous-type leader was himself the *kukerai*, or where the *kukerai* was the true leader's henchman, village affairs went smoothly. But where the true leader and the *kukerai* were rivals, which was quite often the case, the village was continually troubled with factionalism.

Before the Europeans arrived and for some time thereafter, every coastal village and every neighbourhood grouping of hamlets constituted an autonomous political unit, a 'tribe'. There were hundreds of such tribes, and they varied widely in cohesiveness, depending upon the strength of their leadership. Also, there were differences in the routes to leadership. In some tribes the leader inherited his position. This tended to be the mode in tribes dominated, numerically and economically, by a single matrilineal kin group. In other tribes the leader achieved his position by attracting followers through personal attributes of generosity and forcefulness. And finally, some leaders of both types succeeded from time to time in extending their authority over neighbourhoods beyond their own; this was usually accomplished by the exercise of military leadership which was occasionally, but not necessarily, linked with personal military prowess.

After the consolidation of Administration control, indigenous intertribal warfare was brought to a halt, but many of the old tribal groupings persisted. They were deprived of their military aspects but continued to function in other respects, for example as land-clearing and feast-giving teams, and as the unit within which wrongs were righted by informal means. In some cases, such tribal groupings were maintained, and even strengthened, in the form of the new line-villages. In other cases they were subdivided and effectively destroyed by the new line arrangement. In still others they persisted, despite having had their membership consigned to different 'lines'. This persistence of old tribal groupings in the face of the abolition of warfare and of Administration regrouping was mainly the result of the remarkable durability of one particular type of indigenous leadership: the type personified by men who achieved rather than inherited their positions of influence and authority.

This institution merits a closer look. Variations of it were found throughout Bougainville–Buka during the mandate era, the one best known to the author having been practised in the Siwai region of southwest Bougainville:

> Throughout the [Siwai] area, 'renown' was a very concrete concept; it meant, literally, the kind of esteem enjoyed by the man who gives feasts frequently. The concept was most elaborately institutionalized in the northeast quarter of the Motuna-speaking region, which, significantly, was not the richest part of the Plain in material surpluses. Here, at the centre of this complex of ideas and practices was the *mumi*, the man of the neighbourhood who had most renown.
>
> To become a *mumi* a man had to own or control material resources. He could either accumulate these by his own hard labour, plus the labour of members of his household, or he could persuade kinsmen and friends to make him gifts or extend him loans. The usual method of accumulating wealth was by cultivating large gardens and converting the surplus produce into pigs, which could then be either distributed at feasts or sold for shell money. Another method was by making and selling pottery; another by practising magical skills for fees.
>
> Accumulation alone did not however bring renown; wealth had to be distributed in the form of food and other valuables, usually at feasts. It

Kieta, 1915. Government office left, official residence centre, police office and quarters right. *(By courtesy Australian War Memorial from Gash/Whitaker* A Pictorial History of Papua New Guinea*)*

Buka village scene, 1890s. *(By courtesy Mitchell Library from* Sudseebilder, *K. J. Schaffrath, Berlin 1909)*

The Rev. J. F. Goldie. Original caption reads 'of such men did Goldie make Christian gentlemen'. *(By courtesy Mitchell Library from* Isles of Solomon, *Rev. Clarence Luxton, 1955)*

Original Catholic presbytery and chapel at Tunuru. *(By courtesy Mitchell Library from* Blazing the Trail in the Solomons, *Rev. Emmett McHardy S.M., 1935)*

Expatriate-owned plantation, Numa Numa.

The author in 1938 in front of a men's clubhouse.

Line village (1938) as 'encouraged' under both German and Australian administrations.

Appointed village officials were called 'hat men' (1938).

was useful to have had a *mumi* father, to have begun life with some of
the latter's reflected renown, together with what was left of his wealth
after most of it has been given away in the form of mortuary
distributions. Such an initial advantage prompted many informants to
assert that 'Only the son of a *mumi* can become a mumi.' Actually, a
mumi's son was only slightly better off than an orphan. He had to
increase his inheritance many times over, or have very liberal backers,
before he could begin serious feast-giving.

It is little wonder that the feast-giver was highly esteemed. Feast
food—roasted and steamed pork, boiled eel and opossum, tasty vegetable
and nut puddings—provided a welcome break in the everyday monotony
of a vegetarian diet. Also, natives keenly enjoyed the excitement of
milling crowds and the pleasure of dancing and pan-piping. However,
the ambitious man did not merely invite a few of his neighbours and
treat them to a banquet: that would have been a waste of resources. He
usually made the feast the occasion for a house-raising or some other
kind of work-bee; no one esteemed him any less for that.

It was customary to begin a social-climbing, renown-seeking career by
building a club-house. The club-house is a rectangular, shed-like
structure built directly on the ground and without walls. Large wooden
slit-gongs occupy most of the floor space. These are sounded to convene
workers, to announce feasts, etc., and serve as benches for the men who
gather in the club-house to gossip and sleep. Here among these Siwai
only kinfolk visited one another's hamlet dwellings, so that the club-
house was virtually the only public gathering place for men. Hence, the
man who wished to become a *mumi* had to own a club-house; and to
derive most renown for his ownership of one he had to build it rather
than inherit it. The most renown-bringing manner of building a club-
house was to collect all the helpers available, draw out the work as long
as possible, and reward the workers with such a bountiful feast that they
would ever afterward remember the occasion and the club-house with
pleasure.

There were many club-houses in the Siwai region in the thirties—an
average of one to about eight adult males—so that mere possession of
one did not ensure for the owner lasting renown. To serve as a lasting
symbol of the owner's renown, the club-house had to be the scene of
almost continuous activity; and since nothing drew or pleased a crowd
like a feast, the ambitious owner gave as many feasts as he and his
followers could afford. He had men cut down trees, fashion them into
slit-gongs, and install them in the club-house; then he rewarded them
with a pork feast. When no more gongs could be crowded into the
club-house, he caused the roof to be repaired or the floor swept, and
provided food delicacies for each occasion. After a while people would
say of him: 'He is a true *mumi*: he gives large feasts.' And when they
strolled about in search of amusement they would usually end up in his

club-house. They were at pains to ingratiate themselves with the feast-giver, to defer to his judgement, to perform little services for him, laugh at his sallies of wit, praise whom he praised or scorn whom he scorned. In this manner the *mumi* assured himself of a following in his own neighbourhood and even extended his renown and influence beyond.

Some *mumis* stopped there; others were led by their ambition to seek wider acclaim. Their lives turned into continual rivalries for renown. So long as they were active feast-givers they could count on exercising authority over their immediate neighbours and kinsmen, who lined up behind them with patriotic pride in 'our *mumi*' and 'our place'. Those rivalries between neighbouring *mumis* culminated in competitive '*mumi*-honouring' feasts at which the host presented his rival with large quantities of pigs and shell money. The guest of honour then distributed these goods among his own supporters, as rewards for their support, and then set about to accumulate an equivalent or more than equivalent reciprocal gift. If he could not reciprocate within a year or two he forfeited much public esteem and was no longer considered a worthy rival by other *mumis*. If he returned an equivalent amount of goods and no more, that was a sign that he wished to cease competing with the initiating host; thereafter the two usually became 'trade-partners' and assisted one another in competing with other *mumis*. If, however, the debtor-guest reciprocated more than he received, the rivalry continued until one of the principals gave up in defeat. A supernatural sanction supplemented the social sanctions connected with these exchanges: either the club-house demon-familiar or an ancestral spirit of the creditor *mumi* accompanied the 'gift' and tore out the soul of the debtor *mumi* if he did not repay in good time. (In my experience, however, this supernatural sanction provided less motive power than the social ones.)

A successful *mumi* was fortunate in a number of respects. He was singled out among his fellows by means of special beliefs about his personality and destiny—for example, the *mumi* had a better chance than most other mortals of attaining paradise. Also, he had the satisfaction—highly valued among these people—of hearing himself frequently praised. Others showed great respect for his name and person and opinions, and in his own community he exercised considerable authority even in matters not directly concerned with feastgiving. Next to his material resources his most powerful weapon was his ability to focus praise or scorn upon friends or enemies.

There were also certain more material advantages in being a *mumi*. At other men's pork distributions a *mumi* usually received the best cuts. He had little difficulty in raising loans. And he had the tacit right to utilize land belonging to all those persons to whom he regularly distributed pork or other valuables. The term *tuhia* was applied to these persons and, derivatively, to the lands so utilized by them; it was in this connection that natives describe certain *mumis* as having had extensive *tuhias*, which some whites had interpreted as 'kingdoms'.

There were also drawbacks in being a *mumi*. Such a man had to be scrupulous in his everyday conduct and in his commercial or ceremonial transactions. Moreover, he was dangerously situated, in being the potential victim of sorcery aimed at him by envious rivals.

Informants stated that in the days before the German and Australian control *mumis* were primarily war-leaders, their renown having been mainly dependent upon their success in organizing and financing victorious head-hunting forays or pitched battles with rival *mumis*. In those days, informants asserted, a *mumi's* authority was backed by physical force. How closely such assertions corresponded to past actual events it is difficult to judge, but even according to these suppositions, feast-giving was the principle kind of reward given by a *mumi* to his warriors, and feast-giving was the most important factor during the early stages of a man's rise to affluence. Feast-giving rather than personal bravery or martial skill attracted followers who would then fight one's battles.

By 1941 German and Australian control had succeeded in ending inter-tribal warfare, thereby removing much of the basis for the traditional form of tribalism. The Germans and Australians had also imposed a new form of territorial grouping and leadership, which conformed only in part to traditional tribal boundaries and forms of leadership. Yet despite all the coercive authority behind the Administration's *kukerai* system, it was unable in many places to supersede the traditional form.

By 1941 Australians had also attempted to create larger indigenous groupings, but these units, under so-called 'paramount chiefs' or *nambawans*, were in fact more ceremonial than administrative, and seem not to have fostered any sense of wider political boundaries. After several years or decades of contact with whites, the indigenes exercised less control over their own lives than ever before. Despite having become members of a vastly larger and purportedly more democratic 'tribe', they had little or no voice in their own governing.

Nor did they acquire any sizeable stake in the new capitalistic market economy of their islands during the mandate era. By 1937 whites had alienated 28,000 hectares of Bougainvillians' land, nearly all of it of prime agricultural quality and accessible to shipping points. Of this, over 10,500 hectares were planted in coconuts and most of the remainder was designated for coconut planting. Figures on total copra production on Bougainville–Buka are not known, since they were not segregated from Territory totals in published reports. However, it is safe to say that most of the copra exports were produced on white-owned plantations.

Some indigenes in coastal areas sold self-grown and self-processed copra to traders, but this constituted a small percentage of the two islands' exports. Nor did the Administration attempt much in the way of encouraging indigenes in cash-cropping. One or two officials, on their own initiative, required the men in their sub-districts to plant coconuts when their wives gave birth as a means of ensuring 'family welfare', but this was a sporadic enterprise and was not followed up with assistance in processing and marketing.

Locally made handicrafts included the now famous 'Buka' baskets (which were in fact made in southern Bougainville). Such handicrafts provided a few indigenes with some cash, but the volume was small and the prices low (for example a large bowl-shaped basket fetched only a few shillings). Up to 1941 no official effort had been made to encourage this potentially profitable form of indigenous enterprise.

And finally, virtually all retailing remained in white or Chinese hands. Now and then a hopeful Bougainvillian invested his wage savings in a small stock of goods for resale to his neighbours, but such enterprises were usually very short-lived, and received no technical or financial assistance from the Administration. In other words, during the mandate era about the only part played by Bougainvillians in their islands' developing market economy was as wage labourers in white enterprises.

Some Bougainvillians were employed as casual labourers, but by far the largest proportion of wage labourers worked under indenture, principally as unskilled labourers on plantations. In 1939, for example, there were about 3400 working under indenture; about 2500 worked on Bougainville–Buka itself, and the remainder elsewhere in the Territory. (At that time there were also 131 indigenes from elsewhere in the Territory working under indenture on Bougainville–Buka.) It would be pertinent to inquire what these labourers gained, economically or socially, from this, their closest and most sustained contact with whites.

Assuming that the average wage of each labourer was £5 a year (the wage for unskilled labour was between six and ten shillings a month), wage earners would have earned during 1939 about £17,000. By viewing this as income for Bougainvillians as a whole, and subtracting from this figure some £2400 (the approximate amount of head tax collected on Bougainville–Buka that year), this would leave about £14,600 for other purposes, or about six shillings per capita per annum, i.e. enough to buy a strip of calico, a few pounds of rice, and a few sticks of trade tobacco.

But what about other benefits? Mention has already been made of some of the detrimental consequences of the indentured-labour system: the irrelevancy to indigenous life of most plantation-acquired skills, the unsavoury moral atmosphere of labour compounds, the psychological tone of the master–boy relationship, the social disruptions in home life brought about by the workers' long absences, etc. On the other hand, the combination of medical care, muscular regimes and plantation food (monotonous as it was) does seem to have produced somewhat healthier individuals. And, however divisive it may have been in other respects, a term of labour in a white enterprise, especially plantations and mines, served to provide a model for a larger-scale indigenous society. Despite the interethnic conflict that occurred in many labour compounds and colonial towns — Buka Islanders against Buins, Sepiks against Bougainvillians, etc. — the experience of working together, and in the same roles, may have served to dampen some of the intertribal hostility that characterized indigenous life in pre-white times. And finally, Pidgin, which most indentured labourers succeeded in learning, provided a means of communication across old tribal boundaries as well as with the alien Europeans.

Meanwhile, the subsistence economy of Bougainvillians persisted very much as before. Vegetable staples and gardening techniques remained unchanged. Indigenes were introduced to new crops, maize, tomatoes, beans and pawpaw. Some of these they grew on a small scale, but mainly for sale to whites because most indigenes found such food unsatisfactory substitutes for their own taro, sweet potatoes and yams. Many indigenes developed a taste for rice, but the sporadic efforts to grow it did not succeed and few people were willing to expend their precious little hoards of shillings for this kind of luxury.

The change in meat supply was somewhat greater. Interbreeding with new strains considerably increased the size of local pigs, and made pig-breeding a widespread preoccupation. On the other hand the increased production scarcely affected people's daily diets, since pork continued to be reserved for festive occasions. As in the case of rice, many Bougain-villians developed appetites for tinned beef and fish, but few were in a position to afford such luxuries.

In its efforts to introduce white notions of public hygiene, the Administration required pig-proof fences around the new line-villages, and encouraged people to raise their dwellings off the ground on piles. (Pile dwellings were not unknown in pre-European times, but most dwellings were formerly built directly on the ground.) As a result of these efforts pile dwellings did indeed become more numerous, in line-villages that is; in their hamlets most people continued to live in their ground-level houses. As for other architectural innovations of the mandate era, very few indigenes were financially able to emulate whites to the extent of roofing their houses with metal.

By 1941 there were probably no indigenous households on Bougain-ville–Buka without a metal cooking pot or two, although many of them continued to use their own clay vessels. Few adult males were without a steel machete, and many of them owned steel axes or adzes as well. Just as universal were smoking pipes, both wood and clay, for women as well as men. A few Bougainvillians were able to afford an occasional stick of twist tobacco—at threepence apiece—but most of the tobacco used was home-grown (and wholly uncured).

In at least one respect the Administration's policy met with outstanding success. As part of its mandate undertaking, it pledged itself to discourage the consumption of alcohol by indigenes. Even after decades of observing the whites at this congenial pastime, few Bougainvillians followed suit. They were not permitted to purchase alcohol, and drunkenness was sternly penalized. But perhaps more effective was the fact that few Bougainvillians were affluent enough to purchase beer, much less spirits.

White notions of modesty were only partially diffused throughout the islands. Virtually every Bougainvillian past early childhood became accustomed to wearing a *laplap* — a strip of calico reaching from waist to below the knees — but mission efforts to induce the girls and women to cover their breasts met with only sporadic success.

It is possible, but difficult to document, that the Bougainvillians improved somewhat in physical health as a result of their twenty-year wardship. For example, all obvious cases of leprosy, tuberculosis, meningitis, etc. were hospitalized, and both Administration and mission agents conducted campaigns against yaws. Also, as previously noted, a term of years spent on plantations seems to have had a generally good effect upon the indigenous labourers, in terms of physical well-being at least. On the other hand, little progress appears to have been made in eradicating such major killers as malaria and pulmonary diseases. As for mental health, although there is no evidence of increase in extreme forms of mental illness, the period spent under mandate rule can only have increased the psychological stresses occasioned by subordination to incomprehensible authorities and ways of life.

By 1941 nearly all Bougainvillians had become at least nominal adherents to Christianity, of one variety or another, and some had even begun to develop new varieties of Christianity of their own.

At the beginning of the Australian military occupation in 1914 the Marist mission maintained four stations on Bougainville (Kieta, Patapatuai, Koromira and Torokina) and one on Buka Island at Burunotui, and counted some 800 to 900 baptized converts on the two islands. By 1939, the last pre-war year for which there are figures, the number of Marist mission stations had increased to twenty-one and the number of converts to over 30,000, including over 25,000 baptized members and over 5000 catechumens and adherents. Ministering to this large flock were twenty-nine priests, six brothers, twenty-five sisters and five lay nurses, all white, and seven indigenous sisters. To appreciate the great numerical weight of these sixty-five or so white Catholic missionaries, they should be compared with the numbers of other non-indigenous adults living on Bougainville–Buka at that time: six Methodist missionaries, two Seventh Day Adventist missionaries, and some hundred other whites engaged variously in production, commerce, or the Administration. In other words, in terms of numbers alone Catholicism constituted the largest agent of social change in these two islands, and the most widespread manifestation of change. How deep-seated a change it was remained to be seen.

Throughout the mandate era, the primary goal of the Marist mission appears to have been to save indigenes' souls, either from the darkness of paganism or the error of Protestantism. Individual missionaries undoubtedly ministered to the physical and material well-being of their charges, and in due course much mission-wide effort went into health care and education; but evangelism was the overriding objective.

The Marist strategy for carrying out its mission was to set up stations, staffed with priests, nuns and brothers, consisting of churches, schools, living quarters for the missionaries and pupils, vegetable gardens and, at the larger coastal stations, coconut plantations. The gardens were needed to feed the station personnel, and the plantations to help finance the whole mission enterprise. Some financial support came from outside sources, including funds raised by the missionaries themselves in their homelands, but much of the operating costs had to be paid for by profits from the missionaries' own commercial enterprises, a circumstance that added immeasurably to the mission's principal task.

The necessity to be as self-supporting as possible required of many missionaries that they devote too much time and energy to copra production, leaving too little for their evangelical teaching and pastoral duties. For labour they depended heavily upon their schoolboy boarders, who spent more time working than attending their lessons. The mission's acquisition of good agricultural land, however 'voluntarily' tendered, sometimes entailed conflict with its indigenous neighbours and served to identify it with other kinds of *waitman* takeover. And finally, the mission's operation of plantations, in prime agricultural areas and with 'free' student labour, aroused protests from other white planters.

As remarked earlier, the Administration operated only a few schools, and none at all on Bougainville–Buka. In these islands indigenous education was left entirely in the hands of the missions, a task which the Marist mission was pleased to undertake and indeed zealously sought to monopolize. However, during the mandate, the mission's educational goals went no further than their evangelical ones: to save souls. Beyond instructing raw converts enough to prepare them for baptism, the only formal schooling was that given young boys at the station boarding-schools and was designed to turn them into village catechists. It was only towards the end of the era, and largely in response to a growing challenge from the Methodists, that one of the catechist schools (the one at Chabai) was placed in the hands of trained missionary educators. But even this move was intended mainly to increase the pupils' value to the central evangelical purpose of the mission.

The Marist mission also supplemented the Administration's meagre medical services to Bougainvillians. Humanitarian sentiments undoubtedly played a part in this, but as in the case of education it appears that much of the mission's work was specifically undertaken to win converts or to avoid losing them to the Protestants.

How did the mission's primary task—that of conversion—take place, and how were the new converts incorporated into membership of the church? The normal procedure of conversion has been described by the historian, Hugh Laracy; his account is as follows:

> The Marists, like all missionaries, generally found adult pagans—those most committed by habit and interest to old religious allegiances— reluctant to adopt Christianity. . . . therefore, children were regarded as the hope of the mission and the Marists' efforts were mainly

directed to drawing as many as possible into the station schools, where study was a novelty, discipline generally light, calico and tobacco regularly obtained and the [pagan] spirits impotent.

Pupils eventually received baptism almost as a matter of course. Normally, their catechumenate lasted about eighteen months. . . . Pupils usually returned, directly or via the plantations, to their villages. There some acted as teachers and prayer leaders, but most helped diffuse awareness of the *lotu* [Christian religious ritual] simply by their conversation, whetting the interest of their fellows with tales of what they had seen and learned. Infants were baptised whenever the parents approved. The baptism of adults, where there were no matrimonial impediments, was at the priest's discretion. A catechumenate of six months, including a period at the station, might be required to test an adult candidate's sincerity and to extend his knowledge of Christianity; especially once mission influence became established in an area, a request was sufficient to obtain baptism. [Laracy 1976, p. 74]

Perhaps the most obdurate difficulty in winning adult converts was the mission's rule on marriage: only monogamous persons were eligible for baptism. Before a polygamous man — there were no polygamous women on Bougainville–Buka — could be baptized, he was required to put aside all but one of his wives. This ruling was no obstacle to most men: although by indigenous custom any male could acquire as many wives as he wished and could afford, only a few actually did so. But among those few were usually to be found the most affluent and influential men in the indigenous communities, and by excluding them from baptism the mission lost the drawing power of their example. On the other hand many polygamists did cast aside their extra wives in order to be baptized, thereby securing for themselves the promise of a place in heaven. It is not recorded what became of the discarded wives.

Other than the requirement of monogamy, conversion did not demand any radical changes in the converts' lives. It was enough that they attend services and participate on occasion in the sacraments; it was not required that they understand these rites. In fact, the simple cosmologies and ritual practices taught to new converts were similar in many respects to those of the indigenous religions. In each of the native languages encountered on Bougainville–Buka the missionaries were able to find verbal concepts near enough in meaning to Christian ones for their purposes of teaching and preaching.

For example, among the Siwai people their pagan creator spirit, called *Tantanu* (Maker) was appropriated by the Marists and Methodists to designate God; although the divine qualities attributed to the Christian *Tantanu* were considerably greater than those of the indigenous spirit of that name, the identification seems to have been acceptable to both sides. As for the other major Christian figures, Jesus and the Virgin Mary were of course new concepts to the Siwai, but the Holy Ghost (*Mara Mikisa*) fitted easily into indigenous belief in the familiar form of a supernatural bird. A Siwai convert to Methodism once described to me the differences between Catholic and Methodist beliefs: 'The *Popi* [Pope, i.e. Catholics] talk a lot about Jesus' mother and a place called Roma, while we *Taratara* [Methodists] think mostly about Jesus Himself.'

As for mission doctrines about souls (before and after death) and about saints, these accommodated quite easily to indigenous beliefs. The mission's criticism of evil spirits and indigenous magic served only to reinforce what the indigenes already believed: namely, that evil spirits are dangerous and that some magic can be deadly.

The mission concept of sin was more difficult, and few Siwais even bothered to wonder about it. On one occasion this writer succeeded in having three Siwai teachers, two Catholics and a Methodist, discuss the subject together, and they agreed on the following proposition: that sin accumulates within a person as do other evil things, forming a hard round object that lies in the stomach. Catholics can rid themselves of sin through taking communion, but Methodists have no special means of ridding themselves of it and hence have to take special care to avoid doing sinful things.

Except for polygamists then, conversion to the Catholic mission creed and membership in the church did not require a major change in the indigenes' thinking and living. But it did represent a conscious acceptance, however superficial, of something partly new; and it may be asked what persuaded so many people to convert?

Some of the earliest conversions were accomplished through purchase. For example when the Marist mission was becoming established on Shortland Island some indigenes, including a few from Bougainville–Buka, were obtained for work on the station (and of course religious schooling) by payment of goods or money to their relatives. To paraphrase Laracy (p. 75), purchase was a guaranteed means of obtaining an initial following and of creating a core of potential assistance: those who were purchased *belonged* to the mission.

In many other instances individuals who were originally attracted to mission stations for purposes of earning money were converted as a matter of course. Another means whereby individuals were attracted to the mission in the first instance then held within the fold, was by generous hand-outs of tobacco, calico, tools, etc. Religious medals were also passed out in large quantities; the wearing of one seems to have given the recipients a feeling of adherence to the mission, even without baptism. The incentive to conversion supplied by a few specific trade goods was increased for some by their comprehensive belief that Christianity in general, and Catholicism in particular, would somehow provide an unending supply of *waitmans* goods of all kinds, as witnessed by the boat-loads of objects that whites (who were nearly all Christians) continually received.

Medical aid was instrumental in winning many converts. The missionaries were called upon to dispense medicine and first aid as a regular part of their pastorates, and eventually the mission was obliged to establish hospitals in addition to the dispensaries attached to each mission station. Quite apart from any gratitude that may have been elicited by these services, they also served to impress Bougainvillians with the superior 'magical' powers possessed by the missionaries, more powerful even than those of their own magicians and sorcerers. In this connection, there are indications that some missionaries, in their evangelical zeal, did nothing to disabuse them of such beliefs. The motivation behind these endeavours undoubtedly included some genuine humanitarian solicitude, but a need to equal or outdo Methodist medical services also played a part.

Many others became converts in the first instance in order to attend school. In the view of many Melanesians, on Bougainville–Buka and elsewhere, the main basis of whites' superiority in material goods, weapons, etc. was to be found in their knowledge, and schooling was the key to such knowledge. An understanding of English, particularly, came to be regarded as an essential ingredient of that knowledge; this circumstance, reinforced by rivalry with the English-speaking Methodists, moved the Marists to place less emphasis on the use of native vernaculars in schools and services and to augment their French- and German-speaking staff with missionaries from Australia, New Zealand and the United States.

Still another powerful incentive to conversion, especially in the mission's early days in these islands, was the protection, real or imagined, that it afforded against other white attacks or encroachments. On some occasions it was only a missionary's interception that spared some community from punitive action by the Administration. On other occasions, a missionary was the only white willing and able to defend indigenes against other non-official white infringements and greed. An instance of the former occurred in Buin in 1919–20 when mass conversion took place in response to a series of punitive expeditions by the Administration, which included the execution of some feuders for homicide. On this occasion gratitude to the missionaries for having successfully shielded many innocents from Administration retribution was mixed with realistic acknowledgement that Christianized indigenes had wisely remained aloof from the feud.

In some areas whole communities became converted as a consequence of the conversion of an influential kinsman or local leader. And, as will be described, there were instances in which one of the factions of an ancient feud joined the ranks of the Catholics in response to seeing their enemies become Methodists; the reverse also took place.

Another incentive to conversion, individual or en masse, may have been contained in the indigenes' own belief—very widespread in Melanesia— about cargo (*kago*), a millennium in which every wish would be fulfilled, especially those concerning a plentiful supply of good foods and reunion with deceased kin. In some instances the appearance of the affluent and seemingly supernatural whites was viewed as materialization of this prophecy; the missionaries' promises about heaven reinforced such views. Missionary preaching about the terrors of hell may also have influenced some people to convert, although theological arguments were probably less influential than mundane practical considerations in most conversions.

Finally, in attempting to explain why Bougainvillians underwent conversion so readily and in such large numbers, one should not overlook the fact that Christianity was for them not an entirely distinctive institution. It was but one aspect of the whole complex of new — *waitmans* — objects and customs. During the early stages of their encounters with the *waitmans* way of life, the latter must inevitably have appeared overwhelming. It was somewhat later, and then only sporadically, that the idea emerged of selecting only some parts of the new to complement parts of the old.

As noted earlier, the Methodists installed their first mission in 1920, after an abortive attempt in 1916. The 1920 salient was established in Siwai by indigenous teachers from the missionary station on Treasury Island, and this was followed in 1922 by one white and three Fijian missionaries who set up a station on the west coast of Buka. In 1924 and 1926 other Methodist stations were established at Teop and Siwai respectively, and in 1931 Bougainville–Buka became a circuit of its own, separate from the earlier mission headquarters in New Georgia. Meanwhile, responsibility for the mission had been transferred from Australia to New Zealand.

In 1941 the Methodist mission embraced 8,988 people within its pastoral care, of whom 556 were adult members, 288 were junior members, and 135 were 'on trial for membership'. On Bougainville–Buka its strategy resembled the Marists' in some respects and differed in others. Like the Marists, the Methodists based their operations principally on stations somewhat isolated from the indigenous communities, where youthful converts boarded, worked, attended school, and generally led lives quite unlike their lives at home. More so than with the Marists, however, the education of these young converts included training in agriculture and industrial arts. This concept of industrial missions was based partly on the policy that the Methodist missions should be largely self-sustaining economically, and partly on the view that the indigenous Methodist should be industrious (in the white sense) as well as pious.

The Methodists were more exacting than the Marists in their requisites for church membership; they were considerably less tolerant of 'heathen customs', more interested in Westernizing their converts' characters and not only their religious beliefs.

The third Christian mission to become established was that of the Seventh Day Adventists, who began their evangelical work in 1924 in the village of Lavelai on the southeast coast of Bougainville. Their progress was very slow; in 1941 they recorded only about thirty converts, all in the area around Kieta. This is not to be wondered at, because their membership requirements were even more stringent than the Methodists' and included the prohibition of tobacco, betel chewing, and the eating of crustaceans and pork. To forswear pork-eating was an especially onerous test of commitment. Meat in any form was a grand luxury to these islanders, and much of their traditional life revolved around pigs—raising them, exhibiting them, trading them, gift-giving them, and eating them on the most solemn or festive of occasions. (While being impressed with the persuasiveness of a mission able to secure so deep a commitment, one also wonders why such a radical deprivation was demanded.)

Ecumenism was not in fashion in these islands before World War II; in fact, a reading of early missionary reports and correspondence creates the impression that a soul snatched from Methodism (or Adventism, or Catholicism) was felt to be an even greater victory than one lifted out of heathenism. The Marists were first on the scene; while their proselytizing rights were not legally exclusive, as some of them appear to have believed, they were bitter at what they considered Methodist encroachments, and speeded up their evangelistic activities like an urgent military campaign. As for the Methodists (the Adventists being still too meagre in numbers to constitute a threat), although there were several pagan areas in which they could operate, they made so bold as to evangelize mainly in the heart of Catholic strongholds. The rivalry was most intense in southern Bougainville and is here described by Laracy (pp. 63-4):

> Feelings ran highest in Siwai. The Methodists, who eventually attracted half the population, were reinforced in 1928 by an influx of teachers from New Georgia. The Marists were ready for them. The year before Father Boch had equipped a squad of catechists in south Bougainville with bicycles in order that they might more quickly visit threatened villages, challenge Protestant emissaries and report back to their priest. In November 1928 he issued instructions that forceful catechists, 'even insufficiently trained ones', be placed in each village and station work subordinated to visiting, even if it meant making the schoolboys 'a troop of peripatetic scouts accompanying the [priests] . . . from village to village'. Visiting Siwai two months later and observing the bitter sectarian competition, the Government Anthropologist, W. P. Chinnery, suggested to Boch that the missions reach a *modus vivendi*, only to be told, 'If the Protestants wish to have peace with us, let them go where we are not . . . where our influence is established . . . there will be a fight for each individual village if necessary'.
>
> Fighting did break out shortly afterwards; Methodist and Catholic factions destroyed each other's chapels at Osokoli and Hukuha. A judicial commission was appointed to investigate the situation and, though its only official outcome was the restriction in 1930 of the entry of 'foreign' Melanesian and Polynesian missionaries to the mandated territory, it did consider the Marists most to blame for arousing the animosity of their followers. The display of government interest in mission activities (and the threat of further action it was thought to contain) did, however, have a pacifying effect. Rivalry in

Siwai continued into the 1930s but it was more discreet, and decreased as the number of people unconverted to one side or the other declined.

At this point it will be illuminating to go beyond general statements about mission programmes and statistics on conversion to inquire how far these mission activities, singly and in opposition, had affected indigenous life in the off-station Bougainvillian communities themselves. Again the Siwai area, where both Catholic and Methodist missionaries had been long at work and where their converts lived side by side, provides an example for that era:

In matters of belief, then, mission influence has made little impact upon most Siwai. Similarly, Christianity cannot be said to have changed many Siwai practices. It has caused some Methodists to give up productive work on Sundays, and it has encouraged many of the younger men, especially Methodists, to wear cleaner loincloths; but it has had little effect in curbing polygamy or in changing sex mores. It has discouraged the practice of image-sorcery—the carved wooden figures (poripai) used in this sorcery are condemned by the missions as being 'graven images'—but it has probably had little inhibiting effect on other kinds of magical practice. In fact, some zealous converts now use Bibles as magical aids in litigation, and most natives see no difference at all between, say, Catholic . . . christening and Siwai [pagan] baptism. Some Methodists heed their missionary and inhume rather than cremate their dead relatives, but for most natives Christian rites of passage merely reinforce the native rituals.

More significant than their influence upon religious beliefs and practices have been the missions' effects on social structure. As a result of mission rivalry new lines of social cleavage formed or old lines of cleavage crystallized. Catholics and Methodists no longer burned one another's chapels but many tensions continued to exist.

By 1941 Catholics outnumbered Methodists in Siwai, not because of differences in doctrine or practice but mainly because white Catholic missionaries had been at the job there longer and more continuously than their Methodist counterparts. Some villages were entirely Catholic, others entirely Methodist, depending usually upon which missionary began his work there first. Once the missionary had received consent from the highest-ranking local leader to construct a chapel and install a native evangelist, most of the leader's followers moved into his fold. Troubles began only when overzealous native evangelists tried to set up rival chapels in places already nominally affiliated. Many of these efforts were frustrated but some of them succeeded, with the result that there were many villages with both Catholic and Methodist congregations. In such cases the smaller

congregation, usually representing the later mission to have become
established, was nearly always identified with one or two hamlet
units, conversion or transfer of affiliation having followed kinship
lines. In most instances when a whole hamlet unit changed missions
it did so at the behest of one of its more influential members. For
example, one such unit became Methodist when its senior male
member became piqued after being advised by his Catholic priest not
to acquire a second wife. Another unit became Methodist after one of
its brighter young men returned from indenture with glowing tales of
the imagined practical advantages of learning arithmetic and
bookkeeping, which Methodist schooling specializes in. . . . In some
cases the division between Catholics and Methodists corresponded to
long-standing political divisions, with whole neighbourhoods having
purposefully embraced the sect opposite to that of their traditional
enemies. . . .

The tendency towards sect-endogamy had some effect upon the
maintenance of social cleavages between opposing congregations, but
many inter-sect marriages took place, with one of the spouses usually
joining the congregation of the other, depending upon the choice of
residence. There were, however, a few steadfast mission members,
usually men, who did not change sects when they moved to the
places of their spouses, and these accounted for most cases of
scatterings of sect A members in villages belonging predominantly to
sect B . . .

Lines of inter-sect division did not harden so long as a village's
minority sect members did not band together into a definite
congregational unit. Nor, in the case of relations between separate
villages of opposing sects, did the fact of different mission
membership create new cleavages or add significantly to cleavages
already present; active religious hostility between separate villages
seemed to have ended. In 1941, social cleavage between sects was
mainly manifested in villages having opposing congregations, each
with its own chapel and native evangelist. [Oliver 1955, pp. 315-16]

Elsewhere there were whole villages associated exclusively with one or
other of the missions, thereby sparing the local indigenes at least this consequence of whites' eagerness for their land or labour or immortal souls.

To summarize, the three Christian missions at work on Bougainville-
Buka during the mandate era differed from one another in many respects—in
doctrines, goals of conversion, methods of proselytization etc.—but in one
important respect they were alike. They continued to be *waitmans* institutions, as basically colonial as the white-owned plantations and Administration enclaves. All important mission decisions were made by whites,
and nearly all mission material resources reposed in whites' hands. What
indigenization of Christianity there was received its impetus from the

indigenes themselves, and in forms that were more than distressing to the white missionaries and to other whites as well.

Such developments first came to the notice of whites in 1913. They started in a corner of Buka but have since encompassed much of that island and have appeared in many parts of Bougainville as well. In 1913 word reached Europeans on Buka that a resident of Lontis village, a pagan named Muling, was attracting a large following by his claim to be able to acquire *waitmans* goods through magic, all other kinds of effort to do so having failed. His message fell on receptive ears. Before this event, Buka Islanders had been in fairly frequent contact with whites for more than forty years. Large numbers of them had served as labourers, policemen, mission scholars etc., and had come to use and want goods that would permit them to live more like whites, whom they at first greatly admired, and whom one report suggests they first believed to be returned ancestral spirits. In any case the cult swelled to such size that the German Administration, fearful of its potential, arrested Muling and exiled him from the island.

The excitement aroused by Muling's prophecies died down under Australian rule, but his fellow Buka Islanders seemed not to have lost their appetite for *waitmans* goods and ways, so that the Marists were well received when they began active evangelism there a few years later. The missionaries may not have specifically promised the indigenes material wealth in return for conversion, but it is quite likely that the latter read such promises into the missionaries' assurance about spiritual rewards. Thus, Sydney (Australia) was believed by the indigenes to be the future abode of the righteous; it was also known to be the source of most of the goods that reached Buka. Within a few years over 90 per cent of Buka's indigenous population had become Catholic.

Meanwhile, Catholic *lotu* (ecclesiastical services) began to serve purposes which the mission never intended. If the *lotu* worked for the missionaries in bringing what they wanted, some Buka Islanders reasoned, why should it not work for them and bring ships laden with goods? Armed with this argument another Buka Islander, Pako, initiated a new movement to acquire the desired cargo. One of the leaders of this new movement was a Catholic catechist, but the principal one, Pako, along with his associate, the repatriated and durable Muling, were pagans; this form of *lotu* had evidently become disengaged from the mission itself.

Modelling their actions on the mission practices of approaching the divinity through saints, Pako and his associates employed their *lotu* in petitioning their own ancestral spirits for aid in bringing the desired cargo. People refurbished their burial grounds and spent nights there in prayer, and a mood of excitement prevailed in expectation of the arrival of the cargo which had been prophesied by Pako. Gardening and pottery-making ceased (why work when the expected ships would bring all the food and utensils needed?); wharves and storehouses were built to receive the cargo. The movement was a bizarre mixture of new and old. Many old indigenous customs were renounced, new white ones adopted, and equality with whites was proclaimed. At its peak the cult—for such it was with its strong religious emphasis—embraced some 5000 Bougainvillians; this was the largest grouping up to then to unite on these islands, and a sign of things to come.

Subsequently, when some of the cult's members attempted to claim goods landed for whites, the Australian Administration stepped in and exiled the leaders to Madang, where Pako himself subsequently died. After this the popular excitement subsided for a year or so, but was stirred up again by another Buka pagan, Sanop, who moved into Pako's residence in Malasang village and began receiving mysterious messages which he identified as coming from Pako's spirit, again promising cargo, but this time ominously anti-white in tone. Again, however, the cult contained much from Christianity, including the rituals of *lotu*, regular church attendance, and insistence upon monogamy (except for the cult leaders). But even though cult members continued to value the Marist priests' ritual powers (the local priest baptized 200 new converts including some Methodists on one visit to a cult stronghold, and 'Bishop' appeared along with 'Pako' on cult banners), the cult members otherwise distinguished between their brand of Christianity and the European mission itself.

When the movement spread to northern Bougainville it became more militantly anti-white. Then, when talk of 'liberation' was reinforced by a mass desertion of labourers from their plantation jobs, the Administration moved in again. The cult leaders were arrested, along with about 100 followers, and Pako's home was burnt down. Meanwhile no cargo arrived to offset the famine caused by the earlier cessation of gardening, and the disheartened people returned to their ordinary pursuits. With this, the Pako–Sanop episode ended. Faith in cargo was revived a few years later when Japanese ships appeared—but that is another story.

5

World War II

THE SHOCK WAVES PRODUCED BY WORLD WAR II CONVULSED Bougainville at the time. At war's end the principal agents of colonialism — Australian Administration, Christian missions, and white-owned plantation and commercial firms — returned to re-establish their respective controls over indigenous affairs, but neither they nor the islanders were to remain as before.

The events of the war that had the most far-reaching local consequences were the following:

December 1941. Japan attacks Pearl Harbor and moves towards Australia and the South Pacific.

January 1942. The Japanese occupy Rabaul and attack Buka.

March 1942. The Japanese occupy Buka Passage and Shortland Island and begin control of all Buka and of Bougainville's coastline.

May 1942. The Japanese occupy Tulagi, British Solomon Islands.

August 1942. The United States forces land on Guadalcanal.

November 1942. Japanese counter-attacks against United States forces in the central Solomon Islands terminate in failure.

November 1943. United States and New Zealand forces establish beachhead at Torokina and consolidate the isolation of the Japanese forces on Bougainville–Buka.

October–December 1944. Australia takes over occupation of Torokina from the United States, and begins a campaign to destroy or capture all Japanese forces on Bougainville–Buka.

August 1945. The Japanese forces surrender.

March 1946. Australian civil Administration is re-established on Bougainville–Buka.

During the twenty-six months between the German attack on Poland and the Japanese attack on Pearl Harbor, life on Bougainville–Buka went on much as before. A few of the younger Australians in the islands left for home to enlist, and a small detachment of Australian soldiers was posted at Buka Passage to guard the air-strip being built there; but for the rest of the population, indigenous and expatriate, the European war was a very remote and irrelevant affair, or was not thought about at all. Locally, the most important event concerning the approaching Pacific war was the setting up of an intelligence network involving a handful of resident expatriates.

After World War I the Royal Australian Navy devised a plan for reporting the appearance of unusual sea- and aircraft among the islands north and east of Australia, and in September 1939 the plan was put into effect. Coast-watchers, as they came to be called, were appointed at strategic points along the chain of islands, from the Admiralties to the New Hebrides. They were supplied with tele-radios and a simple code, and were linked with naval headquarters in the south through the island centres of Rabaul, Port Moresby and Tulagi. At the beginning most coast-watchers were civilians — chiefly planters, traders, and Administration officials — but in time they were given naval ranks.

The coast-watchers of Bougainville were a remarkable breed of men. For months or years during the war they lived in the island's interior keeping watch over enemy ship and plane movements, cut off from other Europeans except by radio and an occasional air drop of supplies, existing under conditions of extreme physical privation, supported by some indigenes but turned against by others, and in constant danger of capture and death. To a degree that few other individuals can claim, they contributed to Allied victories in the Solomon campaigns. And though they were not there for that express purpose, their presence in enemy-occupied areas for months or years after the flight or evacuation of most other Europeans, provided evidence, if any Bougainvillian wanted it, that Australia had not wholly abandoned the two islands to the Japanese.

At the beginning of the Pacific war, coast-watchers were deployed at several key observation spots on Bougainville–Buka; at Kessa, Buka Passage, Inus plantation, Numanuma plantation, Kieta, and Toimanapu plantation. The heroic exploits of those who managed to survive sickness and capture will be recounted.

Soon after Pearl Harbor, most of the non-official white residents were evacuated to Australia. Of the white missionaries, however, the two Methodists on duty remained, as did all the Marists: priests, brothers and nuns. Then, in late January after the Japanese had captured Rabaul and had begun to reconnoitre Bougainville–Buka by air, the D.O. (District Officer, the top Administrative officer of the Bougainville District), along with his staff and most of the remaining white civilians of east Bougainville, commandeered a small mission ketch and made for Port Moresby. Those still remaining were a few soldiers, coast-watchers and missionaries, and all the indigenes (including some from elsewhere in the Territory then serving indentures on local plantations or employed by the Administration). The cynicism and bitterness provoked among Bougainvillians by this flight and 'abandonment' have endured ever since.

During March 1942 Japanese ships called at Buka and north Bougainville, killed an elderly planter suspected of spying, and captured two other whites — a Marist priest and the sole remaining Methodist missionary. (The latter, along with 1100 other persons, were eventually lost at sea in the *Montevideo Maru*, which was torpedoed by a United States submarine while en route to Japan.) On 30 March Japanese forces landed and occupied Buka Passage and Shortland Island and shortly thereafter the Buin coast as well. Meanwhile the surviving coast-watchers and the small army detachment had established their bases inland and were reporting on enemy dispositions and movements. The army detachment, twenty-five men trained as commandos, was led by Lieutenant J. H. Mackie. The coast-watchers, W. J. Reed, assistant district officer at Buka Passage, and P. E. Mason, the proprietor of Inus plantation, by now held naval rank. From their observation points — Reed overlooking Buka Passage and Mason overlooking first Kieta and then Buin and the Shortland Islands — these two were able to observe and report Japanese movements from Rabaul and Buka towards the north and central Solomon Islands, a contribution of immense value to the Allies' subsequent Solomon campaigns.

Marist Bishop Thomas Wade's decision to keep his mission members at their various posts was based partly on a tradition 'that did not disdain martyrdom', and partly on the hopeful belief that the Japanese would deal tolerantly with noncombatant missionaries. For a while this hope was realized, but as time passed and Japanese reverses in the Solomon Islands increased, more and more restrictions were imposed upon the Marists, including internment and worse.

At the beginning of the occupation the Japanese forces adopted friendly, even fraternal, attitudes toward the Bougainvillians, presenting themselves as their deliverers from white oppression and as ethnic cousins and partners in the glorious new Co-prosperity Sphere. On Buka, where contact was earliest and closest, schools were set up to teach Japanese language, customs and songs. Indigenous officials were presented with Japanese titles and impressive new insignia, and were frequently consulted. The populace was encouraged to revive the ceremonial veneration of their ancestors. There and elsewhere the invaders were at first scrupulous in their bartering for food and payment for labour, and the molesting of indigenous women was firmly and effectively forbidden. Also, on Buka at least, the Japanese deliberately sought to win over the affections of the population at large by a show of friendly egalitarianism. Visitors were hospitably welcomed at the military camps, and friendship between soldiers and indigenes was encouraged.

The initial reactions of the Bougainvillians to these events differed widely. On Buka the Japanese were at first ceremoniously welcomed by much of the populace. For the devotees of the cargo cults described in the last chapter, the flight of the whites and the immense resources of the newcomers spelled fulfilment of the former cult leaders' promises. This was reinforced by the Japanese, who deliberately encouraged ancestor worship, through which it was considered that Christianity would be weakened and a bridge provided to Japanese State Shinto. Even non-cargoists and adamant anti-cargoists were seemingly impressed with the newcomers' military might. Knowing only what they saw, it must have struck them as highly unlikely that whites would ever again return to become their masters.

Belief in continued Japanese mastery was also promoted around Kieta by the statements of some Marist clerics of German nationality, but this was offset somewhat by their (American) bishop; though pro-Allied in sentiment, he insisted for pastoral reasons upon his missionaries' neutrality. Kieta had also been the scene of looting just after the precipitate exodus of the Administration officials, but coast-watcher Reed hurried there from the north and managed to restore Administration authority, which lasted until the Japanese occupied the site.

Elsewhere on Bougainville Island during the first half of 1942, the indigenes had only sporadic contact with the Japanese, or no contact at all, and their attitudes towards whites and Japanese seemed to have varied according to their own individual pre-war relations with the former. Some of them gave heroic proof of their loyalties to a specific missionary pastor or former employer, while others evidently delighted in the discomfiture of some or all whites. And even those without commitment to one side or the other were realists enough to accept the new order, since it appeared destined to remain.

On 7 August 1942 United States forces attacked and captured Tulagi and established a beachhead on Guadalcanal. During the following weeks the Japanese counter-attacked with large numbers of ships and planes from their bases in the north, but by mid-November the counter-attack was called off, in failure. The part played by the Bougainville coast-watchers in this crucial Japanese setback can hardly be overestimated. As wave after wave of Japanese aircraft flew south to destroy the exposed United States bases and the unloading, sitting-duck supply ships, they were spotted and reported by the coast-watchers in time to permit fighter planes to meet and down most of them before they reached their targets. Japanese ships carrying supplies and reinforcements were similarly spotted and met the same fate.

To the coast-watchers and most other remaining whites this turn of events was welcome indication of an eventual Allied victory, but it led the Japanese to the strengthening of their defensive forces on Bougainville and Buka. Buin was built up into a major land base, and several other posts were established or reinforced around the islands' coasts. Whether the ordinary Japanese soldier or sailor viewed these developments as hopeful or otherwise is not reported. For the Bougainvillians themselves—

except possibly those close enough to Europeans to share their perspectives — the Japanese defeat on Guadalcanal was either unknown or uncom-
prehended, and was outweighed by the visible local evidence of increased
Japanese strength.

With their entrenchment on Bougainville–Buka, the Japanese also
adopted a sterner attitude towards the remaining whites. The missionaries were increasingly restricted in their activities, until the point was
reached when some were taken captive and others, including the bishop,
led fugitive lives. Patrols began to harry the coast-watchers, but these jungle-
wise veterans were able to elude them and continue with their work. As
for the islands' Chinese settlers, some had long since been interned and
others sought refuge in the jungle.

The larger Japanese presence, and their more active efforts to turn the
indigenes against all whites, were accompanied by some change in Bougainvillians' behaviour towards the opposing alien sides. Throughout the larger
island the coast-watchers and fugitive whites began to be treated by the
indigenes with indifference or unfriendliness; in some places the latter actively
sided with the Japanese as spies or guides. This mood of hostility was
most evident around Kieta, where bands of coastal villagers calling themselves 'Black Dogs' conducted murderous raids against inland, and presumably neutral, villages, and joined with the Japanese to hunt down fugitive
whites and Chinese. Lest one be disposed to blame the Japanese overmuch
for actively enlisting the Bougainvillians' assistance in that non-indigenous
conflict, it should be pointed out that the coast-watchers did not hesitate
to execute indigenes who endangered their own activities by aiding the
Japanese.

On Buka, Japanese efforts to Nipponize the indigenes continued until
the end of 1943. The schools they established proved attractive to young
people and quite successful in inculcating anti-Allied sentiment. After a
while, when Allied military success began to reduce the flow of supplies,
more and more demands were made by the Japanese on the indigenes for
labour and garden produce, but this was done in the name of shared sacrifice
against a common enemy. These demands took some of the bloom off
the Buka Islanders' earlier enthusiasm for their new masters, but evidently
did not dispose them more favourably towards their former ones.

Under the pressure of Nipponization and general anti-white sentiment,
Christianity undoubtedly suffered a relapse on Buka. But if the pagan leaders
of the local cargo cult had hoped to profit by this, they were to be disappointed. As described earlier, the initial appearance of the Japanese was
greeted by them as partial fulfilment of their prophecies, and for a while
the movement flourished widely and publicly. But for reasons that are
not clear this florescence did not endure. According to some reports it
was actively suppressed by the Japanese, who beheaded three cult leaders
who were about to kill one of their opponents as a human sacrifice to
the cult's guardian spirits. Another more credible interpretation of these

executions — which did in fact occur — is that the victims were suspected
of pro-Allied behaviour at a time when Allied air attacks were increasing.

During the early months of 1943, the position of the Japanese forces
on Bougainville–Buka worsened through shortages and increasing air attacks
from the Allies. The lot of the remaining Europeans and Chinese became
correspondingly perilous, and more and more indigenes turned from friend-
liness to indifference, or from indifference to hostility towards them.

By June nearly all the non-military whites and some of the Chinese had
been evacuated by US submarine. Even the stout-hearted Marist bishop,
Thomas Wade, was finally persuaded, or rather commanded, to depart,
leaving only those members of his order who were unable or unwilling
to escape. Finally, in July, the surviving coast-watchers were evacuated,
their existence having become so imperilled by more aggressive Japanese
patrolling and increasingly hostile indigenes that they were unable to carry
on their reporting; moreover, by this time the Allied forces had moved
so much closer that coast-watching was no longer as useful.

On 1 November 1943, United States forces landed at Torokina, on
Bougainville's west coast, and quickly established there a beachhead of fifty
square kilometres. The purpose of the manoeuvre was not to conquer the
Japanese in direct combat, but to isolate and neutralize them still further,
while using the base for intensified air attacks against the major Japanese
concentration at Rabaul. Somewhat later these measures were reinforced
by the Allied occupation of Nissan Island to the north of Buka, and by
beachheads on New Britain itself, so that the Japanese on Bougainville–Buka
were effectively cut off from all outside help and left to 'die on the vine'.
The dying on that particular vine turned out to be a slow and tragic affair.

When the Americans landed at Torokina there were an estimated 65,000
Japanese on Bougainville–Buka, the major concentrations being at Buin
and around Buka Passage, with smaller detachments at places along the
east coast and strategic points in the interior. In March 1944, the Japanese
mounted a determined counter-attack against the United States base at
Torokina, but when that failed and the Americans gave no sign of
attempting to enlarge the base, the Japanese settled down, the more hopeful
of them to await supplies and reinforcements, the more realistic to stay
alive until the war's end.

In October 1944, Australian forces with some New Zealand reinforcements were sent to Torokina to relieve the Americans, who were destined
for other missions in islands nearer Japan. By this time the Japanese forces
had been reduced in number to about 37,000 to 40,000, some 25,000 to
28,000 having died during the Americans' eleven-month stay there. About
8000 died in combat and twice that number from sickness and starvation.
In fact, by the end of 1944, with large-scale military encounters on the
islands at a virtual standstill, hunger had become the Japanese garrison's
principal foe.

In an effort to replenish their diminishing food supplies, the Japanese
planted large gardens near their bases, and requisitioned more and more

of the indigenes' own food. In addition to depriving the latter of their existing stocks, these measures reduced their already scarce arable land and, through conscription, took away labour needed for growing their own food. To fend off starvation at some of the smaller bases, the Japanese resorted on occasion to raiding indigenous gardens; and instances were reported — difficult to authenticate but circumstantially credible — of desperate Japanese soldiers resorting to cannibalism.

The Bougainvillians' reactions to all these events varied from place to place. In the vicinity of the larger Japanese bases, in Buin and around Buka Passage, they had no recourse but to remain hungry and nurse their resentments in silence. In other places they actively resisted the Japanese, with bloodshed on both sides. Some of those most caught up in this new phase of the war managed to find refuge within the Allied perimeter at Torokina, where they were fed and housed and, if able-bodied, given jobs. In more remote mountain areas, out of reach of both Japanese and Allies, the war continued to be the cause of some deprivation in terms of trade goods and money-earning opportunities; but that was all.

For reasons which appear to have included national pride and military zeal, the Australians decided in December 1944 to break out of the Torokina beachhead and reconquer Bougainville–Buka without waiting for the war to be brought to an inevitably successful end elsewhere. Since this decision was widely criticized at the time and publicly condemned after the war, it requires no further discussion here, except to note that it also surprised and puzzled the Japanese commanders on the islands. Realizing their predicament, and being neither more nor less courageous than soldiers elsewhere, they apparently would have been content to live and let live unless forced to defend themselves. A similar attitude was evidently held by many of the Australians involved in the campaigns, but as the record shows they did not let their doubts and reservations curb their actions when called upon to fight.

The progress of the ensuing major campaigns are indicated in Figure 10. Not shown, however, are the movements and encounters of the various Australian guerrilla parties which reconnoitred and harassed Japanese outposts and patrols from perambulatory inland camps. Leaders of these parties included two experience-hardened pre-war residents of Bougainville, Paul Mason of coast-watching fame and Norman Sandford of Numanuma plantation. The epic adventures of these guerrilla parties and their indigenous supporters provide weighty evidence of the measure of sacrifice and heroism of which Bougainvillians were capable once their loyalties were mobilized. When Japan's capitulation brought an end to these campaigns in August 1945, the Australian forces had captured most of the smaller Japanese bases and were preparing for the final assault on the major base in Buin. By this time Australian casualties numbered 2088: 516 killed or dead of wounds, and 1572 wounded. On the other side, 8500 Japanese were killed during the Australian occupation and another 9800 died of illness. No figures have been published concerning the total number of casualties suffered by the

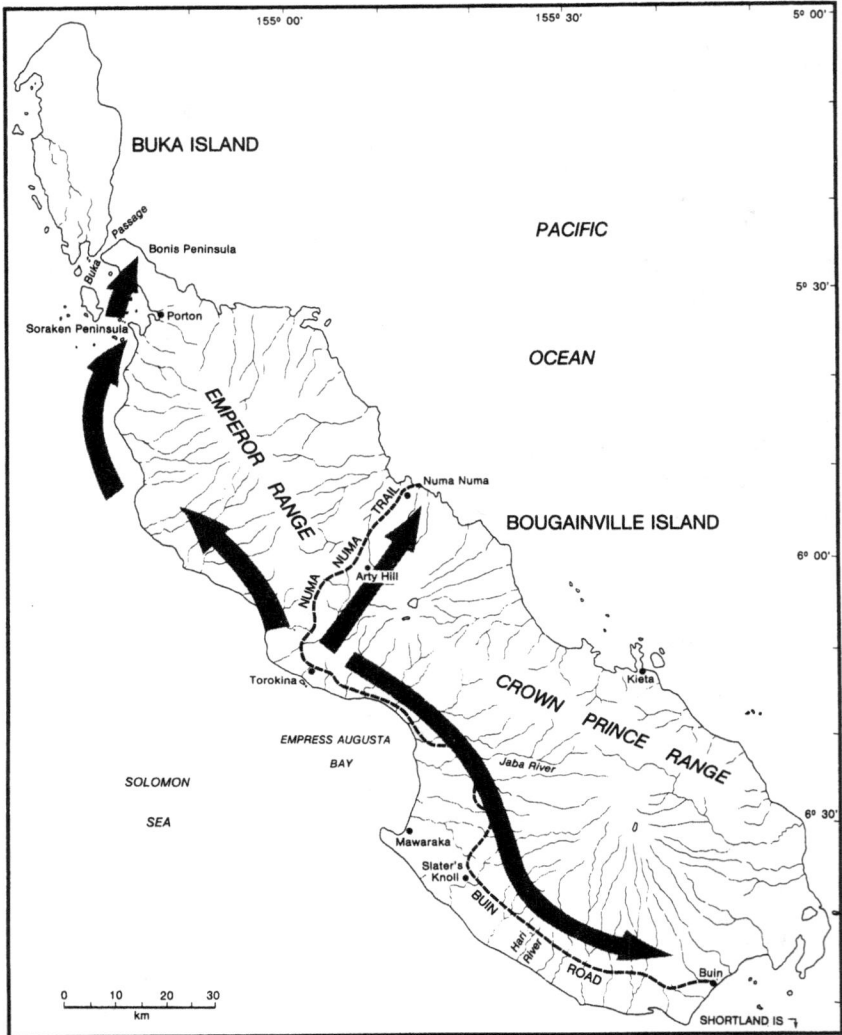

Figure 10   Major wartime campaigns

Bougainvillians themselves in these campaigns, or in direct consequence
of any other phase of this calamitous war, the causes and objectives of
which most of them did not even comprehend.

Full civil government was restored in March 1946; that is to say, the
civilian Australian administrators resumed control of the islanders' lives
after an interim of more than five years of military control.

What effects did the events of the war years have upon the Bougainvillians themselves? As was just noted there are no wholly reliable statistics
available describing the numbers killed or wounded as a direct or indirect
result of the war. They were undoubtedly high, but probably not nearly

A chief (*luluai*) and his principal helper (*tultul*) in 1938.

Australian troops (24th Battalion) advance along the Buin 'road' towards Hiru Hiru, 26 April 1945. *(By courtesy Australian War Memorial, Canberra)*

Rain-sodden camp of 47th Battalion, 20 July 1945. Weeks of torrential rain had brought operations almost to a standstill. *(By courtesy Australian War Memorial, Canberra)*

First contact between Major Otsu, the Japanese peace envoy, and Australian troops on bank of Miro River, 18 August 1945. *(By courtesy Australian War Memorial, Canberra)*

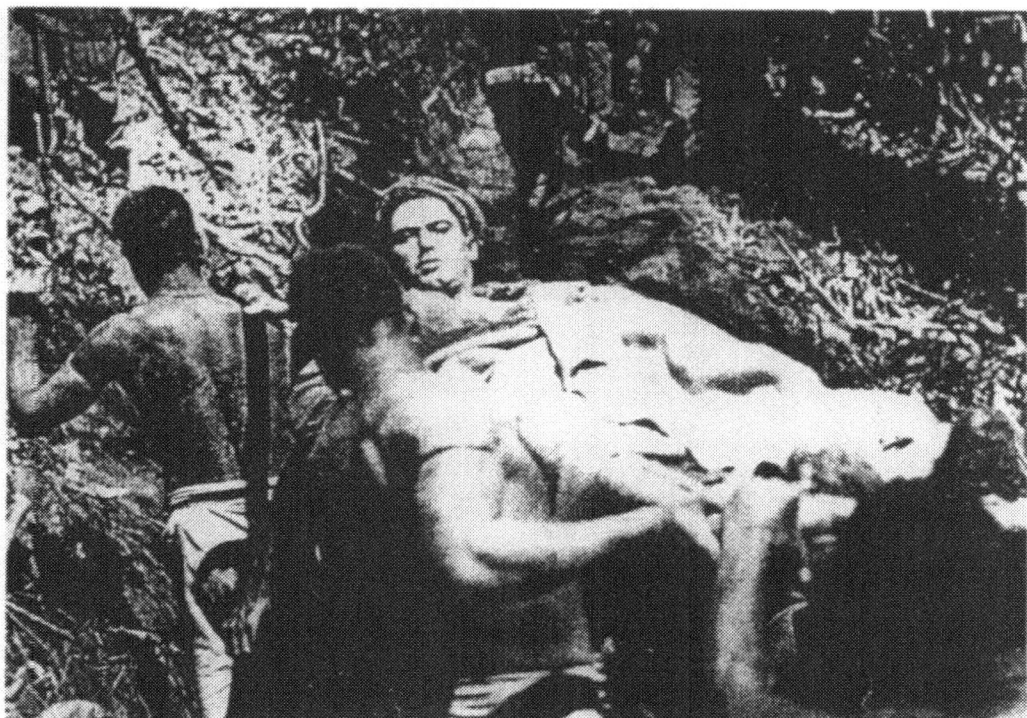

A battle casualty is carried out along a bush track. *(By courtesy Australian War Memorial, Canberra)*

Coast-watchers Paul Mason and Dovai.

Coast-watcher Norman Sandford.

Paul Lapun.

Donatus Mota.

Pig feeding at hamlet house (1938).

Woman at work in taro garden (1938).

Nasioi potters: vessels of local clay could be valuable trade items.

'Shell money' near Panguna: used for ceremonial purposes, payment of bride money, purchase of land rights.

Ceremonial feast. Food-display platform foreground, host's club-house rear. (Siwai, 1939)

Project (but not yet mine) conference in hut at Kobuan, 1966. *From left:* Colin Bishop, Frank Espie, Frank Paholski, Paul Quodling, Dick Spratt, Don Vernon.

Rivalry for converts: Catholics and Methodists engage in a Bible Knowledge contest (1939).

Village elder, 1988.

Ceremonial mask.

Village gardening, 1988.

Sunday outing, 1988.

so high as of those who died from war-induced illness, from starvation, fatigue and lack of medical services. (During the early phase of their occupation, however, the Japanese are reported to have provided fairly good medical care for their indigenous neighbours.)

The effect of the war on the Bougainvillians' material goods — their houses, gardens, groves, house-furnishings, tools, etc. — is difficult to assess, although some estimate may be given by the amount of the war-damage claims which were ultimately paid. In addition, one must include among their losses the almost total destruction suffered by pre-war buildings, both Administration and mission, which housed schools, churches and medical services. Nor can these losses be balanced by pointing to all the camp sites, roads, runways, etc., constructed on the islands during the war, since virtually none of these was of use to the post-war Bougainvillians.

However, despite the damages suffered in terms of life or limb, or physical deprivation or goods, the most far-reaching effects were probably psychological. Demographically, the population quickly resumed its tempo of rapid natural increase, and economically, their homes were quickly rebuilt. But the changes undergone in the mental atittudes of some of them — towards whites and towards themselves — served to ensure that relations between the two would never again be as they were before the war.

The precipitate flight of most of the whites and the collapse of most white institutions at the beginning of the war must have led even the most unsophisticated of Bougainvillians to question the pre-war colonial status quo. The continued presence of the Marists long after most other whites had gone may have tempered somewhat the general disenchantment, but in the end most of the Marists had had to flee as well.

During the first year of the war the Japanese military superiority added disdain to the disenchantment felt by Bougainvillians towards their former Australian masters. In turn, the subsequent Japanese defeat may have led most indigenes to moderate these views somewhat (although the part played by the United States did not escape them); but the whole war experience seems to have nourished a view among them that no colonial regime is necessarily perpetual.

The war also suggested to many that possession of a dark skin does not necessarily and inevitably require one to be treated, at worst as sub-human, and at best as a well-meaning but ignorant child. The early Japanese policy of fraternization encouraged this change in expectations, and it was probably reinforced by the subsequent behaviour of United States and Australian troops, some of whom seemed to have promoted it actively and deliberately.

If these new views of colonialism and ethnic relations had been limited to the Bougainvillians alone, they probably would have been forgotten in time; but they were shared with vast numbers of people elsewhere, and thus began to exert a dominant influence over Bougainvillians' post-war lives.

6

The Post-War Era

For most Bougainvillians the period from the end of World War II to the middle of 1964 was one of gradual and only moderately unfamiliar change. The basis for most of that change lay in the new post-war policies adopted by the Australian government towards its Territories of Papua and northeast New Guinea. In fact, those policies began to be formulated during the war, when the Labor government decided to make native welfare the principal objective in governing those territories, in deed as well as in word. This change of emphasis was of course in line with Labor's traditional ideology and in tune with anti-colonialist stirrings elsewhere. But it was reinforced by a renewed recognition of the strategic military importance of New Guinea to Australia's security, and by the corollary that a more prosperous, better educated, and politically sophisticated indigenous population would provide a stronger shield against future aggression from the north or the west.

In this spirit an agreement was made in 1946 with the newly formed United Nations which gave Australia exclusive trusteeship over the former Mandated Territory of New Guinea (Papua was already an integral dependency of the Commonwealth), subject only to the obligation that administration would be carried out so that 'the customs and usages of the indigenous inhabitants would be protected, their cultural and educational advancement assured, their rights and interests safe-guarded, and an increasingly progressive share in the administrative and other services given to them, as the territory developed'. Meanwhile the Australian government declared its intention of combining Papua and the Trust Territory of New Guinea for administrative purposes, a practical arrangement that had been in effect since the early days of the war. This move was opposed by some members of the United Nations on the grounds that it might slow down the Trust Territory's political development, or even lead to eventual Australian annexation of it. Despite those objections, Australia

continued to administer Papua and the Trust Territory as a single dependency, but also continued to submit its reports to the United Nations covering the latter only.

A whole chapter would be required to trace how the war-born policy
underwent change over the ensuing thirty years. Here in this brief account
will be considered only the highlights of these changes, as background
to the particular concern with Bougainville and Buka.

In terms of official government policy concerning the future political
status of the (combined) Territory of Papua and New Guinea, there were
some major shifts. The first was towards a larger, ethnically balanced,
measure of self-government for all the Territory's residents; then there
was a move towards an emphasis on a government largely of and for
indigenes; and finally towards an independent and mainly indigenous nation.

The move towards implementing the policy of ethnically balanced self-
government began in 1949 with the setting up of a Territorial Legislative
Council. The inclusion of sixteen official members in this advisory body
guaranteed that the Administration would retain firm control of it, but
the addition of twelve non-official members, including three nominated
indigenes and three members elected by the expatriate population, betokened
some broadening of representation. It was the expatriate non-official
members, and particularly the elected ones, who were most vocal in pressing
for a larger share of self-government; the indigenous members tended to
hew to the Administration's line. Of course, 'self-government' meant
different things to different people in those early post-war days. To many
of the Territory's whites it meant more freedom to run their own enterprises without Canberra's bureaucratic control; and while this view of self-
government may have included some sentiment for increased indigenous
participation, it was based largely on concern for the interests of the whites
themselves. As for the official Administration view of self-government,
it was undoubtedly more pro-indigene in sentiment, but was constrained
by the assumption that the indigenes' interest would be better served by
withholding political power from them until they could be educated to
make the 'right' kinds of decisions for themselves.

As for the larger issue of indepedence, the Minister for Territories, Paul
Hasluck, told the Australian Parliament in 1960:

> We are not going out of the Territory in a hurry. In our judgment
> of the situation as it exists today, the Territory will need our help for
> many years to come and the advanced leaders of the indigenous people
> say strongly that they need us for a long time ahead. We are not
> going to abandon them or our own people who are working with
> them.

Nevertheless, events elsewhere moved the Administration to speed up
whatever timetable it may have been following regarding both self-

government and independence. Thus, in June 1960, after his return from
a Commonwealth prime ministers' conference, the (Liberal Party) Prime
Minister, Robert Menzies, made the following statement:

> Whereas at one time many of us might have thought it was better to
> go slowly in granting independence so that all conditions existed for a
> wise exercise of self-government, I think the prevailing school of
> thought today is that if in doubt you go sooner, not later. I belong
> to that school of thought myself now, though I didn't once. But I
> have seen enough in recent years to satisfy me that even though some
> independences may have been premature, they have at least been
> achieved with goodwill . . .

Another such event was a resolution adopted by the United Nations'
Trusteeship Council in June 1960, calling on Australia to set target dates
with respect to political, economic, social and educational development
'so as to create as soon as possible favourable conditions for the attainment of self-government or independence'. The government was also stimulated to faster action by events in adjoining West New Guinea, where an
Indonesian government was moving to take over control from the Dutch.
And finally, many indigenes were themselves beginning to express a desire
for a larger share in their own governance, and this sentiment received
influential support from some Administration officials.

In response to these and other pressures the government set up a new
Legislative Council in 1961 containing a majority of non-official members,
including twelve elected members. The most noteworthy action of this
new body, which contained six elected indigenes, was to foster the
development of a much enlarged and more widely representative successor
to itself. This latter body, the House of Assembly, scheduled for 1964,
was intended to provide the Territory's indigenes with a larger share in
their government, including education in responsible citizenship, while
ensuring the continuation of the Territory's political stability and economic
growth. With this in mind, the new body was to include ten official
members appointed by the Administrator, and fifty-four members elected
by universal adult suffrage (males and females eighteen years of age and
over). To choose the elected members, the Territory was divided into forty-
four open electorates and ten larger special electorates; candidacy in the
former was open to any adult resident regardless of race, but only non-
indigenes, i.e. whites or Chinese, could be candidates in the latter. Equality
of population size was the main criterion in establishing boundaries for
the open electorates, but this was outweighed somewhat in an effort to
conform to ethnic, geographic and administrative boundaries as well. Thus
the smallest open electorate, Manus, contained a population of only 18,000,
while the largest, Bougainville District (Bougainville, Buka, Nissan, etc.),
contained 54,000.

The boundaries of the larger special electorates were set so as to ensure a wide representation from the non-indigenes, and this served to delimit such areas according to degree of capital investment and economic development. With this as a criterion, the districts of Bougainville, New Ireland and Manus were constituted a single 'New Guinea Islands' electorate.

The 1964 election was preceded by an intensive educational campaign, and produced some lively contests. In the end, over 70 per cent of the Territory's eligible voters cast their votes in the open electorates—a remarkable turn-out for a country often characterized as politically underdeveloped. Also, six of the forty-four victors in the open electorates were whites and two were of mixed race, an indication that the indigenes were at that time not yet eager to go it entirely on their own. With the election of another ten whites in the reserved special electorates—most of them businessmen and planters—and the addition of the ten administrative appointees (all white), the ethnic composition of the first House of Assembly was twenty-six whites, two of mixed race, and forty-six indigenes. For the first time in the Territory's colonial history, its indigenes were offered an opportunity to legislate for themselves. They were subject, of course, to the Australian government's ultimate veto power, and limited, in fact if not in theory, by their inexperience in operating within the complex and alien procedures of the Westminster parliamentary system.

The first House of Assembly was in office for four years. An expert evaluation of it concluded that its business was efficiently conducted, in terms of the passage of essential legislation, but that it marked little progress in the political education of its indigenous members. Despite their majority, the latter exercised little initiative, and they tended to follow the lead of the white members, official and elected. Moreover, although ten of the indigenous members were appointed to serve as ministerial under-secretaries in the various departments of the Administration, they took little part in the actual running of them.

Parallel to these moves toward Territory-wide political unification and self-government, and preceding them by several years, were the Administration's efforts to increase the scope and domain of government at the local level. Beginning in 1949, encouragement was given to setting up local government councils, which were intended to replace the previous village or neighbourhood units headed by appointed officials (*kukerai, tultul*, etc.) with much larger units and elected officials. The new councils were empowered to levy (some) taxes, and to carry out (some) public works and developmental projects. These councils were slow to be set up, but by the middle of 1970 there were ninety-three of them, ranging in size from 2500 to 43,000 constituents and encompassing over 1.5 million, or 90 per cent of the Territory's population.

No comprehensive study was made throughout Papua New Guinea of these new units, so it is not possible to appraise their effectiveness. However, it is certain that they varied considerably in realizing the pup05es for which

they were established. As a minimum they appear to have helped to transform strictly local loyalties into wider regional ones. Some of them seem
to have been quite successful in undertaking projects in public works and
co-operative enterprise. A few of them also began to assert their independence of Territorial administrative influence, although some others seem
to have deferred to their Administration 'advisers', no matter how scrupulously the latter attempted to transfer initiative to the council members
themselves. It is probably fair to say that, up to 1965, the new institution
had not progressed very far towards educating the council members, much
less the electorate, in the responsibilities of Australian concepts of citizenship.

Meanwhile much of the pre-war administrative and judicial structure
of the newly combined Territories was retained. The executive functions
were carried out by sixteen departments (health, agriculture, public service,
etc.) headquartered at Port Moresby and under the direction of an Administrator who was appointed by the Governor-General of Australia and was
responsible to the Minister for External Territories. In addition, Papua
New Guinea was divided into eighteen districts, each under a District Commissioner, who was responsible for many administrative matters and for
co-ordinating the activities of the functional departments in his district.

At the summit of the judiciary was a Full Court of the Supreme Court
of Australia whose judges were appointed by the Governor-General, and
against whose decisions there was a right of appeal to the High Court
of Australia. Lower down the judicial hierarchy the administrative officers
continued to exercise *ex officio* judical roles, but official policy seemed to
be working towards the objective of completely divorcing the judiciary
from administration.

The Australian government's policy towards the economy of Papua New
Guinea underwent marked changes after the end of World War II.

In the immediate post-war period the emphasis was upon development
as a strategic bulwark against external aggression, and upon assisting the
indigenes to take a larger part in the development, and a larger share of
its rewards. In the words of E. J. Ward, the Minister for Territories at
the time:

> This Government is not satisfied that sufficient interest had been taken
> in the territories prior to the Japanese invasion, or that adequate funds
> had been provided for their development and the advancement of the
> native inhabitants. Apart from the debt of gratitude that the people of
> Australia owe to the natives . . . the Government regards it as its
> bounden duty to further to the utmost the advancement of the natives
> and considers that can be achieved only by providing better facilities
> for better health, better education, and for a greater participation by
> the natives in the wealth of their country and eventually in its
> Government.

In a more pointed statement issued by the Department of Territories a year later, it was announced that:

> for its own protection Australia cannot afford to permit the territories to remain undeveloped or its native population to remain in a backward stage . . . the limitation of the Mandate so far as fortification and defence measures are concerned will not be repeated under the trusteeship system and the administering authority will, in future, be in a position to take whatever measures it considers necessary for the defence of the Territory.

Indigenes and expatriates (and presumably resident Chinese as well) were to work 'side by side' in developing the Territory's natural resources, which were popularly considered to be immense.

This new concern for the Territory's indigenes was expressed in a liberal war-damages compensation programme; in the cancellation of all war-time labour indentures; and in the phasing out of the penal-sanctioned indenture system itself. Approximately £2 million was eventually paid to the Territory's indigenes as compensation for war-caused deaths, injuries and property damages. The cancellation of war-time indentures was also a well-intentioned move, but it resulted in a mass work-stoppage and may have slowed down reconstruction. Also the war damages paid out to indigenes, much of which were dissipated by the purchase of scarce, high-priced trade-store goods, may have served further to slow down reconstruction by reducing the indigenes' incentive to work for wages.

Slowly, however, the economy started moving again, despite the fears of European residents that the end of the indenture system would spell the end of native labour as well. Indigenes continued to work in expatriate enterprises, with or without contracts; and the higher wages and improved rations now owed them proved commercially feasible because of the high post-war prices paid for the Territory's exports, including copra, rubber, and the relatively new major crops of cocoa and coffee.

Another, perhaps the prinicpal, factor in the Territory's economic progress during the late 1950s and early 1960s was the Australian government's large subsidies in the form of direct grants and departmental budgets. The pre-war policy requiring the Territories to pay for themselves had limited expenditures to about £500,000 per annum in the Mandated Territory and to about £180,000 in Papua. By contrast, the Australian government contributed an average of nearly £23 million per annum to the combined Territories during the period 1958–63.

A large proportion of this expenditure went into projects and services of direct benefit to the indigenes—to make them healthier, better educated, more self-governing, and capable of earning a larger share of the Territory's wealth. Health facilities were greatly expanded, and the large-scale anti-malaria programmes that were mounted from 1960 onwards met with spectacular success. Schooling was also increased by direct government operation and by subsidies to missions.

Thus, during the first eighteen years of the post-war period the official economic objective appears to have been a prosperous multi-racial society, with the indigenes comprising a stable, homogeneous, literate and relatively affluent peasantry. As such, it was thought, the latter would supply a well-treated and well-paid labour force for the larger-scale expatriate enterprises, while residing mainly in their own villages, growing their own foodstuffs and raising more and more cash-crops. Altogether an idyllic picture, made possible by a benevolent government and, it was hoped, a compliant indigenous population which would not demand or be prepared for total economic equality for decades to come.

Unfortunately for this well-meaning policy, events outran the complacent assumptions on which it was based. As already noted, national and international politics helped to induce officials to shorten their timetables for the introduction of self-government, and even to speak of independence. Moreover it became increasingly clear that the political trends envisaged would probably undermine the policy of economic development; an indigenous-dominated New Guinea government could not be expected to tolerate an expatriate-dominated economy for very long. Needless to say, when the implications of changing official views were perceived by expatriate businessmen, many of them became less enthusiastic about long-term investments there.

Turning now to Bougainville–Buka, the three agencies of pre-war colonialism — Administration officials, missionaries and planters — resumed their operations on these islands as soon as the military allowed, and in much the same manner as before. While Administration officials were armed with the new policies described in the preceding pages, these apparently did not come into play for some time. And while the missionaries may have begun to view their tasks in a somewhat broader way than before, the old pre-war mission rivalries reasserted themselves. (In south Bougainville, for example, it was reported that one mission would not provide storage space for another's building supplies.) As for the expatriate planters, the frustrations of rehabilitating overgrown groves with an indifferent and unreliable labour force would have discouraged the most liberal of sentiments among them.

Meanwhile, the Bougainvillians themselves, the subject of all these efforts and sentiments, had undergone some major changes both in life circumstances and in mental attitude. To begin with, many indigenous communities had suffered grievous losses, as, for example, in south Bougainville, where some Methodist teachers reported that the population of their villages had decreased by about 50 per cent as a consequence of exposure, malnutrition, and resulting illnesses. Also, throughout the southern half of the larger island, taro, the principal food crop, had been wiped out by disease, a

blight *Phytopthora colocasiae*. The substitution of another staple, sweet
potatoes, eventually removed the hazard of famine in the blighted areas
but resulted in some major changes in diet, in gardening practices, and
in patterns of land use.

Some £400,000 — about £8 per capita — was distributed among Bougainvillians as compensation for casualties and property damages caused by the
war. At this late date it is impossible to learn what effects this windfall
had on the recipients' daily lives or mental attitudes, but it undoubtedly
relieved many of them of the immediate necessity of resuming wage labour
on plantations or on government projects.

Although the early post-war Administration professed a deeper commitment to the welfare of indigenes, the district headquarters was set up at
Sohano, a small island at the western end of Buka Passage which was
geographically isolated and effectively insulated from all indigenes except
those actually employed by the Administration or those working in officials'
homes. Kieta, the former headquarters, and the historic colonial centre
of the populous east coast, was relegated to sub-district status. From Sohano,
a salubrious, comfortable and thoroughly European enclave, the administrative staff sought to supervise the islands' rehabilitation and to introduce
the new programmes as commanded by Canberra and Port Moresby. Efforts
were made in the early 1950s to transform the pre-war village *kukerai* system
into local government councils, but no progress was made for a number
of years: the District's first, at Teop-Tinputz, was set up in 1958. Eugene
Ogan's account (1972) of a council's beginnings among Nasioi-speakers
in the Kieta area will exemplify how inertia and direct opposition served
to frustrate this well-intentioned move to give the indigenes a larger measure
of self-government:

> The effect of this effort in polarising opinion in the Kieta-Aropa area
> is difficult to believe if not observed at first hand. Local government
> councils became an *idée fixe*, albeit with varying associations, among
> all parties concerned. Administration officials seemed convinced that
> establishment of a council would solve most of the problems of 'cargo
> cult' and general disaffection among the Nasioi. Consequently a great
> proportion of the patrol officer's time and energy went into 'sales
> talks' for the council. Nasioi proponents of the council—almost
> exclusively SDA [Seventh Day Adventist] and Methodist in the early
> stages—became sufficiently enthusiastic to exaggerate further the
> already grandiose Administration claims of what a council would
> accomplish: the favourite theme, in Pidgin, was '*Sapos mipela gat
> kaunsil, mipela stap olsem ol waitman*' ('If we have a council, we will
> live like Europeans'). The patent absurdity of some of these claims
> provided fuel for the anti-council majority.
>
> This dissident majority became equally heated in their denunciations,
> to the point where other efforts at social development—directed towards

goals shared by all Nasioi—were rejected out of hand because they
were tainted as '*samting bilong kaunsil*'. Were the Nasioi encouraged
to plant cash crops? '*Nogat, ol i laik pulim mipela long kaunsil*' ('No,
we won't. They're just trying to draw us into the council'). The
favourite anti-council rumour was that, in order to pay an increased
council tax—and it must be noted that tax in many council areas
is considerably higher than the annual head tax in the [Aropa]
Valley—Valley women would have to prostitute themselves for the
necessary cash.

The Administration's post-war efforts to improve the education of
Bougainvillians began by granting subsidies to mission schools, in return
for some control over the curriculum and, later on, over the qualifications
of teachers. It was not until the 1960s that government high schools as
such began to be set up, the first at Hutjena in 1964. Official post-war
efforts to improve the indigenes' health also got off to a late start, but
in one respect at least met with spectacular success. In the early 1960s
the Administration embarked upon a malaria eradication programme by
mass administration of suppressants and by spraying houses with DDT.
This campaign resulted in a speedy and dramatic reduction in the incidence
of the malady and, probably as a direct consequence, a marked reduction
in infant and child mortality, thereby creating a long-term problem of
another kind; namely, an unprecedented population increase.

The Administration produced impressive results in encouraging Bougainvillians to plant cash-income crops. After several ill-starred beginnings,
allegedly brought about through initial enthusiasm but lack of follow-
through on the part of non-specialist officials, most indigenous enterprises
in this sphere settled down into the growing of coconuts and cocoa. The
missions were also very active in these endeavours.

One of the more visible consequences of the Bougainvillians' turn to
cash-cropping during the late 1950s and early 1960s was their reluctance
to work on the islands' expatriate plantations. A few of the latter still
managed to attract some of their Bougainvillian neighbours, but had to
depend increasingly upon New Guinea mainlanders—from the Highlands
or Sepik Districts—to perform the onerous and monotonous jobs on which
plantation production continued to depend. Most Bougainvillians who
resumed work on expatriate plantations did so on a casual or short-term
contract basis.

Another post-war landmark, for some at least, was the end of the regula-
tion prohibiting purchase of alcoholic beverages. In 1962–3 indigenes
throughout the Territory were permitted full access to beer, wine and
spirits—with consequences that turned out to be more harmful than
beneficial.

The first organized mass reaction to post-war political and economic
changes took place on Buka in the form of the Hahalis Welfare Society.

The full story of this many-faceted movement has not yet been published, and is obscured by the conflicting accounts of its friends and enemies. The more palpable facts are as follows.

During the 1950s two Buka youths, John Teosin and Francis Hagai, discontinued their studies in mission schools and returned home intent upon modernizing their communities through communally organized activities. As Hagai described,

> We held a meeting of all our relatives, and suggested that instead of working separately we should all work together. Thirty people were ready to try and we started in a small way. . . .
>
> We put all our fowls together, instead of keeping them in separate runs, and we grew some peanuts, which we sold to traders. We did not spend this money, but used it to open a small store, which we stocked with bags of rice, cases of tinned meat, shorts, shirts, and women's clothes. Before this, those who wanted to buy any of these things had to travel 20 miles [32 kilometres]. . . .
>
> Soon people saw that our idea was a good one, and they joined us. We gradually increased our activities.

By the mid-1960s the society was officially registered as a private company. By then it consisted of fifty shareholders, and some five thousand other individuals who participated in some way in its activities, that is, over half of the total population of Buka. Its major economic activities were at that time the production of copra and dried cacao beans; it maintained eight co-operative stores, owned its own trucks, built and maintained feeder roads, operated an electric generator, engaged in dressmaking, etc. Productive work was communal and followed a regular weekly schedule. All profits went into a common fund from which members were doled out cash for specific needs. Officers were elected (chairman, village spokesman, president, secretary, treasurer, etc.) and met regularly to plan activities, decide on expenditure and oversee accounts; a philosophy of self-help and hard work prevailed. All very prosy and admirable; just what Administration officials liked to see. But the society, or 'Welfare' as it came to be called, included other practices that pleased officials and missionaries not at all.

What disturbed the missions about Welfare were its alleged cult-like aspects and its sexual 'immorality'. As described earlier, beginning in German times and continuing into the Japanese occupation, large numbers of successive generations of Bukans became devotees of self-styled millenarian prophets who promised white-type material prosperity and an end of white colonial domination, by means of magic and pagan practices. It was probably inevitable that the new movement would acquire some features of the old, or that its opponents would label it cargoistic. According to one undocumented interpretation, the youthful leaders of Welfare, although themselves wholly secular in outlook, deliberately added some cargoistic elements to their programme in order to attract the support of less educated but more influential men.

Like previous manifestations of mass movements on Buka, the latest one included both Christian and pagan (or rather neo-pagan) elements; but this one contained much specifically anti-missionary, principally anti-Catholic, sentiment as well. This centred on the grievance that by providing their converts with catechisms only, in place of the full Bible, the Catholic missionaries were deliberately hiding from them the full scope of Christianity. From their side, the Marist opposition to the society focused mainly on its 'baby gardens', where young unmarried women were domiciled and visited by male members of the society. To its scandalized opponents these baby gardens were regarded as free and communal brothels, which served to attract and hold male members within the society and to express open contempt for mission-imposed morality. To their defenders, who labelled them 'matrimonial clubs', they were a rational system of trial marriage and served as a sensible means for satisfying rational and God-given sexual needs.

Meanwhile marriages continued to be engaged in by some society members, but without the customary bride-price or Christian ceremony. Marist opposition to baby gardens led to excommunication of the society's leaders (some of whom had received training as Catholic teachers), and in 1961 the latter countered by setting up a congregation and *lotu* of their own which included Bible reading, hymn singing, and prayers to God via the members' own ancestral spirits. Most society members took their children away from mission schools and their sick from mission hospitals. By 1962 the Marists had lost hundreds of their parishioners to the new movement. The Methodist mission also lost a large number of its members and adherents to the society, but adopted, publicly, a less uncompromising policy toward the movement's sexual and religious practices, thereby leaving the door open for defectors' eventual return.

Administration officials at first appeared to view the new society as an enterprising self-help effort, and later, with the establishment of baby gardens and a separate *lotu*, as a problem mainly for the missions. But when in January 1962 the society's leaders complained that the government had done nothing for them and advised their followers to refuse to pay the annual head tax, the problem became an Administration one as well. When an Administration official attempted to collect taxes from the Welfare leaders he was opposed, as was a larger party of police sent to arrest the defaulters. More police were then rushed from Bougainville Island and Rabaul to arrest the rebels, but this force, 155 strong, met with heavy resistance and also had to withdraw. Finally, Administration officials and some 400 more police were flown to the scene from other districts. These succeeded in arresting hundreds of the rebels, 256 of whom were sentenced to gaol for periods of three to six months.

This punitive action served to chasten the society's members — for a while at least. Tax-paying was not resumed until some time later, but many of them returned to mission religious services and sent their children back to mission schools. But then, after two months in gaol, all those who had been arrested, except the leaders themselves, were released following a successful court appeal. When the freed men returned home, they and other society members resumed their opposition to missions and Administration, and large numbers of other Buka Islanders flocked to join them, further swelling their ranks.

After a while some of the anti-government feelings of the movement were dissipated when the Adminstration constructed a good all-weather road down the entire length of the island, thereby greatly facilitating the transport of cash-crops and trade goods. Nevertheless, the society remained aloof from the island's Administration-sponsored local government council.

Meanwhile, on Bougainville Island itself there occurred some manifestations that can be more appropriately labelled cargoistic. In 1959, for example, some cultists near Teop were caught up in a plot to murder the local missionary, whose prayers they blamed for delaying the cargo. In 1960 the mission station at Keriaka was looted, and its priest forced to flee for his life, by indigenes who blamed him for their lack of success in obtaining cargo by means of their cult practices.

Faced with these reverses, the Marists adopted a more positive attitude towards the economic side of their people's lives. For example, they helped to organize several timber-milling and house-building co-operatives, one of them in collaboration with the Methodists (in itself a radical change from the mission's anti-ecumenical past). At Torokina they assisted in the replanting of war-devastated coconut groves, and in Nagovisi they engaged in road-making. But perhaps the most far-reaching of the Marists' secular enterprises took place along Bougainville's northwest coast where over 3000 mountaineers were encouraged to resettle and then assisted in planting over a million coconut palms and cacao trees.

These kinds of projects sponsored by the missions, and evidence of Administration concern such as was shown in the building of the Buka road and in the anti-malaria campaign, may have served for a while to reduce anti-white sentiment in parts of Bougainville, but other events elsewhere fed the discontent. One of the most widely publicized of these latter concerned a timber project in southeast Bougainville, a succinct account of which was written by the Marist mission priest, Father Wally Fingleton:

> . . . some 10 years ago the Forestry Department negotiated for a large area of timber surrounding Tonolei Harbor. The Administration assured the people that all it wanted was a 40-year lease on the timber. They were assured that their land was sacrosanct. At the same time, they were promised that, if they agreed to the timber sale, they would be given roads leading from the back of the timber lease to the

harbour, thus giving them an outlet for their products. The sum of
£30,000 was paid to the land-owners for what was estimated at more
than 500 million super feet [1.2 million cubic metres] of standing
timber. Subsequently, the Administration announced that it required
200 acres [80 hectares] of harbour frontage at Tonolei. The owners
agreed that the Administration could have the use of the land for the
period of the lease. They were not willing to sell it. Canberra
usurped the 200 acres by 'right of eminent domain'. One of the
owners has some ground at Samiai, a swampy area on the coast
outside Tonolei Harbour. In 1958 a two-year rental for one acre at
Samiai was arranged by the Administration between a Japanese salvage
company and the owner. He was paid $14 for the two-year rental of
his acre of land. When his Tonolei Harbour frontage, on good
ground, was completely alienated by Canberra, he was offered $2 per
acre. He, along with the other owners of the 200 acres, refused to
accept the money. The ground was taken, anyway. The Japanese
occupied Tonolei Harbour during the war. They were pushed out by
the Allies who claimed that they were doing so to hand the land back
to the native owners.

Some 20 years ago, the Tonolei owners in question set out with
bows and arrows to shoot the Qantas Catalina that had initiated a
service to Buin. They looked upon its advent as the start of a new
invasion. The warriors were assured that their land was not in
jeopardy. They took their bows and arrows home. Now, they are
told that by the stroke of a Canberra pen, without any armed
invasion, they are no longer the owners.

They can scarcely believe their ears. The timber company,
Bougainville Development Corporation, has, to date, lost more than a
million dollars, and is in recess. Contributing reasons for failure were
wrong techniques and a vicious squeeze by the Japanese market. There
is not one road in the area. So much for promises. Disillusionment
and distrust continues among the natives.

Another well-publicized expression of Bougainvillians' dissatisfaction
occurred in April 1962 when a United Nations mission visited Kieta on
an inspection tour of the Trust Territory. At a general meeting at Kieta
several Nasioi, all Catholics, complained of their treatment by Australians
('like dogs'), and requested that the administration of Bougainville be turned
over to the United States—probably a residue of war-time experience with
the powerfully armed, richly equipped, more egalitarian and generous GIs.

The first Territory-wide elections for the new House of Assembly in
1964 provided Bougainvillians with an unprecedented opportunity to express their sentiments about their governance in particular and their lot
in general. It will be recalled that the whole Bougainville District (Buka,
Bougainville, Nissan, etc.) constituted one open electorate—the most

populous one in the Territory — and that the district was joined with New
Ireland and Manus to form a single special electorate, in which only non-
indigenous candidates were eligible.

Nine candidates, all of them indigenes, competed in the open electorate.
Most of these were, or had been, Catholics, and all were comparatively
well educated. The winner by a wide margin was Paul Lapun, a member
of the Banoni language-community of southwest Bougainville and conversant in all the other languages of south Bougainville. Lapun was forty-
one years old at the time and a successful farmer and cash-cropper. He
had received secondary education in a Catholic seminary, and was fluent
in English. Although popularly believed to be a devout Catholic, he was
not as closely identified with any Christian mission as were some other
candidates. Nor was he as vocally for or against the Administration as
were some other candidates. Although he was believed by some of his
backers to possess the esoteric knowledge essential for acquiring cargo,
he himself appears to have made no such claims (in comparison with at
least one other candidate who was a leader in the Hahalis Welfare Society).
In any case, whatever the basis of his strong support at the time, Lapun's
appeal was so magnetic that in one of the district's twenty-three census
divisions (the only one for which such information is readily available)
most of the large majority who cast their votes for him declined to voice
any other preferences, despite their right to do so.

This tendency to support only one candidate showed again in the special electorate, in which the votes constituted only 58 per cent of those
cast in its three component open electorates. In this case the circumstance
that all the special electorate candidates were whites must have reinforced
the Bougainvillians' tendency to abstain.

The 1964 elections were a fitting culmination of Bougainville–Buka's
early post-war period of gradual and familiar change. The next twenty
years were to witness the creation on the larger island of a huge mining
enterprise, and with it an accelerated rate of economic, social and political
change. That enterprise was only one of the forces involved in that change
but it was a very powerful one; however, before describing it we shall
take a closer look at the ways Bougainvillians lived just before the mining
began.

7

Bougainville Before the Mine

A HISTORIAN WRITING ABOUT BOUGAINVILLE (AND BUKA) CAN with some justification divide the description into periods or eras, as has been done in this book; the focus, as in most 'colonial' histories, has been on the actions of the outsiders, including what they did to the 'natives'. ('Natives' also have 'histories', but their lack in most cases of written records renders them accessible only through the meagreness of archaeology or the ambiguities of oral traditions.) For the anthropologist attempting to place his descriptions within the time frames delimited by colonial historians the task is difficult, to say the least. In the first place, in order to learn what effects the earliest outsiders had on the natives it is essential to know what the natives were doing (and thinking, etc.) before the outsiders appeared. In nearly every case, however, the first authoritative and comprehensive study of an illiterate native people has taken place years, even decades, after first contact with the outsiders had resulted in changes in their ways of doing and thinking.

And such was the case with Bougainville. Before World War II only three of Bougainville's societies had been studied comprehensively by trained anthropologists. One of those was Buin, which was studied by the German anthropologist, Richard Thurnwald in 1912–13, and again by him and his wife in 1934–5. A second was Halia (just southeast of Buka Passage), which was studied by the English anthropologist, Beatrice Blackwood, in 1930. The third was carried out by myself in 1938–9. (A German ethnologist, Ernst Frizzi, visited and wrote about the Nasioi and the Kongara before World War I, but his reports are relatively brief and piecemeal.) At the time of the three larger studies just listed, none of the societies described in them was 'pristinely' non-Europeanized, but enough of their respective aboriginal cultures was still current, or fresh in the memories of their elders, to permit credible reconstructions of their pre-European patterns.

Because of war-time interruptions and other circumstances, no other Bougainvillian societies were studied until 1963, when the American anthropologist, Eugene Ogan, commenced his long-term research among the Nasioi. After that and throughout the 1970s, competent ethnographic studies were made of the Nagovisi, by Jill Nash and Donald Mitchell; the northeast Buin, by Jared Keil; the Eivo-Simiko, by Michael Hamnett; the Aita, by John Rutherford; the Teop, by Robert Shoffner; the Buka, by Max Rimoldi; and the modern-day Siwai, by John Connell. While the focus of all these more recent researches was on current practices and beliefs, it is possible to extrapolate from them some clues about certain aspects of those societies' pre-European pasts, including those beliefs and practices most directly affected by outsiders. But first, some inferences about the numbers and the health of the islands' pre-European peoples.

The earliest effort to enumerate the indigenous populations of Buka and Bougainville was made by German officials, but the figures arrived at pertained only to people dwelling along or near the coasts, in the areas under more or less continuous police control; they are of little value even for the areas in question. Not only were they based on very superficial nose-counting, but during that period, and indeed for many years thereafter, intertribal fighting resulted in some shifting of people between 'controlled' and 'uncontrolled' areas.

Later, when Australian officials superseded the Germans during World War I, they instituted annual census patrols into the known, controlled areas, and began more systematically to bring the other areas under administrative control. By World War II there were only a few small isolated areas, mainly in the north central mountains of the larger island, that had not been visited by Australian patrols, so that there were actual nose-counts or honest estimates available for both islands. (See Figure 4.)

Census-taking is difficult and more or less inaccurate under the best of conditions, and under the kinds of conditions that prevailed on Bougainville even as recently as 1939 the figures are not very dependable. But they are the best available and, for better or worse, will have to serve as a basis for calculations of population trends.

What can be said about the numbers of Bougainvillians a half-century earlier than 1939, before the activities of Europeans had begun to exert influences upon their birth and death rates? These influences were both constructive, through medical services and the discouragement of murder and warfare; and destructive, through the introduction of alien diseases and the employment of adult males on plantations far from home. Lacking precise information from Bougainville–Buka concerning the relative importance of these opposing influences, one might be led to turn to better-known Pacific island populations and inquire how their numbers have tended to change under similar influences. Such inquiries, however, can lead to widely varying conclusions. In some places the first few decades of European contact were unmitigatedly disastrous; in others an initial period of population decline was followed by a gradual, continual rise in numbers. Generally speaking, the populations that were met with earlier were the ones to suffer most in the early decades of contact, and in this respect the indigenes of Bougainville–Buka were perhaps more fortunate than those of, say, Australia and eastern Polynesia. This kind of reasoning is straw-clutching at best but the population figure it leads to, approximately 45,000, is perhaps better than an outright guess.

Guesses about where these 45,000 or so indigenous people lived prior to European contact are no better founded. According to Parkinson, the German planter referred to earlier, the coasts were only sparsely populated in his day (1882–1906), but he noted that such had not always been the case. Coastal settlements had become depleted by warfare or kidnapping, or whole communities had moved inland to escape European labour recruiters, or other islanders. On the other hand, the type of lands well suited to indigenous gardening is mostly far enough inland to discourage the owners from living directly on the coasts. Some day archaeologists may provide better answers to this question of where people lived in pre-European times, but until then we must rest content with indications such as these.

What was the state of health of the Bougainvillians just prior to their direct contact with Europeans?

An authoritative answer to this question is of course not possible. The first Europeans to see these indigenes at close hand were interested in other matters, and were not trained for making observations of this kind. Decades later, during the Mandate era, medical patrols occasionally toured the more accessible parts of the islands, but the records of their observations have either been destroyed or disappeared into the depths of some archives. In fact, all that one can say about the health in that and previous eras is derived from generalizations made about New Guinea as a whole, namely that the principal physical ailments affecting the indigenes were (in order of prevalence): yaws, tropical ulcer, scabies, malaria, pneumonia, injuries from accidental causes, skin diseases, infections of the upper respiratory tract, leprosy, pulmonary tuberculosis and spinal and cerebro-meningitis. This information is based on records of patients actually treated in hospitals or medical centres; how accurately it represented conditions among indigenes unable or unwilling to reach the medical centres is impossible to say. Nor do these data indicate which of these ailments had been introduced or exacerbated through contacts with Europeans, although it is safe to conclude that many of them (e.g. malaria, yaws, tropical ulcers) were prevalent in New Guinea in general, and on Bougainville–Buka in particular, prior to European contact. Later observations have recorded a high incidence of goitre among present-day Bougainvillians residing in iodine-poor volcanic soils heavily leached by rain. (Friedlaender 1975, p. 19) It is highly likely that such was also the case in pre-European times.

Turning now to the pre-European native institutions most directly affected by outsiders, including especially the mining enterprise, it is essential to note that the subject is *cultures*, in the plural. Whatever cultural uniformity may have prevailed on Bougainville–Buka at the beginning of human settlement there, 28,000 years ago, had diversified greatly over time: first, through wide dispersal of the bands of pioneers followed by millennia of sequestered isolation; and much later, through immigrations from elsewhere. The multiplicity of the islands' languages offers the most conspicuous aspect of that diversity, but this chapter will focus on some other, less conspicuous, ones, beginning with these islanders' ways of satisfying their material needs and wants.

Indigenously, Bougainvillians appear to have resided in two forms of settlement. Those on very small offshore islands, or immediately along the coasts, resided in clusters of houses which were sometimes numerous enough and close enough together for us to call them villages; whereas inlanders tended to live in smaller and more widely scattered hamlets. Some villages may have contained as many as fifty or so dwelling houses (i.e. about 250 to 300 people), but most of them were probably nearer to twenty houses in size. As for hamlets, they ranged in size from one dwelling to ten or so, and probably averaged four or five. As we have seen, Europeans have for decades encouraged or compelled Bougainvillians to build their dwellings in larger, village-size settlements for purposes of closer administrative supervision, etc., but their desire to live in small and scattered settlements still seems to prevail, in inland areas at least. (There are sound economic reasons for this, in addition to some impelling social and sentimental ones.)

The dwellings were made entirely of plant materials — wooden poles, leaf-thatching, woven rattan, fibre-strip bindings. Throughout both islands they usually had closed sides and were built mostly on the ground, but there were some local variations in architectural style. In some instances cooking sheds were associated with dwellings, but more usually the occupants cooked, ate and slept inside the dwelling itself.

In many villages and sizeable hamlets there was to be found another type of building usually larger than most dwellings. These were the men's houses, places where youths and adult males congregated and where some of them slept from time to time. Such buildings were conventionally reserved for males, sometimes so exclusively that a female would have been killed had she entered one.

Bougainvillians were, and most continue to be, farmers — by tradition, by necessity, and by choice. In pre-European days they obtained most of their foodstuffs from their own gardens, supplemented by fishing, pig raising, collection of tree crops and wild plants, and a little hunting. With the use of a few simple, self-made tools they managed to subsist in a natural environment where few Europeans so equipped would be able to survive.

The basic tools were made of stone, shell and wood. Stones were ground into numerous sizes and shapes of axe and adze blades, or used as hammers. Marine shells were used for cutting and scraping; bone for piercing and splitting; and wood for digging, piercing, hafting, beating, cutting and numerous other purposes. Unlike most modern European tools, these were limited in variety but versatile in use. For example, the same stone adze was used for felling a tree, hollowing out a canoe, thinning down a plank, carving an image—and attacking an enemy.

In other words, the craftsmen (which is to say nearly every adult male) depended more on his versatility in skills than in variety and refinement of tools. But let us attempt to reconstruct the purposes for which these tools were used.

Food-getting consumed most of the time and energy of most Bougain-villians prior to 1965. This was even more so in pre-European times, before the acquisition of European cutting tools made gardening easier, and alter-native ways of earning a living were introduced. There are no detailed records available concerning the food-getting activities of these islanders in pre-European days. An eye-witness account (by this writer) of what they were like just before World War II (i.e. before commercial cash-cropping began to be important) may be taken as not significantly different from what they had been six decades previously. This account concerns the Siwai, a people speaking the non-Austronesian Motuna language and living in the plains and foothills of southwestern Bougainville.

More than half of an adult Siwai's working hours were taken up with food—producing, collecting, processing, and consuming it. This was an average, for women spent about ten hours daily at it and men generally less. They ate a wide variety of comestibles; but when they spoke of 'food' (*pao*) they were usually referring to taro. In month-long records I kept of meals eaten by representative households, taro constituted some 80 per cent by weight of everything eaten. Informants told of having eaten a few 'wild' taro during straitened times; otherwise all taro consumed was grown in their gardens.

Taro gardens were laid out on well-drained land where the soil was deep and free of sand. Another technical requirement was that gardens be located in areas of secondary growth; I did not see a single instance of primary forest being cleared for gardens. My informants said that it was too difficult, even with metal axes, to cut down and remove the huge trees; but there were other reasons why new gardens were located on old garden sites.

Taro gardens were laid out in patches fenced in to keep out pigs. These patches were rectangular; a single patch varied from 1000 to 4000 square metres in area. Very rarely did one see isolated patches; they were gener-ally arranged in sequence, as shown in Figure 11. Taro was ready for harvesting after about four months' growth, and planting was a continu-ous process without perceptible seasonal differences in growth and yield. Added to this was the fact that the Siwai did not know how to preserve taro and hence had to consume it within a few days after harvesting. As soon as the plants were harvested from a matured garden the corms were cut off just below the stalks, for eating, and the stalks were replanted in another site after their leaves had begun to rot. The gardener's ideal was to have several contiguous patches in various stages of growth. Figure 12 illustrates the technique. Patch A was completely harvested and overgrown with reeds and small trees; most of its fence timbers had already been used for firewood and those remaining were scattered and rotting. Patch B had been completely harvested of taro and contained only a few banana and plantain trees still bearing. Patch C contained growing taro, some of which was ready for harvest. Patch D contained unripe taro in various stages of growth. Patch E had been wholly prepared for planting and contained a few plots of new taro shoots. Patch F was in process of being cleared and fenced; some trees had been felled and split for fence timbers, others had been strip-barked and left standing. The remaining plant rubbish was being piled and burned and the ashes were scattered over the patch to enrich the soil.

The ideal technique, in the hilly northeast region, was to progress in one direction for about six years and then return to the starting point and begin again. For such a process each family would require a continuous strip of fertile land about 30 metres wide and about 1300 metres long.

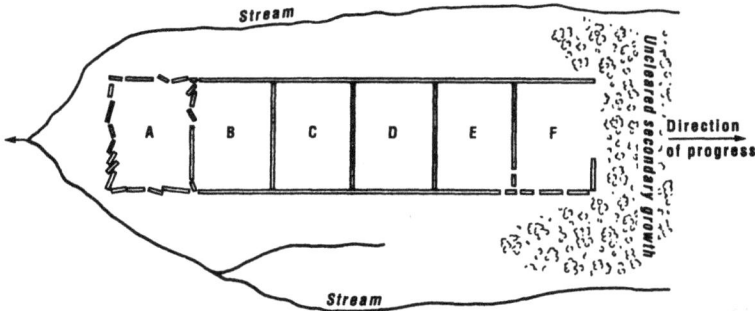

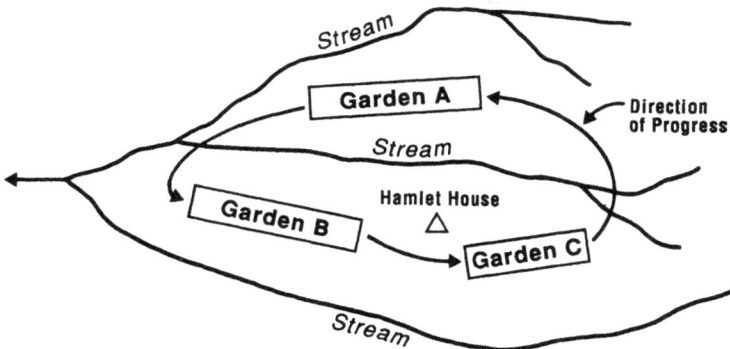

Figures 11–12. Siwai garden cycles

A mapping survey of northeast Siwai indicated that such an ideal was seldom realized. Some kind of barrier, either another household's garden or an effective natural or cultural boundary, usually got in the way. What usually happened was that each household had three or four gardens, as in Figure 12. Garden A had been deserted for three or four years; it would be replanted when Garden C was completely cultivated. Garden B was partly harvested (at the southwest), and there was a stand of ripe taro at the northeastern end. Garden C was being cleared and fenced in at the time; the gardeners planted the stalks from Garden B in the southern end of Garden C. Firewood was collected from the southwestern end of Garden B. In most taro patches they also grew some tobacco, a dozen or so plantain and banana plants, and several yam and gourd vines trained along the fences.

In 1939 sweet potatoes were second only to taro in Siwai diets. As their local name, *peteita*, indicates, they were of recent introduction, assertedly from Alu, but their use was expanding and even during my stay potato acreages increased. Older people expressed some contempt for them, calling them pap: 'children's food; not solid, strength-giving food like taro'.

No Siwai meal was complete without a few slivers of coconut meat or a portion of coconut oil poured over the vegetables. For special feast dishes it was indispensable; it was shredded and baked with sago or mixed with boiled taro. It was used to 'grease' food generally, and in other ways. Most important was the coconut's everyday use as a light repast or thirst-quencher; except for water and broth, it was a Siwai's only drink. There was no cash trade in coconuts: the only potential purchaser was a Chinese trader located on the coast one day's walk away. But judging by the frequency with which ambitious people had to taboo their groves to accumulate nuts for feasts, there was still an unfulfilled subsistence need for coconuts.

The breadfruit season lasted from early May to the middle of June and while it was in progress people roasted the fruit and ate it as a staple. Canarium almonds left a deeper impress on Siwai life. When they ripened in late July and August, even garden work was forsaken for a few days to allow time for collecting nuts and extracting the kernels.

Compared with the vast areas of sago palms found in New Guinea, the few small groves scattered around the marshy places of southern Bougainville appeared quite insignificant. Nevertheless these antediluvian, giant fern-like plants did have a number of important uses for the Siwai. The starch obtained from the pith of the trunk was a substitute food staple, fronds were the principal thatching material, the broad bases of branches were used as troughs, and rotting palm stumps crawled with choice edible grubs. Sago flour was obtained by felling the palm, stripping off its outer hull, shredding the pithy centre, and washing out its starch into a wooden standing-trough. After the starch had settled, the water was drained off and the flour packed into cylindrical-shaped leaf containers for storage.

In addition to the plants the Siwai grew or encouraged to grow for food, they collected *hari* nuts, and many kinds of edible leaves, nuts, ferns, mushrooms and fungi from the forest. Some of these wild foods were ordinary fare, others had special uses: one kind of wild yam was a rare delicacy; pig's-wife fern was eaten only with pork; the leaf of the *surasia* tree was a delicacy used in invalids' broth. All such foods were used as relishes, because they were not 'solid' like taro; they constituted only a small proportion (about 2 per cent by weight) of any ordinary meal.

When food staples became scarce as a result of crop failure and sago shortage, people warded off starvation by collecting wild sago, yams and taro. Such times were infrequent and only one such 'big hunger' was recalled by my informants, who attributed it to a prolonged drought.

Salty condiments were used extensively and were considered indispensable for the domestic meal. People nearer the coast made a salt by evaporating salt water; those farther inland utilized the ashes of several plants. Salt ash was kept on the smoke rack over the hearth to increase its pungency.

Collecting edible insects was only an incidental activity. If people saw them they sometimes tried to catch them; there were no special implements involved. In addition to the sago grubs, which were regarded as a delicacy, the Siwai liked to eat beetles, white ants — large and small — and certain kinds of spiders.

Hunting was less haphazard. Hunting for wild pigs was in fact a serious undertaking requiring skill, persistence, and some courage. Many male Siwai took sporting pleasure in hunting, even though the booty was usually small. Wild pigs and possums were the chief quarries, but people would shoot arrows at tree-rats, flying foxes, and birds if they saw them reasonably close.

Fishing was also an occasional pastime with many Siwai, but did not add significantly to their diets. One often saw a solitary man searching for fish and eels with his bow and arrow up and down a stream, hopeful but rarely successful. Once a fortnight or so, women and girls would spend a few hours wading in a stream searching by hand for prawns that hid under ledges and roots. Almost everywhere large and small basket-traps were manufactured and left in likely spots — every 60 to 90 metres of a stream's course would harbour one; but these remained empty for weeks at a time, and a catch became a lively conversation topic in the neighbourhood. In addition to these individual and spasmodic attempts to catch fish, people sometimes — as often as two or three times a year in communities bordering large creeks — engaged in co-operative fish drives.

Pig-raising was vastly more important than hunting and fishing in Siwai economy. Nearly every household owned at least one pig, and most averaged three or four. Domesticated pigs were fed once a day, during the late afternoon, and the rest of the time they were allowed to run free and forage. According to local belief, based apparently on experience, a full-grown pig must be given about 2.5 kilograms of food daily in order for it to remain properly domesticated. If fed less, it would break through the strongest of fences and devour garden produce, or wander farther from home and invade the less-protected potato gardens, or, worse still, it would run wild altogether in the forest. Before the introduction of European strains, wild pigs were probably not very different from domestic ones — thinner and tougher, perhaps, but probably the same breed.

Every settlement contained a few fowl which fended for themselves by stealing food leavings and catching insects. They hid their eggs in the scrub, but most people did not care for eggs anyway. Only on rare occasions did the Siwai kill and eat one of these athletic birds.

Dogs were also to be found in every settlement — almost in every house. A few of these wretched rail-thin creatures were used in hunting but the rest served only as pets, if that is the proper word to describe the lot of those half-starved, often kicked animals.

The Siwai domestic meal, the one full meal of the day, seldom varied in its main outlines: each individual past early childhood had a basic portion of about 1.5 kilograms of taro or sweet potato along with a helping of cooked greens flavoured with salt or coconut oil, and now and then some small taste of relish — meat, fish, grubs or other delicacy.

The Siwai chewed the betel mixture frequently and smoked almost continually. The betel mixture consisted of areca nuts, collected from purposely planted palms, along with catkins taken from a pepper tree, and lime obtained by burning shells collected on the southern beaches. The Siwai said the betel chewing staved off hunger, but they also chewed it immediately after meals just for its pleasant taste. Local tobacco was smoked in trade-store pipes. It was grown in the gardens along the fences, and although some of it was briefly 'dried' before use, those inveterate and hardy smokers often picked a green leaf, 'cured' it over a smoky fire for a minute or two, and smoked it forthwith.

Comparatively speaking, the Siwai just described occupied a more favourable setting for gardening than did many other peoples of these islands. Like others living on the higher plains and lower foothills, they were able to find ample land that was both well drained and not too steeply sloping. Their settlements were larger, closer together, and more permanent, since they did not have to travel far from houses to garden sites. In contrast, peoples living higher up the mountain ridges — such as those in what was to become the mining area — had to eke out their existence from smaller and more widely scattered garden plots perched on steeper slopes. These latter undoubtedly worked harder for their food than did the people of the plains, and lived more isolated lives.

In even sharper contrast were the lives of the coast-dwellers; these spent much time fishing, partly for their own consumption and partly for trade with inlanders. Some coastal settlements had access to good farming land but others, hedged in by swamps or coralline ground, or located on tiny offshore islands, had virtually none.

In all of their subsistence activies the Bougainvillians made use of magic to make crops grow, to induce fish to be caught, etc. On the other hand, I was impressed with the amount of experimental knowledge gained from

Geologist Ken Phillips with Dateo of Pakia village, October 1964.

Shares in Bougainville Copper Limited were available to local residents and employees. Colin Bishop listens to an explanation in the local language (*tok ples*), 1971.

Fijian Prime Minister visits Panguna, 1974. *From left, foreground:* then PNG Chief Ministe[r]
Michael Somare, Deputy Chief Minister Albert Maori Kiki, Don Vernon,
Rato Sir Kamisese Mara.

Arawa township, 1988. North Solomons Provincial Government offices in foregroun[d]

trial and error, which evidently lay at the basis of so many of their subsistence activities. There was nothing particularly Melanesian about the application of a mixture of supernaturalism and science to activities as crucial as food-getting; like humans everywhere, these islanders depended first of all upon their practical knowledge and labour and skills to do what was needed. Then—and usually only then—they resorted to other means to help ensure the desired results. Like prayer, magic helps humanity to relieve the anxiety that accompanies uncertainty in human enterprise. It is probably as old as humanity itself, and in some form will probably endure just as long.

Another characteristic feature of pre-European Bougainvillian life was trade. Nearly every community seems to have produced something lacking in and wanted by some other: taro was traded for fish, clay pots for colouring matter, pigs for spears, etc. Parkinson described one episode of trading between inlanders and coast dwellers as follows:

> Each district is in a state of constant war with its neighbours, though, as I have already mentioned, a more peaceful state of affairs is gaining ground year by year as the result of the efforts of individual members of various tribes labouring together away from their homes [as labourers on European plantations]. This is more especially the case in the coastal districts; the beach dwellers are nearly everywhere still on a war footing with the mountain tribes. If these mountain people come down to the beach, it is always in sufficient numbers to guard against attack. The principle is carried so far that when they enter into negotiations for barter, the exchange of goods is always accompanied by a display of force on both sides. In the year 1902, I witnessed such an event in Bougainville. The natives met six kilometres south of Keaop [Teop] (Cape l'Averdy) to trade their produce; the mountain people usually bring taro and exchange it for fish. The Keaop natives, mostly women with their loads of fresh and baked fish, came partly in canoes, and partly on foot along the beach. Armed men formed a kind of advance guard. Shortly after, the hill natives appeared; the armed men first, followed by the women with their burdens. The groups camped about five hundred metres apart, and started to chant a loud song.
>   During this a detachment of men came forward from each camp; an old man strode out from each detachment holding a bamboo full of water; following them came a dozen or so armed warriers. As they approached, the old men stepped up to each other, exchanged a few words, and flung the water in the containers in all directions. The escorts then came up and exchanged and chewed betel. The women then commenced to sing again, this time only for a short while. That finished, the women came forward with their loads of taro and retired

again. The Keaop women then performed their part and laid their fish beside the exchange, probably to see that nobody was cheated. The business over, the Keaop women fetched the taro, and the mountain women the fish. A short song then followed, and women departed with the produce they had brought. The men conversed for a little while, and then followed the women.

The whole affair went off in such a quiet and orderly manner, without bargaining or haggling, or useless, idle talk, that one would have thought the greatest harmony existed between the parties. But the old people assured me that in everyday life they regarded each other as the bitterest enemies; during the exchange of commodities however, a temporary peace reigned. This peace is observed not only so that the bartering shall go off without a hitch, but also to ensure the safety of the women. As soon as they are safely out of the way, the men often enough start fighting as soon as either party finds itself in the majority.

One can see in the varieties of this exchange a kind of developmental sequence from direct barter of consumable objects involving haggling, to commerce as the modern world knows it. In some instances people from one place took their wares—taro or spears or clay pots—to a likely meeting place, and if the local residents were interested and not too hostile, would proceed to barter for local products, say fish or shell necklaces, with wary bargaining on both sides. Elsewhere, based on longer relationships, people from two places would meet at an agreed time in some neutral place to barter objects at tacitly agreed-upon exchange rates, i.e. without haggling. Still another stage was reached when certain kinds of objects, such as clay pots or shell necklaces, were used for buying more immediate consumables. In fact, the shell necklaces used in this form of barter came to have such standardized exchange values over such wide areas that one might call them 'money' were it not for the fact that they were also 'consumed' or used by their owners as ornaments on occasion. Thus Bougainvillians, unlike some other Pacific islanders, were not novices in commerce when European traders appeared on their beaches, and they were well prepared in some respects for the introduction of European currency.

A second noteworthy thing about their trade was its political implications. As noted earlier, these islands' populations were divided into innumerable small tribes, some perhaps of no more than a hundred or so people, and all more or less hostile to each other some of the time. Ties of kinship and common cult practices sometimes served to bind separate tribes into somewhat larger unities, but usually not for long at a time. In fact, it appears that trade was the one activity that kept many otherwise warring tribes at peace with one another for lengthy periods. Not that trading disposed such tribes to be characteristically friendly with one another—even such commerce was usually carried on in an atmosphere of suspicion and wariness — but so long as active trade relations persisted, outright fighting was sometimes postponed.

In all the foregoing, attention has been focused on the practical utility of the activities. But what about their aesthetics? Some casual observers might be led to assert that the Bougainvillians have no art; only crafts. (Such observers would perhaps also agree, tacitly with one definition of 'primitive art' as being "those things Europeans call 'art' made or done by people Europeans call 'primitive'!") It is of course true that most (but not all) of the designs executed by these islanders were associated with objects of practical utility — house posts, canoe prows and paddles, bamboo containers, spear shafts — or with items of ceremonial or religious use — wooden trumpets, images. But this does not render such objects unartistic. As in all human societies, there were skilled and unskilled designers and carvers and weavers, and those that were skilled should be called artists even though they were also, or principally, gardeners and fishers.

The kinds of decorative work produced are better shown than described, but some words should be written about their musical art. A description of Siwai music will serve as an example.

The Siwai devoted more time to music than to graphic and plastic art. Every club-house contained several wooden slit-gongs, ranging in size from 450 to 90 centimetres in length, and from 150 to 30 centimetres in diameter. These gongs were sounded with vertically held wooden beaters struck against the lip of the slit, and they were used both for signalling and for making music. The gong played by women consisted of a single wooden board laid in a waist-deep hole dug in the ground. Two or three females stood in the hole and jumped on the board in unison to mark tempo for their companions' dancing and singing around the hole.

Flute-playing in concert by large numbers of men and boys accompanied many feasts and ceremonies, but individuals also played the flute by themselves for their own diversion. Flutes were of many sizes and were made of bamboo; they varied in number of units from one to ten. The Siwai also played large trumpets made by attaching a mouthpiece made from half of a coconut-shell to one end of a hollow wooden cylinder. And children sometimes amused themselves by playing on small musical bows.

Singing was highly popular and accompanied many kinds of occasions. Parents sang lullabies to their children, and used songs to teach them new words. Lovers composed boastful songs about their adventures, and relatives expressed their grief in moving laments. Even wailing followed a conventional musical pattern. Most dramatic of all was the harmonized singing of large numbers of men and boys, which took place at some feasts.

Attempts to reconstruct the pre-European social structure of a non-literate people are fraught with uncertainties; this was particularly true of Bougainville–Buka, where a lapse of several decades intervened between the first arrival of Europeans and the first studies carried out by trained anthropologists. By the time the latter appeared on the scene, warfare and indigenous forms of coercive social control had been effectively outlawed by Europeans, thus undermining some of the props of the indigenous political regimes. The process had been carried a step further through the institution of a new system of Administration-appointed chiefs. Also, the indigenes' economies were undergoing changes resulting from their use of steel tools and their employment for wages on European plantations. And finally, many of their religious beliefs and practices, as well as certain of their social institutions, were giving way to missionary doctrines.

Certain aspects of life that seem not to have changed before World War II were the basic territorial forms of social grouping: households, hamlets and neighbourhoods. Throughout most of Bougainville-Buka households were—and have remained up to the present time—the most important kind of indigenous social grouping, socially and economically; people sleeping together under the same roof also spent most of their waking hours together and pooled most of their energies and goods for common consumption. Most households consisted of a single monogamous family, although some men had more than one wife, and housed them either together under the same roof or in closely adjoining houses. All the older males of a household usually worked together, as did their womenfolk, and all partook of the commonly prepared meals.

The households were grouped into hamlets or villages which comprised up to about 300 residents, which was about the upper limit to durable territorial groupings. Cutting across these territorial units were social units made up of kinsfolk interrelated by *matrilineal* ties, in which kinship is traced through women rather than through men. Every one of Bougainville's societies was divided into such units, which were everywhere alike in being exogamous (i.e., marriage between members of the same unit was prohibited, although the force of the proscription differed from one society to another). Such units were also alike, everywhere, in being totemic: each one was associated exclusively with one or more species of animal or bird, whose members were respected in some way and whose precursor was believed to have been kin-related to the human ancestress of the unit. Beyond those similarities, however, Bougainville's matrilineal units differed from one society to another in several respects. One extreme was represented by people residing in northeast Buin, where such units were large, non-segmented and non-localized; they had no religious functions, other than totemic beliefs; and they owned no property in common. At the other extreme were the Nagovisi, who were comprehensively divided into two society-wide exogamous and totemic matrilineal moieties (halves), which were divided into numerous land-owning matri-clans, which were themselves subdivided into localized, land-using matrilineages. The latter retained their localized character through the preponderance of *uxorilocal* residence (the practice wherein a man, upon marriage, goes to live in the home community of his wife; the opposite is virilocal residence). In addition, the clans and lineages of the Nagovisi were the owners of other kinds of valuables, including shell money, and they were the focal points of most religious rituals and of much political influence.

Bougainvillian societies also differed—in some cases, widely—with respect to the political side of life.

In pre-European times every durable neighbourhood grouping constituted what may be called a tribe, in that all its members, its citizens, were usually at peace with one another and enjoyed access to one another's land, and usually fought as a unified group under a single leader when occasion demanded. Beyond this, however, tribes varied widely in their degree of cohesiveness. In some instances they were made up of very loose and unstable coalitions of hamlets, and drew together only when attacked by outsiders. In other instances local tradition or a particularly commanding leader succeeded in welding a neighbourhood-tribe into a unit that was closely knit—economically, administratively and judicially. Generally speaking, the more cohesive tribes were to be found among the coastal Austronesian-speaking peoples, and in the Buin area of southeastern Bougainville. Elsewhere tribes appear to have been smaller and more loosely organized. Thus regional differences in tribal size and cohesiveness may have been associated with differences in the ways men became leaders.

At one extreme were those tribal neighbourhoods dominated numerically, or in terms of land-holdings, by one particular matrilineage. In such cases the members of the principal matrilineage constituted an aristocracy, and their senior member a hereditary chief, to be succeeded in time by the eldest son of his eldest sister (not by his own son, who would of course have been a member of a different matrilineage). So far had this process gone in some coastal tribes, and in Buin, that these societies reached the point of clear-cut stratification, having been divided into two or even three hereditary classes: aristocrats, commoners and intermediates.

At the other extreme were those tribes whose leaders earned their positions of authority solely by exercising military or political skill. Usually, actual fighting prowess was less important than the ability to gain and inspire followers, which was exemplified by forcefulness of personality and by shrewd distribution of favours and hospitality. In some instances it appears that certain leaders, either hereditary or self-made, were skillful enough to extend their sway over neighbourhoods beyond their own, but such enlarged tribal coalitions rarely endured past the demise of the individual leader who had created them.

As warfare was ubiquitous among these islanders in pre-European times, a few words should be said about its character. Weapons consisted of spears, bows and arrows, and clubs; tactics can best be described as hit-and-run. Frontal encounters of massed forces occurred only rarely; the favoured method of attack was ambush, including the killing of solitary enemies—male or female, adult or child. Trickery seems to have been standard procedure; one hears much about one tribe visiting another to trade or attend a feast and then treacherously attacking. When at all possible the enemy

victims were carried away and dismembered and their skulls kept as trophies.
In some places the victors also indulged in cannibalism. This was certainly
the case on Buka and northern Bougainville, but seems not to have prevailed
in the south.

But for all their preoccupation with fighting, and for all their martial
airs, the casualties resulting from warfare seem to have been fairly few,
except in instances like the one when well-organized Shortland Islanders
in a fleet of war canoes destroyed a whole village. Nevertheless, although
warfare may not have served to reduce the populations very markedly,
it did cause many people to shift their residences to less exposed locations.
Also, the constant possibility of attack must have affected all indigenous
life to some extent.

From the vantage point of 1991, we can see that nothing has provoked
more discord between Bougainvillians and outsiders than disagreement about
land ownership—specifically, about principles of land tenure. In pre-
European times it is likely that no two Bougainvillian societies were exactly
alike with regard to those principles, nor have any two of them undergone identical impacts upon those principles since Europeans appeared on
the scene. Nevertheless, the now familiar Siwai can serve as a starting point
in this discussion—not only because they are known best to this writer,
but also because the Siwais' land principles included some that prevailed
throughout Bougainville–Buka. The following account is fairly detailed,
as indeed it must be in order to avoid oversimplification of this complex
topic.

The Siwai people thought of themselves as being different from the neighbouring speakers of different languages, but as such they did not own
any land in common. Those Siwai residing in settlements bordering those
of the Buin, the Banoni and the Nagovisi did recognize boundaries between
lands owned by themselves and those owned by specific social units—
mainly, matrilineages—of persons speaking Buin, etc. As for those Siwai
residing along the northern borders of their geographic area, they asserted
a kind of pioneering privilege over the huge uninhabited forested and mountainous space that separated them from the distant Nasioi, but recognized
that the Nasioi possessed a similar privilege over the regions nearer to themselves. Thus, although one could not accurately speak of a Siwai 'tribal'
domain, it was possible to refer to a Siwai geographic area. A similar
distinction applied to the area occupied by the Nasioi.

Within the geographic area of the Siwai, much of the land was actually
utilized for residing, gardening, hunting, fishing and collecting, and was
'owned' by one or another set of persons—in many cases by universal
consent, in many other cases against conflicting claims. In addition, there
were portions of the Siwai geographic area that were neither used nor
claimed, because they were, for example, swampy (although some swampy
lands were claimed by one or another set of persons, based on some mythical

episode in their ancestors' past). And thirdly, there were portions of the Siwai geographic area consisting mainly of virgin rainforest considered too far from any settlement for hunting or collecting, and too difficult to convert to gardens. Such areas were not 'property' in the sense used in this resumé; they were not 'owned'. Given the more or less static economic needs of the (non-expanding) Siwai population of the 1930s, there was no need to extend cultivation into the virgin forest (a very laborious exercise even with the use of steel machetes and axes). Nevertheless, if some individual had the wish and the energy to convert a portion of virgin forest into gardens he could have done so; he could have claimed personal ownership of the part first cleared, along with enough of the contiguous forest to enable him to extend the clearing (in line with the common practice of long-fallow gardening). Thus could a Siwai 'create' property in land—a practice that was seldom undertaken during my sojourn in Siwai, but that was universally acknowledged to be legitimate.

Once such property in land had been created, however, it seldom remained for long in the exclusive ownership of the individual creator. Even during their lifetimes most such creators widened the property's 'ownership' to include other individuals. Some other kinds of property— tools, garments, ornaments, magical formulas and materials—were owned by individuals, but that was seldom the case with land.

If individuals rarely owned land, what combinations of persons served as co-owners of any particular plot?

Before answering that question it is necessary to point out that in Siwai practice any unit of land could be owned in two ways: full ownership or divided ownership. The former prevailed when the same set of owners possessed access to it or use of it exclusive of all other persons. A plot held in divided ownership was identified with both its residual owners (landlords) and its provisional owners (tenants). In our Western societies the titles (use-rights, obligations, etc.) of the *residual* and *provisional* owners of a real property are usually spelled out in writing and in fine detail; among the Siwai of the 1930s, the respective titles of residual and provisional owners of a real property were somewhat less explicit, but were nevertheless widely agreed upon.

Take the common case of a unit of land lightly covered in secondary forest (and therefore suitable for gardening) and held in undivided owner- ship by a combination of persons interrelated by close matrilineal ties of kinship—say, the descendants, through females, of a pair of sisters four generations back. Anyone of this combination possessed the incontestable right to garden on that land. If, however, the land unit contained more cultivable land than its co-owners felt they needed, it was customary for them to lend use of some of it to less favoured kinfolk or mere neigh- bours. There was a tacit understanding that the rights of the latter, the tenants—i.e., the provisional owners, would terminate after the garden had been harvested (usually, a period of about two to three years).

In a similar way it was customary for the full owners of land to permit non-owners temporary use of their land and its resources (e.g., for hunting or collecting building materials or for traversing), provided of course the latter were perceived to be friendly. (And provided, in the case of successful hunting, that some of the kill were given to the residual owners, as a kind of rent.)

Returning to the question of what persons or combinations of persons in the Siwai of that era owned land in full or residual title, the answer is fairly simple: all units of owned land were named and most of them were identified, by full or residual title, with one or another of the society's hundreds of matrilineages.

In Siwai belief, most matrilineage-owned land—i.e., most of the real property in the Siwai geographic area—had been acquired by pioneer settlement. An individual pioneer had transformed a previously unowned tract of land into 'property' by using it (for house site, gardens, hunting grounds, etc.), and that property eventually became identified with his or her matrilineage. Doubtless, in many cases a pioneer male settler (the individual who had converted the land into property) passed it on to his own offspring, who in turn combined it with their mother's, hence their own, matrilineage estates, or made it the property base of a new matrilineage. In some cases the male heirs of a property-creating man may have passed their interests in that property to their own children, thereby initiating the formation of a *patrilineage* estate, but during my sojourn in Siwai nearly all owned land was held in the form of matrilineage estates. The exceptions to that arrangement will be described below.

With respect to landownership Siwai matrilineages were similar to a modern corporation, which may be defined as 'a body formed and authorized by law to act as a single person'. Decision-making regarding the use or disposal of such estates varied among the members of any matrilineage. In general, most decisive authority rested in the hands of a matrilineage's *Simiri* (Old-ones), its oldest male and female non-senile members. ('Mental senility' had about the same connotations for the Siwai as it does in European societies—which of course are not very precise!) In some cases a somewhat younger male member possessed as much or more actual authority than his matrilineage's Old-ones if he were an outstanding leader, or Big Man (*mumi*), but most Big Men were respectful of their Old-ones inasmuch as they depended upon them, as well as on other matrilineage mates, for support in achieving and retaining Big Man status.

In terms of use, any member of a matrilineage had an uncontestable right to garden on his or her matrilineage estate, but it was customary to obtain the consent of the Old-ones concerning the location of the proposed garden—mainly in order to avoid conflict with the location choices of other matrilineage mates. It was also customary for an individual to consult the Old-ones before planting semi-permanent tree crops (in those days, coconuts and areca palms). The Siwai, like most other Bougainvillians of that era, recognized a distinction in terms of ownership between a deliberately planted tree and the land on which it was planted. (In this connection, however, no distinction was made between a land's surface and its lower layers: the concept of mineral rights was absent and indeed irrelevant, since there was no mining!) Moreover, it was not only customary but obligatory for non-members, including the husband of a member, to obtain permission from a matrilineage's Old-ones even to garden on its land, although such a husband or widower was usually accorded as much privilege as a member if he resided nearby and was congenial with his wife's matrilineage mates.

Congenial non-member neighbours were also granted provisional use-rights to garden on a matrilineage's estate, if space was available and if the estate's Old-ones were agreeable. However, that privilege was seldom granted, nor indeed requested, with respect to the planting of semi-permanent trees.

Turning now to the disposal of a matrilineage's land: although the Siwai were given to stating that 'matrilineage land cannot be alienated', the fact is that it sometimes was alienated. This could occur in three ways.

First, when all of a matrilineage's members died their land was taken over by the membership of a neighbouring matrilineage — preferably a related one. In some cases, however, if a matrilineage included no females capable of childbearing, they sought to ward off extinction by 'adopting' a young and usually unmarried woman from a matrilineage belonging to their matriclan, hoping thereby to renew the membership of their own matrilineage.

There were also occasions when a matrilineage deliberately and willingly alienated some of their land. On some rare occasions land was actually sold (for native shell money). Much more frequently, the transfer was carried out in the mode of gift-exchange, as follows.

When a person died it was customary for kinfolk and neighbours to attend the wake and cremation — the more important the deceased, the greater the number of mourners. It was also customary, indeed obligatory, for the deceased's close relatives to reward other mourners with a meal, which usually included pork. If that pork, or shell money to purchase that pork, was provided mainly or entirely by members of the deceased's matrilineage, then that ended the exchange, of mourning in return for a pork meal. If on the other hand some persons not belonging to the deceased's matrilineage provided a large proportion of the funeral pork (or money to purchase it) — it was usually some male neighbour who did so — the latter was customarily 'given' a part of the land belonging to the deceased's matrilineage. Most transactions of this kind took place when the adult sons of a deceased man paid for his funeral feast out of their own assets; not being members of their father's matrilineage, they had no residual rights in that matrilineage's land. In most such cases the transferred land was simply merged with the sons' own matrilineage estate, but I recorded a few instances in which the process had been repeated, which marked the beginning of *patrilineage* estates.

Finally, I recorded several accounts of seemingly historical authenticity of land having been obtained by capture in times prior to the *Pax Australianis*. In the minds and hearts of some of the Siwai whose land had been alienated in this way within this century, their claims to it had not been extinguished, and were voiced on many occasions.

Except in parts of Buin, where land was owned corporately by localized groups of individuals owing allegiance to one or another patrilineally hereditary chiefs, most land estates on Bougainville–Buka were held, in either full or residual ownership, by units made up of matrilineally related kinsfolk. In some societies the landowning units were, as in Siwai, matrilineages; in others they were matri-clans, combinations of two or more matrilineages tracing descent from a more remote ancestress, either actual or mythical. In the case of the Nasioi, in whose geographic area most Europeans have resided—and acquired land—the pre-European land tenure system was closely similar to that of the Siwai, but beginning as early as German times the Nasioi have been subjected to much more frequent, more transforming, and in some respects more destructive contact with Europeans than other Bougainvillians—events that will be described in the next chapter.

In the foregoing discussion the focus has been on the economic uses of land—for gardening, hunting, collecting and fishing. Certain places also figured importantly in the lives of Bougainvillians as sites of religious, or religiously-coloured, events. In all the well-studied societies in these islands, each has socially significant matrilineal units: in some societies matrilineages, in others matri-clans, in others matri-moieties, in still others two of these or all three. Such matrilineal units owned, exclusively, one or more sacred shrines—typically, a cave, a bizarre rock outcrop, a dense grove of bamboo—associated with some event in the unit's mythical, or possibly historical, past. This might be the spot where a legendary ancestress had tarried and died, and turned into stone; or where one of its totemic animals had 'joined' the unit. In several societies such a shrine served as a setting for the unit's own religious rituals, such as the naming of a newborn member; in others it was the place where the unit's heirlooms (such as strings of shell money) were hidden. Another kind of place deemed sacred was the one, usually secluded, where bodies of relatives were buried, or where their charred bones were deposited. (Before missionaries discouraged the practice, cremation had been very widespread.)

Thus, most Bougainvillians had some *religious* devotion to places— shrines—occupied by the spirits of actual or mythical kin. In addition, of course, they had a sentimental attachment to certain locales, growing out of long-time familiarity with particular landscapes and with the living relatives and friendly neighbours residing there—a real oasis in an otherwise hostile world.

With all the above ties between particular individuals and particular places, one would expect that most Bougainvillians would not have wished to move to distant places. Yet move they did, some of them quite frequently and to locations relatively far away, and without displaying symptoms of disabling *mal du pays*.

In listing those aspects of Bougainvillian cultures that have served to promote most discord between indigenes and outsiders, one must add the differences between their respective native languages. By 1964, when the (English-speaking) agents of the future mining enterprise first appeared on the scene, most of the adult Bougainvillians with whom they—and their Administration sponsors—were required to communicate were fluent in Pidgin, and some were partly conversant in English as well. But for all of those Bougainvillians, Pidgin and English were second languages, acquired long after birth. These tongues were so different from Pidgin and English as to form barriers in translation, and therefore in mutual communication. Let us consider some items in these islanders' vocabularies.

In a large number of instances, a European and a Bougainvillian would have no difficulty in finding precise and adequate translations for words in each other's languages, as in the case of such concrete items as 'adze', 'pig', 'pot', and the like. The European would experience somewhat more difficulty in arriving at a concise translation of certain indigenous terms that denote, in a single word, such things as 'an old fallen coconut that has begun to put out roots and leaves', or 'a condition of the ocean surface characterized by shallow and parallel ripples'; but with some circumlocution (as just exemplified) it can be done. Similarly a Bougainvillian motor mechanic would be capable of inventing informative circumlocutions when discussing in his own language the workings of an internal combustion machine, or even an ore-concentration plant. On the other hand, whole paragraphs would be required to express in any European language the meanings summed up in some important indigenous terms, such as those that have been translated, with confusing and even perilous inaccuracy, as 'chief', 'law', 'ownership' and 'clan'. Conversely, much more than schoolboy command of English would be required to express in a Bougainvillian language the profundities and ramifications summed up in such English words as 'sovereignty' and 'independence', or to unravel the ambiguities contained in English homonyms.

The grammars of Bougainvillian languages are if anything more alien to the uninitiated English-speakers than are their most exotic words. Consider, for example, the structure that has to be employed in translating two fairly simple English phrases into Motuna (Siwai), a non-Austronesian language of southwest Bougainville:

Niñgi monosim
First-person singular acting, verb 'to see', third-person/thing singular,
past definite tense
[i.e., I saw him/her/it]

Mono-roro-piti-henute
Verb 'to see', second-person dual, first-person dual, future tense
[i.e., We two shall see you two]

Thus in Motuna (the non-Austronesian language I know best), a word
denoting transitive action—a 'verb'—will include particles specifying: the
kind of action; the person, sex and number (single, dual, plural) of both
actors and those acted-upon; and the tense (remote past, recent past, past
habitual, present, present habitual, future, future habitual, etc.). In terms
of 'adjectives', this same language divides all its 'common nouns' into some
forty categories, for each of which there is a distinctive set of numerical
adjectives. Thus the limbs of persons, spirits, pigs, fowls are counted in
one way, while the limbs of all other living things are counted in another.
Other such categories are: coconuts, stones, spears, banana plants, leaves
of taro and sago palms, shelves and benches, settlements, forests, water
and streams, day and night, moon and lunar cycle, etc. (Perhaps the nearest
thing in a Western European language would be the convention in German
of classifying all nouns as either masculine, feminine, or neuter.)

These differences may strike the reader as entertaining, or tedious,
according to intellectual tastes, but no one who troubles to ponder the
matter will dismiss them as inconsequential. For a person's language serves
not only to express thoughts, but acts in some very fundamental ways
to help shape those thoughts. One does not have to subscribe to the theory
that thinking is nothing more or less than silent talking, but there is a
correspondence between the two. And when languages differ as much as
do those of Europe and Bougainville–Buka in both word meanings and
grammar, it is quite certain that the thoughts of their respective native
speakers will also differ in many profound ways.

Just consider, for example, the English phrase, 'thank you', the verbal
acknowledgment expected and usually offered by a native English-speaker
in return for the bestowal of an object or service, however large or small.
In some such interactions the native English-speaking donor expects an
equivalent return for the gift some day; in others no such return is ever
expected. In both instances, however, one usually expects a verbal 'thank-
you' as well, and would be resentful if it were not forthcoming. In contrast,
in most Bougainvillian languages there is no word or phrase like 'thank-
you': this fact prompts some English-speakers to say that 'they have no
way of expressing gratitude', which often leads to the further conclusion
that 'they do not *feel* gratitude'.

Whether any particular Bougainvillians ever 'feel' what English-speakers label 'gratitude' is of course impossible to discover. It is quite apparent, however, that most Bougainvillians have been inculcated with an attitude towards giving and receiving that differs markedly from that of English-speakers. For most of them perhaps there are two kinds of giving: to close kinsfolk, and perhaps to a few intimate friends, one gives with little or no desire for, or expectation of, return; but with respect to all other persons the bestowal of any service or object carries with it the clear and mutual understanding of eventual reciprocation, even though the initial presentation may be ceremoniously described as 'free'. Under such circumstances the giver does not expect, nor even wish for, a mere verbal reciprocation in the form of a 'thank-you'. Before I came to comprehend this matter, I was greatly puzzled by the disappointed looks that used to greet my 'thank-yous'; my benefactors were evidently disquieted by the suspicion that, according to the obscure customs of Europeans, such verbiage itself was reckoned to be a sufficient return. To people reared in English-speaking societies a distinction is also made between 'free' (i.e., altruistic) giving on the one hand, and the kind of reciprocal exchange that is also labelled as a 'gift'. But with English-speakers the distinction is more or less tacit; to talk about it would be considered unseemly, even crass. To the Bougainvillian, however, words would be less important than actions in this kind of situation; any unseemliness would lie in the inadequacy of the eventual return 'gift'.

In anticipation of the next chapter: To relate these and other exotic features of Bougainvillian languages to the presence of the mining enterprise, it should be kept in mind that that involved myriads of transactions — exchanges — between mine staff and Bougainvillians, sometimes with and sometimes without the mediation of Administration officials. The process of communication and mutual comprehension was rendered difficult — in many instances, impossible — by at least two factors. One was mutual incomprehension about the 'official' identity (including the decision-making authority) of the individuals engaged in the transaction. For example, how decisive and how durable was the authority of the 'clan elder' who 'agreed' to long-term lease of a piece of land? or of the mine's Village Relations official (in many cases a very junior staff officer) who agreed to the 'agreement'? or, of the (usually junior) Administration official who sanctioned, or in some cases, compelled it? The other obstacle to mutual comprehension was the nature of the transactions themselves. Needless to say, for Bougainvillians (or other non-English speakers) inculcated in their own exchange practices and vocabularies, the legally-defined English practices and vocabularies for transactions like 'lease' and 'compensation' were, initially, at best ambiguous and at worst totally incomprehensible.

Complicating all that has been the different meanings attached to certain key English words by different English-speakers — e.g., by Administration officials in contrast to missionaries with respect to, say, 'law' and 'justice'.

Or to reverse the direction of ambiguity, one can sympathize with the white trying to comprehend the deeper and different meanings which *rot* in Pidgin has beyond that of *road* in English.

Turning now to Pidgin: languages so labelled occur around the world. They are generally defined as being reduced in structure and vocabulary in comparison with the language or languages from which they derive. In most cases they originated when people speaking different languages interacted fairly regularly but for somewhat limited purposes, such as trading, working together, governing, or religious proselytizing. The several varieties or dialects of Pidgin spoken in Melanesia originated during the nineteenth century through contacts between native Melanesians and the Europeans who recruited ('blackbirded') them or employed them on plantations and boats. Their vocabularies came mainly from English (with a few French words in the New Hebrides, and some German words in northeast New Guinea), but their grammars derived from the Melanesian (i.e., mostly Austronesian) languages spoken by the native labourers and traders, most of which are fairly similar in word-structure and syntax.

Such Pidgins came to serve not only as contact languages between Europeans and natives, but also among natives themselves, many of whom were brought into frequent contact with speakers of mutually unintelligible native languages. Then, after a while one or another of the pidgins spoken in Melanesia came to be the 'national' and semi-official language of each of the region's emerging nations. As for the variety spoken in Papua New Guinea, while it consists of several dialects (Highlands, Madang, eastern New Britain, etc.) the differences among them are slight. In 1970 about 15 per cent of the Territory's native population spoke Pidgin; by 1991 the percentage is estimated to be much larger. Without it, it is doubtful that the nation could have held together as well as it has.

Today, few resident whites in the nation are fluent in Pidgin—partly from laziness and partly from their condescending opinion that it is a 'corrupt' and debased form of English and that it is incapable of expressing 'advanced' thought. Needless to say, this opinion is egregiously incorrect in both respects. English also contains many 'incorrectly pronounced' French words, but it is not 'debased French' on that account. And even concepts such as those of nuclear physics or corporate takeovers can be expressed in New Guinea Pidgin with the invention or adaptation of some new words.

Finally, in drawing up a list of the factors that have served most to obstruct mutual comprehension and concord between whites and Bougainvillians, one must add their discordant concepts of 'time'. Individual whites may have different individual sensations concerning the durations of this or that episode of their lives (e.g., 'time passes slowly when there's nothing to do', 'it seems only yesterday since . . .'), but they do agree upon and generally reckon and live in synchrony with conventional fixed intervals— centuries, years, months, days, hours, minutes and seconds.

In the Bougainville language I was most familiar with, there were labels
for several phases of the solar day, e.g., dawn, midmorning ('time for
walking about'), midday ('sun overhead'), sunset ('the spirit Huru is drying
his hat'), and darkness or night. Here is what I wrote about them in 1938–9:

> The introduction of night may have helped the Siwai to fix upon a
> time to eat and sleep but it did not succeed in regulating all their
> other activities so strictly! It is true that most natives spend at least
> part of the forenoon at some constructive labor, but the time for
> beginning (*kongkongno*) ranges from seven to eleven, depending upon
> the task and the individual. Also, throughout the day they nap
> whenever opportunity arises and make haste almost never at all.
> Times for the beginning of work-bees or feasts or ceremonies may
> sometimes be announced as 'very early' (*kiakia meng*, dawn true) or
> 'at noon itself' (*ra rera meng*, sun is exactly overhead), etc. but this
> attempt at precision has little effect on times of arrival, which usually
> straggle out over three or four hours. Either the language as
> constituted cannot be made explicit with respect to time, or the Siwai
> have so little awareness of the passage of time, and place so little
> value on time regularity, that no amount of precise comprehension
> could make them 'punctual': I suspect that all three of these factors
> enter in. With respect to their lack of time-awareness, I noted many
> occasions of individuals setting out on enterprises without rational
> consideration of the time element involved. For example, men would
> set out on journeys to known places too late in the day and would be
> placed in the fearful circumstances of having to walk in darkness. Or
> again, in connection with cremations the ideal is that the last embers
> of the burning pyre will die just as dawn appears; but even using the
> morning star as a guide for lighting the pyre, according to my
> observations they were never able to cause the two events to coincide.
> [Oliver 1955, pp. 97–8]

The Siwai greeted the appearance of the new moon with gong-beating,
to discourage that sharp-pointed demon from piercing growing coconuts,
but they did not reckon in lunar months. Christianized natives recognized
Sundays as days for attending church services, but events were not measured
in terms of weeks. As for years, the annual ripening of the tasty *galip*
nut was recognized as a cyclical regularity, and the interval between ripen-
ings came to be labelled a *karisimasi* (Christmas), but time was not reckoned
in 'so many *karisimasi*'. There were verbal ways for expressing 'long ago',
'very long ago', 'present', and the near and distant 'future' — but no verbal
criteria for measuring them.
    As I said then:

'History' provides a sense of time, a sequential frame of reference, for

many peoples, but the Siwais' concept of history has little resemblance
to that of the Western historian. Their cosmogonic myths refer to an
epoch when the *kupunas* (creator spirits) walked about creating things,
but there is no conceptualization of how long ago these events took
place beyond dating them as *u'kisa*, 'a long time ago'. In one sense
this epoch may be thought of as extending into the present, inasmuch
as acts of creation are still taking place. On the other hand, most
Siwai conceive of the age of creation as having had a definite ending,
if not a beginning; and after the close of that epoch things were more
or less 'straight'—for a while, at least. The present epoch, that of
mankind, followed the age of creation although some human beings
were already present when the *kupunas* were walking about. Some
missionized individuals now speak of a Judgment Day, which will
mark the end of the present epoch, but there is no concept of this
sort in native Siwai belief.

With respect to actual events Siwai memories are quite short. A
few exceptional old people are able to recall the names of lineal and
collateral relatives six generations back, but most natives can extend
their genealogies only four generations and even then they remember
only matrilineal kin that far removed in time. Each *maru* (growth-
magic) formula contains a long list of former [clan] matrilineage
members who used to perform the ritual, but these names are not
considered to be linked in any specific genealogical manner.

Many elderly Siwai are able to describe events said to have taken
place as far back as the childhood of their parents. The kinds of
incidents best remembered had to do with warfare, epidemics, and
contacts with aliens. Beyond that, however, the historic past is a
blank.

A third kind of sequential frame of reference which might
conceivably have led the Siwai to adopt some time scale is the
individual's life cycle. But here again there is no regular measure,
since age is phrased in terms of activity and physical criteria and not
of years. Moreover, even physically evaluated age-progression is not
formalized in terms of age-grade institutions, nor is socio-political
rank closely correlated with age.

Turning now to language, the Siwai express temporal sequence by
the use of adverbs and verb tense. *U'kisa* means 'a long time ago',
while 'later' or 'in the future' is expressed by *romokisa*; these, then,
express relative sequence only. *Rirokisa*, on the other hand, specifically
means 'present' (or 'new'). There are also special terms for 'yesterday',
'today', 'tomorrow'; for 'two (or three, four, etc.) days hence' and
'two (or three, four, etc.) days ago'—however, I did not once hear a
Siwai date events in terms of moons, and only rarely in terms of the
new concept of 'year'. [Oliver 1955, p. 98–9]

By 1965, as result of Westernized schooling, the Siwai had become more accustomed to clocks and calendars, but it is my impression (based on occasional contacts with them in the 1970s) that many of their traditional ways of thinking and acting regarding 'time' still prevailed.

In this and previous chapters I have been at pains to point out that there were — still are — many differences among the fifteen or so cultural units of Bougainville–Buka. Nevertheless, it is my (only partly supported) hunch that the Siwai were and remain typical of most other Bougainvillians with respect to 'time'.

As a postscript to this chapter's characterizations of the Bougainvillians' social institutions, one feature calls for additional explanation. This has to do with the size of their social units whose members shared feelings of identity and loyalty. In most cases such units did not extend beyond a dozen or two close kinfolk and near neighbours. Most whites are accustomed to thinking of themselves as, say, Australians or Melburnians, or Liberals or Collingwood supporters; in comparison, those 'we' units of the pre-colonial Bougainvillians seem very small.

8

The Mine

THE GOLD FEVER THAT GRIPPED MAINLAND NEW GUINEA IN THE 1920s and 1930s spread to Bougainville as well, where small quantities of gold and traces of copper were discovered in the mountains behind Kieta. Beginning in 1934 a gold mine was operated at Kupei, but on such a small scale that it was regarded by most European residents as a pathetic pipe-dream: by the time the Japanese invasion ended its operations, only about 1789 ounces of gold and 80 ounces of silver had been extracted from it. After World War II traces of alluvial gold were discovered in the mountains west of Kupei, but only a few attempts were made to extract it.

In 1960 a geologist of the Australian Bureau of Mine Resources, Geology and Geophysics re-examined the country around Kupei and confirmed the presence of intensive low-grade copper mineralization. Again, this would have been dismissed as nothing more than a scientifically 'interesting' discovery but for certain current developments elsewhere.

At the start of the 1960s, disparate developments around the world combined to produce the possibility of profitable large-scale mining on the island. Each by itself was unlikely to have achieved that; together, and with the foresight of some remarkable individuals, they changed low-grade mineralization on the island from a matter of academic geological interest to one of major risk investment.

A layman hesitates to assign priorities to these factors, which included: new theories of ore genesis and novel techniques in geological exploration; new prospecting techniques, ranging from stream sampling for dissolved metal ions to innovative use of helicopters for both human and equipment transport in heavy jungle; technological breakthroughs in the scale of machinery, both for materials handling and for minerals concentration; and a worldwide market uncertainty about the future assured availability of copper as a vital industrial commodity.

In Australia, the vision of one mine executive from Broken Hill had already led to a series of new discoveries undreamed of twenty years earlier. Maurice

Mawby had spent heavily on diversified risk exploration of the bauxite resources at Weipa in north Queensland, and for a token fee had operated for the Australian government the uranium mine at Rum Jungle in the Northern Territory. Another visionary, Canadian geologist Haddon King, began to envisage new theories of ore genesis: if you were right about how minerals became concentrated, the likelihood of finding economic mineralization was greatly improved.

Meanwhile, in the United Kingdom, an entrepreneurial financier, Val Duncan, began to create a world-size mining company on the tiny initial base of Rio Tinto, a rundown Spanish mine operated since the days of the Romans; Australia, and particularly uranium, became his chosen fields of opportunity. Australia had a stable and pro-development government, unlike those of mineral-rich countries in Africa and Latin America which were then in the process of decolonization and nationalization. In due course Duncan's Rio Tinto merged with the London-based mining house Consolidated Zinc (created by L. B. Robinson) to form RTZ in London and CRA in Melbourne, with Mawby at its head.

Would entrepreneurial flair, geological foresight and a pro-development Australian overseer have automatically led to mineral development in PNG? I have been told it was 'at best highly improbable' without the other factors of technology and world demand.

World demand went through a turbulent decade, as forced nationalizations of major producers such as those of Chile and Zambia caused fears of involuntary or deliberate restrictions of supply. In addition, prices on metal exchanges for copper were fluctuating; as a result, several developing countries formed the Conseil Internationale des Pays d'Exportation de Cuivre, or CIPEC, the attempted copper equivalent of oil's OPEC.

While all those events were taking place, new technology in material handling and scale of metallurgical concentration made remarkable advances in economies of scale. The visible growth of ore trucks from a carrying capacity of 50 tonnes to 170 and then 210 tonnes was only the outward manifestation of this. New techniques and skills were also developed in planning of open pits, in size of ball mills, and in production scheduling.

The consequence of those technological innovations was that, given a large enough ore resource to justify the investment over years, huge quantities of mineralized rock at 1 per cent copper or less became mineable, where a generation earlier the 'cut-off grade' which divides economic ore and waste had been perhaps 4 per cent. In a true sense, technology created new reserves. In Australia and elsewhere, this dramatic increase in the ability to mine and process low-grade ore profitably extended across the whole field of minerals commodities, not simply to copper ore.

So, were technology and demand the extra determining factors? Again, I have been told 'not so'. In 1960, RTZ and CRA were entrepreneurial but not cash-rich miners. The Panguna deposit had no 'cap' or 'pod' of high-grade gold which might have paid for a modest self-financing start. Neither the Australian government nor Australian banks were willing financial backers of larger minerals projects in their own homeland, let alone in a UN Trust Territory. International banks, sensitive to the disruptive nation-alizations taking place elsewhere in what was coming to be called the 'third world', were also unenthusiastic about risk in a not yet independent nation and in the fluctuations of copper prices.

The final necessary element was the ability to have in place — long before even final production details were firmly settled — a system which would make both the high capital cost and the proposed output from a probable mine securely 'bankable' in the eyes of lenders.

That might not have been achieved at all except for independent feasibility studies and for innovatory international marketing of a new financial concept, namely, that long-term contracts for supply of premium copper concentrate to smelter firms in Japan, Germany and elsewhere in Europe, could be a financeable prudent risk.

Many of this book's readers are likely to possess only vague general knowledge of the massive scale of much modern mining, of huge bulk ore carriers, and so on. As I write elsewhere, neither Bougainvillians nor Administration officials (nor this writer) could then do more than try to imagine the scale of development which was to occur.

We were not alone. In 1960 almost no Australian mining engineers, even the most experienced, had themselves much more than *seen* operations on this scale. The Australian tradition, of which Mawby was a part, was then centred on underground mining of relatively high-value ore. The experts in the new skills of large-scale open-pit mining and large-scale metallurgy were mostly North American, as were the bulk of the relevant lending banks. In consequence, American expatriates joined Australians in the initial top management level of these new ventures in Papua New Guinea, just as they did in similar ventures in Australia and elsewhere. It was transfer of tech-nology on a giant scale.

When I resumed visits to Bougainville in 1968, exploratory drilling and road-making were already on a huge scale by South Pacific standards. It was as hard for me as for other non-miners to conceive that the activity was only a 'project' supported by risk capital, and neither a proved deposit nor as yet a financeable mine.

The new economies of scale imposed their own stern disciplines. Both the exploration companies and the potential lenders crunched numbers over many months — a 'small' mine on low-grade ore was simply not possible. Indeed, even the mining and treating of 30,000 tonnes a day, which the original plan envisaged, was calculated to be insufficient for profitable operations. A very costly and extensive exploratory drilling programme had to be undertaken in order to produce forecasts large enough (80,000 tonnes a day) to repay the borrowings needed to finance the operation, given the expected fluctuations in the world price of copper over a ten-year period. In the course of that exploration the most promising area was found to be in the Panguna Valley.

After eight years of exploration, evaluation, and construction, mining began there in April 1972 and continued, profitably, until May 1989, when hostile actions by some local residents forced it to shut down.

During its seventeen years of operation this mine was the largest industrial enterprise in Papua New Guinea in a number of respects, including especially its production of national revenue.

It produced and shipped to overseas buyers concentrate containing nearly three million tonnes of copper, 304,412 kilograms of gold, and 780,875 kilograms of silver, for a net sales revenue of about 1900 million kina. Up to the end of 1988 a total of some K685 million had been paid to the Territorial, later national, Papua New Guinea governments in direct taxation, customs duties, retail sales taxes, and withholding taxes on dividends; plus another K166 million in dividends on shares owned by the government. Altogether, these sums constituted an average of about 17 per cent of the government's internal revenues, and the overseas sales of the ore concentrates amounted to about 40–50 per cent of the nation's foreign earnings.

In addition, the new country received as early as 1967 (long before leases were granted) an unparalleled 'free ride' into what was to become its largest single industry, source of finance, and exporter.

The miners offered the Administration 20 per cent of equity in the development at par (rather than the much higher price later offered to other investors). Australia took up the offer much later when the project's viability was established. This acquisition passed to the new country almost unnoticed with the other transfers of decolonization. As D. S. Carruthers and D. C. Vernon wrote in 1990:

> Given that CRA was taking all of the risks on exploration and feasibility, this was an offer unprecedented in mining projects in under-developed countries, and to the authors' knowledge, has still not been matched voluntarily by any other company anywhere in the world.

Moreover, up to the closing of the mine the company had paid about K75 million to the government of the North Solomons Province in royalties and taxes, plus another K22 million to Bougainvillian landowners in the form of royalties, rent and compensation. These figures do not include the monies paid by the mining company to the Administration in the form of rents, etc., nor the amounts paid by the company to Papua New Guinea residents in dividends, nor the income taxes paid to the Administration by company employees.

The number of individuals employed by the company and its contractors varied over the years. During the exploration and evaluation phase it was in the hundreds; during the construction phase it reached more than ten thousand. Since the beginning of mining the number averaged about four thousand, of which about 80 per cent were indigenous Papua New Guineans. How much of the salaries and wages of those employees remained in Papua

New Guinea is impossible to discover, but it cannot be ignored when judging the economic impact of the mine, particularly on Bougainville.

Now to describe the physical stage on which the mining operations took place, the principal participants, and the rules governing their actions.

Figure 13 shows the places eventually occupied by the mine and its associated facilities, along with the areas leased to the mine company. Comparison of this map with the others will indicate roughly how many residents had to be shifted to make way for the mining enterprise, and how much economically useful land was reserved for actual or potential use of the enterprise. The latter includes land in actual use by the local residents and land capable of economic use by them. Some statistics bearing on those curtailments and dispossessions will be provided later in this chapter.

The principal participants in the mining enterprise were the residents, mainly indigenous, of the region demarcated; the officials and agents of the mine company; and the governmental bodies and their particular agents which regulated relations between the residents and the company. In addition, one Australian planter and several members of the Catholic clergy played roles.

Members of five distinct Bougainville ethnic (i.e. language-cultural) units resided in the region directly affected by the mine. Of them a large majority were Nasioi, and most of these were rural villagers engaged in subsistence farming and some cash-cropping of cocoa. Although mainly 'rural' in lifestyle, most of these Nasioi were well acquainted with the local colonial institutions and with many individual Europeans. As will be recalled, it was along this region's eastern coast where Europeans first settled: first the Catholic mission, in 1904; next, colonial officials, German and then Australian; then Australian planters and Chinese merchants; then a detachment of Japanese; and finally a return of all of the pre-war types of outsiders — officials, planters, merchants and missionaries. Individuals among the region's Nasioi doubtless differed in the content and degree of their Westernization and in their attitudes towards outsiders, but their contacts with persons and things Western were on the average far more frequent than those experienced by most other Bougainvillians, including most other Nasioi.

The second of Bougainville's native peoples to experience the direct impact of mining activities were the Austronesian-speaking Rorovanans, who resided along the coast a few miles north of Kieta. The Rorovanans, it will be recalled, were descendants of fleets of Shortland Islanders who had 'colonized' — seized and settled on — this coastal area in the nineteenth century and had remained ethnically 'intact' ever since. They appear to have adjusted more comfortably to the immediately adjacent Europeans than had their Nasioi neighbours.

By the mid-1960s the nearby village of Uruava (Arawa) was more Nasioi than Uruavan. As related earlier, when this writer was on Bougainville in 1938–9 the residents of this village still spoke the Austronesian language of their ancestors, who like the Rorovanans had migrated here from Short-

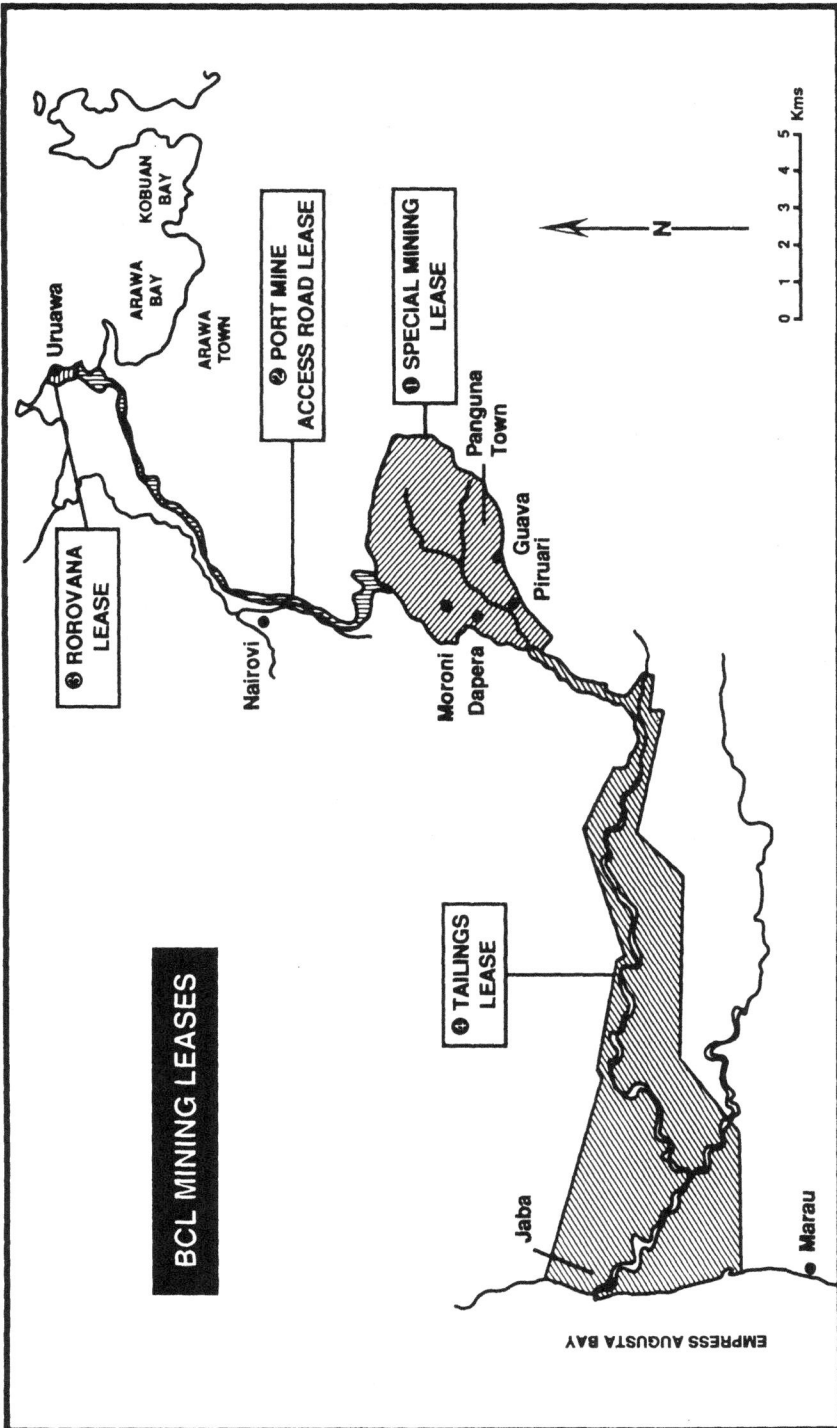

BCL MINING LEASES

① SPECIAL MINING LEASE
② PORT MINE ACCESS ROAD LEASE
③ ROROVANA LEASE
④ TAILINGS LEASE

Uruawa
KOBUAN BAY
ARAWA BAY
ARAWA TOWN
Nairovi
Panguna Town
Guava
Piruari
Moroni
Dapera
Jaba
Marau
EMPRESS AUGUSTA BAY

N

0 1 2 3 4 5 Kms

land Island, but a century or so earlier. By the mid-1960s, however, inter-marriage with the neighbouring Nasioi had led to the near-extinction of the Uruava language, along with many other Uruavan customs.

The fourth set of Bougainvillians to be directly affected by the mine were those of the Nagovisi people who resided west of Panguna along the southern borders of what was to become the mine's tailings. The fifth were a community of Austronesian-speaking Banoni, whose small coastal village near the outlet of the Jaba River had to be moved to avoid flooding by that tailings-widened stream.

Now to unravel the corporate transformations undergone by the company that discovered, developed, and operated the mine.

The Consolidated Zinc Company Ltd of London (CZC) owned the Zinc Corporation Ltd of Broken Hill, Australia, and had a managing interest in New Broken Hill Consolidated. In 1962 it merged with the Rio Tinto Company Ltd to form the Rio Tinto-Zinc Corporation Ltd (RTZ). At the same time, in Australia, Consolidated Zinc Proprietary, a wholly owned subsidiary of CZC, merged with Rio Tinto Mining Company of Australia Ltd to form Conzinc Riotinto of Australia (CRA). To reduce this (incestuous!) genealogy to its essentials: London-based RTZ owned 85 per cent of Australia-based CRA in April 1965. When it was decided that CRA would undertake a systematic search for copper deposits in the Southwest Pacific, New Broken Hill Consolidated Ltd joined with it as a junior partner, and it was this combination that launched the exploration on Bougainville in the form of an entity, named CRA Exploration (CRAE), which was created for that and other mineral searches in the southwest Pacific. (Among other facili-ties, CRAE possessed the *Craestar*, a former Japanese tuna-fishing vessel, fitted out with a laboratory, a helicopter landing deck, and accommodation for an exploration party of ten.)

After CRAE's exploration had turned up promising prospects on Bougain-ville, the work of evaluation was handed over to yet another new entity, Bougainville Copper Pty Ltd, which was a subsidiary of Bougainville Mining Ltd. Then, after the commencement of mining, in 1972, the company was registered in the Territory of Papua and New Guinea and renamed Bougain-ville Copper Ltd (BCL) — which, I am relieved to report, it has remained ever since. In the account that follows, all these transformations will be simplified by referring to the organization as either the company or BCL.

The governmental entities which initially regulated the exploration and mining on Bougainville were diverse in both ends and means. Final authority over all Papua and New Guinea matters, including those of Bougainville, rested with the Australian government in Canberra, and specifically the Minister for Territories. The Bureau of Mineral Resources, Geology and Geophysics played a role in copper search throughout Australia, and its regional geologist stationed in Port Moresby had visited Bougainville in 1960 and found porphyry-type copper mineralization there, but the bureau did not follow up this lead.

The Territory's Administrator in Port Moresby was of course subordinate to Canberra, but could, and often did, propose and administer minor changes in policy, including changes regulating mining. In Port Moresby the mine company's principal contacts were with the Assistant Administrator, Economic Affairs, and the Department of Lands, Surveys, and Mines. On Bougainville, the company's main governmental contact for the first few years was with the Assistant District Officer (later, Commissioner), stationed at Kieta, the District Officer being headquartered at Buka Passage at that time.

In 1966 a Mining Warden's Court was set up in Kieta to adjudicate complaints ('plaints') against the company for damage to property. In 1969 a new office was set up, the Chief Liaison Office in the Bougainville District, charged with overseeing all relations between the company and residents of Bougainville. Also in 1969, the Public Solicitor's Office based in Port Moresby began to be more deeply involved in relations between the company and Bougainvillians; it actively championed the latter's definitions of their rights and claims, taking out writs against the Administration itself, and pursuing litigation up to Australia's High Court.

Another governmental personality closely concerned with the mine in its early years was the Hon. (later Sir) Paul Lapun, Bougainville District's first member in the Territory's House of Assembly. Lapun's background and election were described in Chapter 6.

In addition to the entities just listed, during the early years of its operations the company had some relations with members of some of Bougainville District's local government councils, including those of the Kieta Council, but those elected officials had very limited functions and authorities, and the institutions themselves did not last long.

Individual Catholic clergymen have played influential roles in the company's relations with Bougainvillians from start to present. A few have been silent or neutral, but many have been vocally against the company, some loudly and effectively so. They have not only championed Bougainvillians' claims for larger shares of company revenues, but also denounced the mine's effects on the physical environment and on public morality. Moreover, some of the most influential Bougainvillians with whom the company dealt over the years had been trained in Catholic schools, some of whom had reached, and later left, the priesthood.

Other players on the Bougainville scene during the early years of the company were the island's plantation managers and merchants—the former mostly European, the latter mostly Chinese. By the late 1980s nearly all of the former had left, after selling off their properties. Even while still there, few of them had much to do with the company, except to experience increasing difficulties in recruiting and keeping their workers in the face of the company's higher wages and more 'interesting' jobs. As for the merchants, the increase in the sales receipts in the early days of mining diminished when company-sponsored supermarkets opened in Arawa and Panguna.

To complete this resumé of the physical and institutional setting in which the company operated, we turn to the official rules that governed many of its actions.

The basic legislation governing prospecting and mining in the Territory was Australia's Mining Ordinance of 1928–40, which, as one writer observed, 'was adapted to the needs of gold mining and small-scale operations as they were in the latter part of the previous and the early part of this century.' (King 1978, p. 77) The ordinance did not anticipate mining of the type and scale of that carried out on Bougainville. Nor could it have envisaged operation in a cultural setting that differed so widely from that of Australia, particularly in the domain of land ownership. Those parts of the ordinance most relevant to this narrative had to do with royalty payments and land leases.

According to the ordinance, all mineral rights in the Territory belonged to the Crown — i.e., the Administration; hence all royalties derived from mining had to be paid to the Administration for the benefit of the Territory as a whole. Obviously, this rule was entirely alien to Bougainvillians' traditional ways of defining land and its ownership. By this time Bougainvillians were accustomed to many of the differences between their own rules and those of their German, then Australian, rulers — but to none as oppressively alien as this. Some of their ancestors had been deprived of their lands by their enemies in time of war. The fathers and grandfathers of some of them had 'sold' land to European planters — although it is not unlikely that most of those 'sales' had been thought of by the sellers as long-term leases. But to be acknowledged as owner of the land's surface and not of the things under it, was difficult to comprehend and virtually impossible to accept. (The Panguna landowners of the 1960s doubtless knew about the earlier gold-mining in nearby Kupei, but seem not to have learned anything from that about the ordinance's sub-surface definitions.)

The company operated under several types of leases at various stages of the enterprise.

First was a Prospecting Authority, which was valid for up to two years. Under its provisions the company obtained seven leases, covering an area of more than 500 square kilometres (as shown in Figure 13), for which it paid the Administration — not the landowners — an annual rent of $1 per square mile (2.59 square kilometres). The company was required to compensate landowners for damage or disturbance to buildings, trees and gardens, but not to pay 'rent'. In 1966, after having been denied access to numerous places by suspicious owners, the company persuaded the Administration to amend this part of the Prospecting Authority to permit payment of some rent (i.e., an occupation fee) of $A1 an acre per annum of land in continuous use by the prospectors, 25 cents an acre for 25 per cent occupancy, and 10 cents an acre for 10 per cent occupancy.

In 1969, after test-drilling had indicated the presence of a mineable ore body, the company applied for and received Special Mining Leases comprising

3770 hectares in the Panguna area, good for forty-two years and thereafter renewable. Annual occupation fees of $A2.47 per hectare were payable to the Administration, and $A16 per hectare to the landowners.

'Leases for Mining Purposes' are those obtained for purposes ancillary to the actual mining and initial processing performed on a Special Mining Lease. In the case of Bougainville these included areas for tailings disposal, roads, gravel pits, and water supply. In addition, the company has over the years obtained 'special agreement leases' for parts of the port area, townsite, etc., as will be described.

Frequent reference has been made to 'landowners', specifically to those residing in the Panguna and other mine-lease areas of north Nasioi. Up to the time the leases were granted, no officially sanctioned, expertly supervised cadastral survey of those areas had been made, and no authoritative list of 'owners' had been recorded. Nor did the hardpressed prospectors and their accompanying Administration officials have the time, or the kind of expertise, required to draw up such a list. They lacked knowledge about the intricacies of Nasioi social organization and property ownership, and did not even know what questions to ask. Under the circumstances the best that could be done was to ask 'the natives' who owned what, and to assume that unchallenged statements made by what appeared to be a village's most authoritative (or most vocal!) 'leader' were true. Unfortunately, among the Nasioi (as in most other Bougainville native societies) leaders had several roles. Some men spoke for their land-owning social unit because of their seniority in years, others because of their current eminence as feast-giving, managerial 'Big Men'. There were cases in which the two types were combined in one man — at least for a time.

Another complexity of Nasioi land tenure had to do with different types of ownership. In societies like that of the Nasioi, where customary rules of land tenure were not codified in writing, they were highly susceptible to changes and to circumstances such as the presence of a vigorously persuasive co-owner. In 1966–8 anthropologists made brief visits to Panguna to advise the company in its relations with Bougainvillians, but the only intensive and long-term study of Nasioi land tenure ever made was carried out by an anthropologist, Eugene Ogan, who had no association with the company and whose field researches among the Nasioi began in 1962 and continued, with breaks, for several years. Ogan's description of Nasioi land tenure was not published until 1971 — too late to serve the company and the Administration as a guide in their dealings with Panguna's self-professed 'landowners', but still useful for an understanding of the causes of some of the company's subsequent difficulties, including the proximate causes of the Bougainville Crisis. Ogan prefaces his published study as follows:

> until about 1955 there was neither physical nor social pressure on land anywhere. Before World War I high infant mortality . . . and a *post-partum* tabu on sex relations kept population density low. World War II

caused a sharp population decline . . . and only in the last 15 years have
Western medical techniques—especially a malaria eradication
programme—brought about a potentially disastrous population
expansion . . . Cash cropping and consequent increased demands for land
were not major factors until the late 1950s. In short, until relatively
recently there were few occasions for the kind of intensive conflict over
land rights which could have created precedents for a neatly defined
system of land tenure . . . Needless to add, the subsequent appearance of
prospectors in the area and the possibility of greatly enhanced land values
did provide an occasion for 'intensive conflict over land rights'. [Ogan
1971B, p. 84]

As described by Ogan, the Nasioi land tenure system resembled in many
respects that of the Siwai summarized earlier, although the Nasioi system
of the 1960s was more explicit than the Siwai one of the 1930s in the way
it distinguished various types of ownership. Continuing Ogan's analysis:

Nasioi land tenure is best described in terms of a hierarchy of rights of
usage. This hierarchy is sufficiently complex that merely to talk of
'owning land' is to obscure important aspects of the situation. One is
better advised to employ such terms as 'primary', 'subsidiary' and
'derivative' rights . . . Primary rights to a tract of land were normally
established by a man who cleared the area of primary forest. These
rights included the right to make gardens and to erect houses; the right
to plant tree crops of longer life than the ordinary garden; and the right
to allot sections of garden land to others. Garden land seems to have
been allotted to women according to the following priority: the primary
right-holder's wife or wives; his close uterine kinswomen; other women
of his [matrilineal] clan; wives of his clansmen. An ambitious man
might also allot use rights to any other household or individual willing
to help him clear a particularly large tract. His assistants thus
acknowledged and at the same time increased his prestige . . . Primary
rights did not include the right to alienate the tract by transferring
rights in freehold or fee simple to a complete stranger outside [his
politically autonomous community] . . . subsidiary rights included the
right to request, with reasonable expectation of success, garden plots of
the primary right-holder and to claim primary rights to the tract upon
the death of the primary right-holder in the event that no formalized
transfer to [the latter's] children had taken place . . .
   Subsidiary rights were held in the first instance by the primary right-
holder's immediate uterine kin and secondarily by his co-resident
clansmen regardless of more exact genealogical connection. These rights
thus effectively precluded alienation of land beyond the [autonomous
community].
   The common instance wherein rights were formally transferred
involved the children of a primary right-holder as opposed to his

subsidiary right-holders. If the children provided a sufficiently large feast [without an equivalent reciprocation] to their father's close uterine kin and other clansmen, they thereby obtained primary rights which would otherwise have reverted on [the father's] death to the subsidiary right-holders. The [above gift-feast by a man's children] might be made during the father's lifetime, or in connection with his funeral at a so-called 'head feast' . . .

Further complicating the system were what might be called derivative rights. These essentially consisted, on the one hand, of claims to some share in the product of a tract of land, and on the other, claims to primary or subsidiary rights in extraordinary circumstances. Thus, a clansman both geographically and genealogically distant from a primary right-holder might none the less claim primary rights upon the latter's death in the absence of a stronger claim. Derivatal rights were obtained through a variety of relationships — affinal, cognatic [i.e., consanguineal ties of any kind], residential, and other — and cannot be reduced to any simple formula. In the modern situation, where cash is being given by Europeans in what they believe are straightforward freehold or fee simple transactions, the claims of derivative right holders . . . create particularly thorny problems.

In addition to the above, Ogan noted three other general points about Nasioi land tenure, two of which are relevant here:

First, despite the importance of uterine kinship in the transmission of land rights — as symbolized by frequent Nasioi statements to the effect that 'the land belongs to the women' or 'the land is the women's affair' — land rights were normally exercised by men . . . This is consistent with the sharp sexual division of labour, in which men clear land and plant perennial trees while women garden. Second . . . perennial trees such as coconuts or areca [betel] nuts were individually owned by the planter or finder and could be inherited by his children without further transaction.

The remainder of this chapter is in two parts. The first will outline the development of the mine project up to the commencement of mining, in April 1972, including the direct and visible effects it had on nearby Bougain-villians. The second part will follow the same procedure, but much more briefly, for the period from 1972 to November 1988, when the active attempts to shut down the mine began.

As we have seen, in 1963 CRA Exploration was granted a Prospecting Authority over an area of 630 square kilometres in central Bougainville (see Figure 14). The company was required to compensate landowners for damages to their trees, crops and buildings, and to pay the Administration a small annual rent — but not to pay rent to the landowners. In fact the

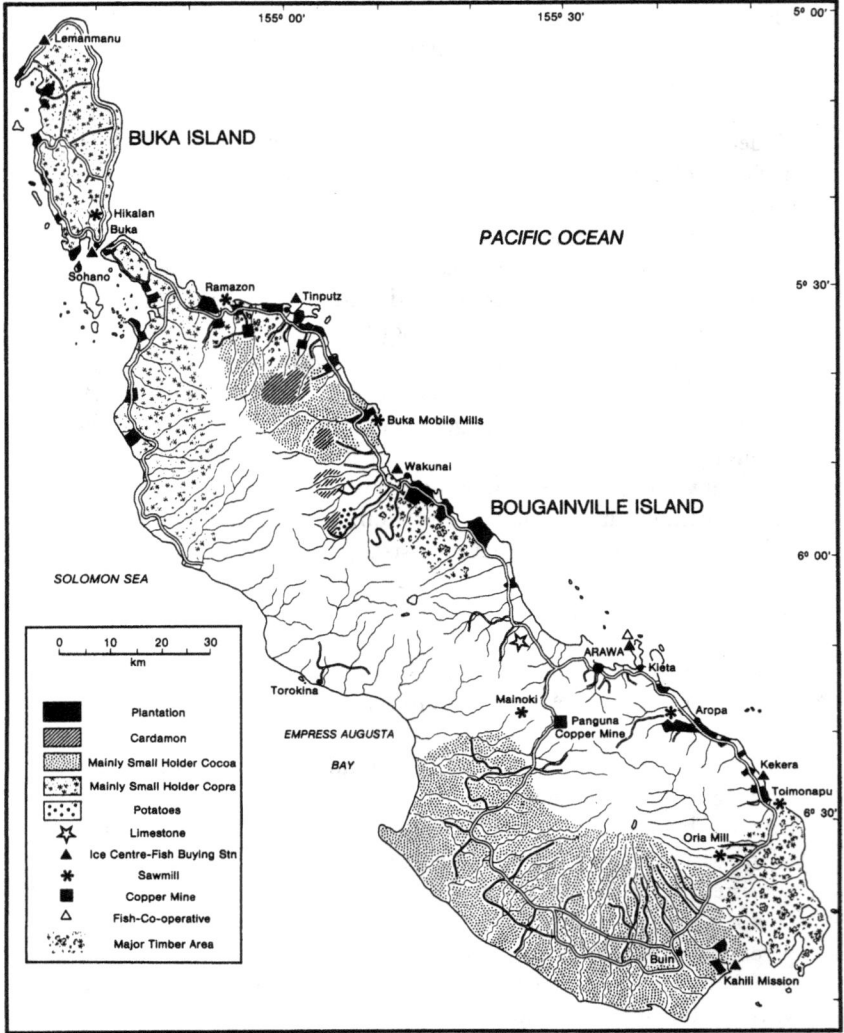

Figure 14   Other economic activity

company was not even required to obtain permission from landowners to prospect or to conduct test-drilling on their land. The small party of CRAE geologists walked into the remote Panguna area in mid-1964. They had had little experience in dealing with Niuginians. They were accompanied by a member of the Territory's Mines Department, designated to act as liaison, but that official had had no previous experience on Bougainville, and in any case did not remain with the party very long. On arrival in the Panguna area, the party was met by a large number of local residents, curious, and evidently anxious, about the party's purposes. Two days of talks succeeded in making clear to the local residents at least two things: first, that there might be valuable minerals under the surface of their lands but, secondly, those minerals belonged not to themselves but to the Administration in faraway Port Moresby. Despite that, however, a few of the residents agreed to allow the party to proceed (probably in a mood of helpless inevitability); most of the others were hostile to the 'trespass' and were to remain so for a long time.

Nevertheless, prospecting continued, including the employment of helicopter-transported drilling equipment to obtain cores of ore for laboratory analysis. This was accompanied by increasing numbers of outsiders: many Europeans and hundreds of non-local Bougainvillians and of Niuginians from other districts of the Territory. Some of the local landowners continued to react with resignation; others responded by posting 'keep out' signs, or by physically blocking entry onto their lands. Rumour spread that the outsiders were not merely searching for minerals but were actually mining them — that the cores removed in test-drilling were being sold for profit, with no shares for themselves. Another rumour was that the wooden pegs hammered into the ground for surveying and for the demarcation of lease boundaries were in fact designations of expropriated land. More alarmingly, the story spread that the drilling was having the effect of creating a huge cavern into which all this part of the island would collapse.

Meanwhile, the company along with several Administration officials, were co-operating in an attempt to temper the Mining Ordinance somewhat, by permitting the company to pay occupation fees to owners of the land within the Prospecting Authority. The motives behind this attempt included a wish to reduce local opposition to the prospecting, along with an evidently sincere desire to compensate the affected owners for use of their lands. But before this amendment had been formalized and publicized, a visit from the Australian Minister for Territories turned that clock back. In a whirlwind visit to Bougainville, with a one-hour stop-over in Port Moresby for 'advice', he announced to a large assembly of Bougainvillians that the fruits, if any, of the Panguna mine would go to the Territory as a whole and not the Panguna landowners in particular, nor to Bougainvillians in general, except as residents of the Territory.

In the words of one company official:

> The effects [of the Minister's declaration] were dramatic . . . trees were felled across the helicopter pad . . . surveyors met opposition . . . landowners [demanded that] mining activity [be] confined to the area of ground of present activities (about one mile [2.59 kilometres] square) . . . until they had an opportunity to see the effects of the mining on the land, and also what they could get out of it. A few days later our present area of activities was surrounded by 'tambu' ('keep out') signs. [King 1978, p. 88]

The amendment to the Mining Ordinance called for a change in the designation of lands within the Prospecting Authority, from 'native' to 'private', and a payment to their owners of $1 an acre ($2.47 per hectare) per annum during prospecting, and $2 per annum for the duration of any mining lease. What the amendment did not take into account was the certainty that much of the land 'rented' would eventually be wholly destroyed if mining took place. Nor did it provide for the precise kind of cadastral survey needed to determine who owned what — thereby prolonging an uncertainty that would bedevil the company to the end. And thirdly, it did not respond to the company's wish that some of the Administration's revenues be put aside into a development fund that would benefit Bougainvillians in general and those directly affected by the mining in particular.

Despite the payments made to them by terms of the amendment, the locals' opposition to surveying and test-drilling persisted. Most physical expression of that opposition was stifled by the establishment of a police post near Panguna, but the attitudes of most landowners did not change. They were exemplified by the action of the residents of Guava village, who returned most of the money they had been paid in compensation for trees felled by the company's agents. Many residents in the area refused to accept compensation, evidently acting on the assumption that acceptance would signify that they had willingly 'sold' their lands.

As described earlier, Bougainvillians in general, and the Nasioi in particular, had fairly clear-cut concepts regarding the full or residual ownership of land, but were normally quite generous about allowing others temporary use of their land. Why, then, were the residents of the Panguna area so opposed to the mere presence of the prospectors on their lands — a presence that would be (as was repeatedly 'explained' to them) only temporary? The answer to that, almost certainly, is that most of them disbelieved those explanations, so the question becomes: why were they so adamantly opposed to the long-term lease, or permanent alienation, of land, which at that time and in that area was not in short supply?

One answer to that question is that the company prospectors were totally outside the circle of people to whom usage rights were normally extended. Added to that was the fact that they were white, and it was through whites that the Nasioi, as they said, had lost much of their land for plantations, roads, etc.; had been 'treated like dogs' as plantation labourers and Administration subjects; had been abandoned to the Japanese; had been employed at low wages; and so on.

A reason for opposing the company's prospectors in particular was their close association with Administration officials. While Bougainvillians tended to distrust whites in general, many had learned to differentiate among them. They perceived that some whites, including most missionaries, a few planters, and even some officials, were sympathetic and sincerely beneficent. They had also learned that there were institutional frictions between Administration and the missions, and between plantations and the other two. Unfortunately for the company, it was perceived by most Bougainvillians to be closely allied with the Administration from the start, thereby sharing in the hostility engendered by some Administration policies and personnel. Some association with the Administration was essential at the beginning of the project, but as time passed the company needed to distance itself from it and act with more generosity towards its Bougainvillian neighbours. This was necessary not only in order to get on with developing the mine, but in anticipation of the time, not long off, when Bougainvillians and other Niuginians would be governing themselves, including deciding the fates of companies.

Another reason for opposing this particular *waitmans* enterprise was the Administration's insistence on enforcing the Ordinance's doctrine about the state's ownership of minerals. It was reported at the time that the landowners' rejection of this alien, and to them totally unfair, doctrine was supported and encouraged by 'certain' Catholic missionaries, mostly American. They are said to have declared that the doctrine was peculiar to Australia, and did not apply to 'civilized' nations. (For the record, there was, and is, considerable public ownership of subsurface rights in the United States, and some private ownership of them in Australia.)

Finally, local opposition to the company's activities was heightened by the desire, expressed by many residents in the prospecting area, that the treasures, if any, be left in the ground until they could be extracted by their children or grandchildren. In 1968 a few of them told this writer that they had the necessary shovels, etc. and were eager to work at it themselves. Other hopefuls, although aware of their own limitations in knowledge and technical skills, proposed that 'education' would qualify their descendants to operate the mine—which of course might someday become true. At that time even the company geologists and Administration officials had not seen an open-pit mine as large as Panguna was to become, and consequently were unable to dissipate the Bougainvillians' shovel-size illusions. In fact, even after watching the mine grow to its 1989 size, many Bougainvillians were still cherishing the 'cargoistic' hope that each of their communities could someday operate a mine of its own.

Despite such opposition, the results of early test-drilling were promising enough to encourage company officials to define the dimensions and quality of the ore body. The final decision to mine was not made until 1969 (after an expenditure of $21.43 million), and actual mining did not begin until April 1972; meanwhile a number of other measures had to be taken.

First, guaranteed markets had to be found for the future mine's future products. To secure such markets several years ahead of production was no easy undertaking, but was eventually accomplished by letters of intent exchanged with companies in Japan, Germany and Spain. The company, and the Territory, were fortunate in that the growing proof from thousands of drill cores promised not simply copper concentrate, but a significant 'sweetener' of minor quantities of gold and silver. Japanese companies agreed to buy the largest share: 1,025,000 tons (1,000,000 tonnes) of copper concentrate over a period of fifteen years. For the same period West Germany's Norddeutsche Affinerie agreed to purchase 787,500 tonnes and Spain's Rio Tinto Patino 180,000. Over the years these buyers remained the

largest of the company's customers, although revisions were made from time
to time in their original contracts, in line with their changing needs. In
addition, substantial sales were made from time to time to the People's
Republic of China, South Korea and the Philippines; and smaller ones
elsewhere.

Secondly, financing had to be secured for construction of the project's
mining and support facilities. This was accomplished by obtaining loans and
issuing shares. The Bank of America acted as manager for the raising of a
$200 million loan through the Euro-dollar market; of this sum $125 million
was repayable within three to four years, the remainder by 1978. In the event,
these initial borrowings, with interest, were repaid on schedule. During the
years of the mine's operation several other loans were obtained from various
sources in order to finance new equipment and facilities — including, for
example, trucks for ore haulage, some of which cost $220,000 each.

Initially, a total of 18,917,013 ordinary (50-cent) shares were issued at
$1.55 each. Of the 1,114,000 reserved for sale within the Territory, 893,000
were sold to individual indigenes or to largely indigenous groups, the
remainder to missions or to long-resident expatriate individuals or companies.
No breakdown is available giving the number of shares held by individual
Bougainvillians, but the company made a special effort to publicize the
offering locally.

A third requirement for mining was to conclude an agreement with the
Administration. This, among other objectives, would rectify aspects of the
Mining Ordinance shown to be unworkable in the Territory, and, of utmost
importance, would be acceptable to the independent nation of Papua New
Guinea, whose establishment was imminent. After year-long negotiations,
the Bougainville Copper Agreement (BCA) was passed by the Territory's
House of Assembly in August 1967. Its principal provisions included:

• Approval of the current leases.
• An undertaking by the company to complete current investigations at a
  cost of $4 million, in addition to the $4.5 million already spent, and to
  complete within five years all construction required for mining, processing and shipment (at a cost of up to $30 million, including investigation
  costs).
• An undertaking by the company to provide the necessary port and roads,
  and enough housing in an off-site (public) town to accommodate employees
  not housed at Panguna. In addition, the company would build and temporarily finance a hospital in the public town on behalf of the Administration. For its part, the Administration would provide for the extra
  schools, police, and postal services required by the company.
• An undertaking by the Administration — subsequently cancelled — to continue the licensing arrangements for a total of eighty-four years, during
  the first half of which there would be no renegotiation of the terms of the
  agreement.

The Bougainville Copper Agreement also included the following taxation arrangements:

- A three-year tax holiday, and a long-term guarantee that 20 per cent of net revenues from copper sales would be exempt from taxation.
- A special additional tax, when the company became taxable, to bring its total tax to 50 per cent of taxable income.
- A ceiling of 50 per cent on total imposts until the twenty-sixth year of operations, and thereafter increasing by 1 per cent every year up to a final ceiling of 66 per cent.

And, of critical importance to everyone concerned:

- A royalty, payable entirely to the Administration, of 1.25 per cent of f.o.b. value of concentrate produced.

As noted earlier, the royalty provision of the BCA derived from the Mining Ordinance; subsequent events showed it to be the most controversial financial part of the agreement.

The principal physical components of the mining project were the mine itself and its crushing and concentrating facilities, all at Panguna; a port for storing and shipping the concentrate; a pipeline for conveying the concentrate from mine site to port; an area in which to dispose of the mine's overburden and tailings; housing for mine employees and their families; a road linking the mine with the port and off-site housing; and a source of electric power for mine, port and housing.

By 1972, when mining commenced, the ore body at Panguna was estimated to contain about 900 million tonnes of ore averaging 0.48 per cent copper and 0.55 grams of gold per long ton (1016 kilograms), along with some silver. The whole roughly elliptical body measured about 1500 by 1000 metres wide at the rim and about 600 metres deep. Like all known porphyry copper deposits of economic grade (i.e., by this time, 0.4 per cent or more) it is geologically young, being in the range of two to five million years old. It is made up of mineralized quartz diorites intruded into an older sequence of andesitic volcanic flows. Similar but much smaller ore bodies have been discovered nearby, including the one at Kupei. Bougainville is part of an ancient volcanic chain of islands, and one of its two still-active volcanoes, Mount Bagana, is only about eight kilometres north of Panguna.

Most of the Panguna ore body lay under a thick layer of soft overburden and rock, which had to be removed to expose it to mining. Once exposed, the ore had to be loosened — mostly by explosives — and hauled to the nearby processing facilities. There it was crushed, ground, and then subjected to flotation. The large amount of water required for this came from the Jaba River eleven kilometres to the west, where a pumping station (capable of supplying 140 million litres of water a day) sent it up to the mine site. (The writer's main purpose in reporting these figures is to convey the magnitude of this vast industrial complex located, as it is, on a remote and economically undeveloped island.)

Loloho, the company's port and concentrate-storage facility, was built on the island's east coast at Anewa Bay, which had to be dredged to provide berthing for ships up to 40,000 deadweight tonnes. There also was located the 135-megawatt oil-fueled steam power station designed to supply electricity for all of the company's installations, as well as for other nearby non-company needs. At the time of its completion, in 1971, its output equalled nearly twice the amount of electricity generated throughout the rest of the Territory at that time.

By the end of 1971 Panguna contained 323 family residences, a number of two-bedroom flats, and ten three-storey blocks of rooms for single employees, along with dining halls, a supermarket, extensive sports facilities (including an olympic-size pool), a cinema, and an inter-denominational Christian chapel. In the absence of level ground, however, housing for the rest of the company's employees had to be located elsewhere. For reasons of proximity and salubrity the company preferred a site partway to the eastern coast and about 600 metres above sea level. However, the Administration opposed the location on the grounds that it would displace about 120 Bougainvillians living there, and instead acquired the expatriate-owned coconut plantation located on the coast at Arawa, about half-way between Kieta and Loloho, and about thirty kilometres by road from Panguna.

The Administration purchased the plantation and in co-operation with the company built the town, which was designed from the beginning to be a public one. For its part, the company constructed 399 houses for its employees and their families, and assisted in the construction of a hospital and other civic facilities. In addition to its company houses, the town came to include residences for non-company people (including those employed by the Administration), commercial stores and offices, etc. Inevitably, there was also a large colony of squatters' shacks. By late 1972 Arawa's population was about 5000. By 1988 it had grown to 15,000 and had become the third-largest urban centre in Papua New Guinea.

The road was eventually built to connect Panguna with the east coast and Loloho and Arawa. Completed in December 1970, it was second only to Panguna itself in scale and cost of construction. Part of it lay parallel to the earlier jeep road leading to the mine, but its cost was 15 million kina more (i.e. K55 million at 1991 values). The jeep road continued to have a useful purpose as the route along which was built the power line from Loloho to Panguna.

The 26-kilometre stretch of the new road from coast to mine site was bitumen-surfaced and 8 metres wide. The first 14 kilometres of it from the coast was fairly straight and averaged only 2 per cent in grade. For the next 6 kilometres, however, it climbed 1000 metres to the crest before descending to Panguna. In all, some eleven million cubic metres of rock and earth had to be moved in building it, with labour supplied mostly by Niuginians from districts other than Bougainville.

Aside from perennial grievances about apportionment of its earnings, the disposal of the mine's tailings has proved to be the company's most persistent political headache; and not without reason. Initially about 26 million tonnes of waste rock and 6.8 million cubic metres of soft overburden were removed and piled along the side of the valley of the Kawerong River, which flows westwards from Panguna; and that was only the beginning. During the mine's seventeen years of operation, another 556 million tonnes of waste material were separated from the ore and disposed of, along with 595 million tonnes of silt material left over from ore processing: altogether an additional 1225 million tonnes were either stacked up alongside the Kawerong's banks or dumped into its stream. A few miles downstream the Kawerong and several other streams flow into the Jaba River and thence to the west coast. By mid-1971 those naturally clear streams were already silted, aggraded, and widened. By 1988 the flow and spread of the tailings had raised (and thereby widened) stream beds by up to 20 metres in the deepest parts, and had blocked stream flows in many places, creating more new swampland while infilling other areas.

Before the widening of the Jaba River, a small village of Banoni people resided on the coast near its mouth; not only did they have to be resettled, but the widened river blocked all pedestrian movement north and south until a road and pedestrian bridge were built over the Jaba at Balo in the 1970s. Prior to the mine's operation only about 750 people resided in the inland area now covered by the tailings and its impounded swamps, but hundreds of others from adjacent areas used to fish, hunt and swim there. Moreover, the tailings sediment flowing from the Jaba into the bay increased the size of the delta from 65 to 900 hectares and spread silted water far out from the shore, with some known or suspected lethal effects on marine life there and up and down the coast.

In direct response to these palpable effects on the physical environment of the tailings area, the company undertook three measures. One was to set up procedures for monitoring environmental changes. Beginning in 1969 and continuing into 1989, systematic sampling of stream flow and content was undertaken, inland and in the bay. In 1976 marine and freshwater biological research was added to the monitoring. The whole programme was carried out by a separate staff of scientists and engineers. The second measure consisted of a programme, established in 1973 and staffed by agronomists, to discover how best to revegetate the areas covered by waste rock and tailings with plants which were economically useful. The third measure was to construct a 31-kilometre pipeline for conveying tailings from the mine to the bay. This project, which was estimated to cost K65 million and to be completed in late 1989, was intended to eliminate use of the river system for tailings disposal and thereby allow the company to commence the rehabilitation of the area during the life of the mine. Such a pipeline had been considered when the mine was under construction, but technology was not available at that time to render it invulnerable to the region's chronic seismic movements.

Now to consider the direct, large-scale and more or less detrimental effects that the mining has had on Bougainvillians, and the company's efforts to ameliorate them. Considered here are those effects which were widely perceptible and which the affected Bougainvillians characterized as harmful. Other, more indirect effects, both harmful and beneficial, will be considered later on, as will some of the more general efforts of the company to make its presence on the island more tolerable and useful.

Without question the most palpably hurtful effects of the mining project were experienced by residents of the mining area who had to be resettled, some of them twice, to make way for mining. Between 1969 and 1989 nearly 200 households were relocated. The total cost to the company was about $1.64 million; its cost to the relocated people, in terms of physical and psychological hardship, cannot be expressed in figures.

The first people to be resettled were the residents of Moroni, whose village was located within the planned open-pit mine site. The new location was on an exposed ridge overlooking the pit. On it the company was required by the Mining Warden, under terms of the Mining Ordinance, to construct houses in permanent materials of the same size as the ones destroyed; in the event, they consisted of one- and two-room fibrolite houses with iron roofs. In addition, the company was required to build latrines, water tanks, a chapel, and an access road; to clear and plant gardens equivalent to those destroyed; and to provide rations for six months while the crops matured. In an effort to compensate for other losses, tangible and intangible, each one of the village's fifty-four residents was paid $300, including $200 for the 'hardship which will follow the enforced change from a traditional village environment to a European way of life, and provide for the additional cost of maintaining a European-style residence in an urbanized community'. (Bedford and Mamak 1977)

At the time some Europeans considered the change to be highly beneficial: 'paying the natives to do what they had been trying to do, namely to live like Europeans.' At the beginning of the resettlement a few of the villagers may indeed have been pleased to live in the new European-style dwellings, and to eat the proffered, mostly canned, rations—but not for long. In Bougainville's climate all buildings require constant and expensive maintenance and repair; iron roofs develop expensive leaks; and water tanks clog with sediment. Also, the continuous, round-the-clock mining noises to which the resettled villagers were exposed, proved to be a painful change from the quiet of their former village site.

Few other resettlements proved to be quite as painful as that of Moroni, but none of them was lacking in hardships of some kind. In some cases, people underwent two moves, on account of changes in the company's expansion plans; many had lengthy waits for houses to be built, or access roads put through.

All in all, the resettlement required by the mine project has brought about the most palpable of its harmful effects. The number of Bougainvillians affected was relatively small—195 households, or about 0.1 per cent of all residents of the North Solomons Province—but their sufferings stand out as a grim reminder of the human costs of operating an open-cut mine in a place like Bougainville.

One measure to reduce that cost was 'nuisance compensation', which is described in the following excerpt from a study by Bedford and Mamak (1977):

> in most areas where resettlement of all or part of a community was required, another class of compensation was also commonly demanded [and granted]. Some recompense for the inconvenience caused by blasting, deforestation and subsequent flooding, dumping of waste, noise and dust generated by heavy machinery was given to most villages in the upper Pinei and Kawerong valleys. Again precedents were set early in the Mining Warden's Court hearings and, although the company frequently challenged claims for 'nuisance' compensation, the Warden was generally favourably disposed towards granting something to villagers who were exposed to the extremely destructive processes adopted to get the mine into production . . . For example:
> * In 1969 Pakia villagers received $A17 per head on the basis of the following claims: flooding of the tributaries of the Pinei river as result of earthworks; school children endangered by heavy construction machinery on the only access route to their school; the road rendered virtually impassable by mud and dust; a general loss of aesthetic appeal in the village neighbourhood as a result of construction activities.
> * In 1971 the Moroni villagers received $A1,460 for the following damages: disturbance to sleep due to continuous 'round the clock' working of bulldozers; disturbance, annoyance and inconvenience from blasting; annoyance and general unpleasantness in the village caused by rotting vegetation and dieseline; destruction and loss of water supply and sources of firewood; disturbance of burial grounds.

The company sometimes challenged claims because of their potential for endless ramifications, but most company officials known to the author acknowledged and expressed regret for these and other palpably harmful consequences of their operations. This of course did nothing to reduce the victims' immediate hurts and long-term dissatisfactions.

Not all villagers in the mine-lease areas had to be resettled, but none of them escaped deprivations of one kind or another, particularly loss of gardening land. However, since most of the deprivations originated or were exacerbated after the commencement of mining, description of them will be deferred.

Some resettlement had to be undertaken in the areas adjacent to the highway from the port to the mine. Apart from that, the most visibly disturbing aspect of the road's construction (and indeed of the construction phase in general) was the presence of thousands of labourers. These included strangers from other parts of Bougainville and, even more disquieting, thousands of 'redskins' (lighter-skinned indigenes of other districts of the Territory, especially of the Highlands and Sepik regions). While on their jobs and in their huge camps, built mostly along the new highway, they were only potentially menacing, but on their excursions on off-days (sometimes to places as far away as Buin and Siwai) they posed threats, real or imaginary, to villagers' security, especially that of nubile females. The company issued strong warnings against such excursions, and many villages posted 'keep out' signs — but with what success is not known.

Many Bougainvillians earned unprecedently high and steady wages from the construction, but the only other mitigating benefit obtained from that immense activity was the money they earned from selling native food to the labour camps. From the very beginning of exploration the company's search parties had purchased sweet potatoes and other garden and tree crops from local residents to help feed themselves, and that trade continued throughout the operation of the mine. A peak in it occurred during 1969–71. In that period company trucks made scheduled visits to several produce-assembly villages in northern Nasioi and bought large quantities of food. These were often larger than needed, but paid for anyway, in order 'to maintain good will'. In 1971, for example, some $330,000 was spent on these purchases. For the same political purpose the company constructed a vehicular road to link the Jaba River pump station with the network of roads of south Bougainville, to enable the residents of that region to transport their vegetables and fruit to the Panguna market, and their cash-crop, cacao, to Kieta for export.

As mentioned earlier the Administration over-ruled the company's choice of a second inland site for housing employees, and chose instead to acquire the coastal plantation of Arawa, owned by an Australian. It was expected that the more or less coercive acquisition of land owned by a European would arouse less opposition. (Under terms of the 1967 BCA, the Administration was responsible for acquiring land for facilities that included 'public' purposes, as was the case with the proposed town.) Faced with the alternative of condemnation and compulsory acquisition, the plantation's owner agreed to sell.

Much to the surprise of many whites, the transaction aroused heated opposition from the owner's Bougainvillian neighbours. This was partly because of his personal popularity (he provided them with facilities for drying and selling their copra and cacao); it was also part of a growing opposition to forced sale of any Bougainville land. In fact, opposition was so strong that the Administration postponed for a time the acquisition of an additional 150 hectares of adjoining land needed for future expansion of the prospective town. Meanwhile, out of that opposition was formed an organization, named Napidakoe Navitu, which was to play a leading part in promoting Bougainvillian nationalism.

Turning now to Loloho, the company's port on Anewa Bay, the nucleus for it had already been acquired in 1964 by purchase of a small expatriate-owned plantation, which served as a shore base for the company's survey ship, *Craestar*. More land was needed to accommodate the port's proposed ore storage, wharf and power station, and for that the company asked the

Administration to acquire an adjoining 70-hectare grove of coconut palms owned by the residents of nearby Rorovana village.

The first offer made by the Administration was to purchase at $430 per hectare or to lease for forty-two years at $13 per hectare per annum. It was flatly refused. While the Administration-established value for land elsewhere in the Territory was only $260 per hectare, the owners knew that more than $432 per hectare had been paid for the Arawa land. The Administration temporized; after two months it instructed the company's surveyors to commence work, having warned the Rorovanan owners that the land would be leased by compulsion if they did not 'voluntarily' sell or lease it. In the face of continued opposition, the Administration sent in a police escort with the company's work crew. In the ensuing confrontation, the Rorovanans (including women barebreasted for the occasion) obstructed a bulldozer and were dispersed with tear gas. The event was witnessed by a large crowd of onlookers (including many camera-ready reporters) and was so widely criticized, throughout the Territory and abroad, that the chastened Administration had the company postpone construction until some agreement could be reached.

Meetings took place in Bougainville, Melbourne and Canberra over a three-month period. The Rorovanans held out for a higher price; but the Administration did not want to allow a precedent for land acquisition elsewhere. Finally, in order to avoid very costly delays in the mining project — estimated by the company to be around $30,000 a day — the Administration gave in to the company's urgings. It negotiated an agreement with the landowners that included a leasehold occupation fee of $7000 per annum for 56 hectares (based on a land valuation of $2470 per hectare) plus a large lump-sum payment to compensate the owners for destruction of buildings already on the land. In addition, the owners were give an option of purchasing 7000 shares in the company when the first issue was made.

Armed with this precedent, the owners of the 150 hectares needed for expansion of the town willingly negotiated a 99-year lease with the Administration and the company, at $18,600 per annum (subject to periodic review). In addition, the Administration agreed to provide the owners' village with electricity and an access road, and the company agreed to offer them an option to purchase 7000 shares.

Various measures were undertaken to compensate Bougainvillians for actual or potential damages to their crops, their forest products, and their usual supplies of fish.

In the early days of prospecting the company's agents offered payments to landowners for timber used in building camps and drilling rigs, but in most cases the offers were rejected because of suspicions that acceptance might have been viewed as selling the land. In 1966 a Mining Warden's Court was set up in the prospecting area to regularize damage claims, and by the end of 1969, 350 cases had been heard. The procedure was for the plaintiff to present a claim to an Administration officer, who, after verifying the claim, filed it in the court. The following excerpt from Bedford and Mamak (1977) is typical of the kinds of claims lodged and compensation awarded during that period.

> Okiong-Amentat of Korpei village who had gardens on the land known as 'Damara (No. 2)' owned by Simon Poraka sued the company for the following damages: Destruction of 20 small pitpit clumps, 6 mature banana clumps, 100 roots of tapioca, 1 Kaukau (sweet potato) plot 63 feet by 140 feet [19 by 42 metres]. Amentat had no other gardens at the time and sought assistance in the preparation of a new garden as well as cash compensation for destruction of his crops. The judgement handed down at the Mining Warden's Court on 9 October 1969 was:
>
> Compensation for destruction of crops, $86
> Cash payment in lieu of rations for 22 months while a
>   new garden was established, $84
>
> In addition the company was to prepare a garden site covering an area of one-fifth of an acre [809 square metres] within three months of the court hearing.

By 1979 a schedule of compensation for subsistence crops, animals and trees had been standardized to the satisfaction of the Administration, the company, and, evidently, most claimants. The procedures were simplified by having the claimant deal directly with a company employee (a member of the company's Village Relations staff). Some examples are:

- $5.00 each: breadfruit tree, mango tree, coastal sago palm
- $2.00 each: banana plant, pepper plant, pandanus tree
- $1.00 each: pawpaw plant, passionfruit vine, limbum tree
- $0.50 each: ipika leaves (a green vegetable), tomato plant
- $0.20 each: ginger plant, pineapple, tapioca, clump of pitpit
- $0.10 each: choko vine, kumu leaves, sugar cane, thick bamboo
- To clear ground and plant an acre of garden crops, $200

And last, but certainly not least, to compensate for death of a pig, $10 to $80, depending on size.

Claimants who were dissatisfied with the standard rates could lodge claims in the Mining Warden's Court. One who did was Benggong, many of whose cacao trees had been destroyed by construction of the port–mine road. Benggong demanded compensation of $30 a tree per annum during the life of the mine (cacao prices were high at the time—but not that high!) The company offered to pay at the rate paid in the purchase of Arawa plantation, and in a lump sum; while that was much lower than Benggong's figure, it was higher than the $5 recommended by the Administration, who were fearful of setting such a costly precedent for compensation elsewhere in the Territory. The Mining Warden, who was sympathetic to Benggong's claim, awarded him $35 per tree. The company objected to both the amount of the award and its proposed method of payment, and took the matter to the

Territory's Supreme Court — not on the amount of the compensation (which could not be appealed) but on the grounds that the Mining Warden had exceeded his jurisdiction in making the award. The Supreme Court found for the company, but that was not the end. The Territory's Public Solicitor thereupon took the case to Australia's High Court, who decided that the matter *had been* within the Mining Warden's jurisdiction. By this time both company and Public Solicitor were agreed on the desirability of a compromise. The company wished to avoid the adverse public image of a huge and wealthy multinational company engaging in a legal battle with a poor native villager; the Public Solicitor wished to dampen the expectations which other potential claimants might have as result of this unrealistically high award. In the end they agreed to a formula for this and further claims of this kind, namely, lump-sum payments of $15 for each coconut palm destroyed in a lease area and $23.50 for each cacao or coffee plant. Even those rates were 500 per cent higher than those paid by the Administration elsewhere, and of course much lower than Benggong and his compatriots had hoped to receive, but in the end they were accepted. The rates were higher than the company wished to pay, but it wanted to put the matter behind it, and avoid more adverse publicity. It not only agreed to the compromise but decided to pay compensation for all cash crops within a lease area as soon as its boundaries had been determined, and whether or not the crop had been destroyed. (Doubtless, the distinguished engineers, financiers and industrialists who gathered in the company's board rooms in Melbourne to authorize $100 million expenditures or to seek sources for $200 million loans, were surprised to be dealing with damaged $15 coconut palms and $20 dead pigs.)

As the company and Administration had feared, the very high rates set for Benggong's compensation encouraged other cash-crop growers to expect the same. That, however, was not to be; after a few similar claims before the Mining Warden's Court had resulted in smaller awards, the claimants were persuaded to accept less. This of course did not signify that they were satisfied with the smaller amounts paid them. They became increasingly discontented with both the company and the Administration, including their alleged champions, the Mining Warden's Court and the Public Solicitor. Less discontented were those cash-croppers who, in lieu of cash, accepted the company's offer to replace their plantations with larger ones elsewhere. Under this agreement, the company undertook to pay for the labour to make and maintain for three to four years a new garden one-third larger than the one requisitioned or destroyed.

In addition to demands for compensation for individual wild-growing trees and other plants, some residents of the lease areas submitted claims for 'bush compensation' for loss of use of forested areas. The company surveyed samples of such places and valued them in terms of their useful plants, their hunting possibilities and their aesthetic qualities (by *whose* 'aesthetic' criteria is not reported). It arrived at values ranging from $62 to $288 per hectare, to be paid in lump sum. After the owners responded with a demand for a flat $1000 per acre ($2470 per hectare) for all such 'bush', discussions between company and Public Solicitor arrived at a lower, but equally arbitrary, figure of $124 per hectare, to be paid in annual instalments over the forty remaining years of the lease.

Among Bougainvillians, rights to fish in certain sections of a stream were owned no less than rights to dry land. It was necessary to assign a monetary value to the loss of the possibility of catching and eating fish — either as result of the diversion or pollution of a customary fishing place, or of the impossibility of fish reaching it any more. Discussions between company and Public Solicitor arrived at a formula that included estimates of the amount of protein supplied by fish in the claimant's diet (in the mine tailings area this was estimated to be 80 per cent); the amount of protein issued to labourers under the Territory's Labour Ordinance; and the projected retail price of tinned fish over the next five years. In 1971 representatives of the people having fishing rights in one of the mine-polluted streams accepted the proposed compensation of about $18 per person per annum, on condition that the amount be reviewed at the end of five years. In this same part of the tailings area the company also agreed to pay a total of $3000 per annum in settlement of claims for 'nuisance' — specifically, those based on loss of customary places of recreation, of bathing, of laundering, and of 'aesthetic appeal'.

Not all of the streams polluted by mining activities were affected to the same degree or in the same way, and claims for damages varied both in amount and in kind. Regarding the latter, I cannot resist quoting one claims report submitted to the company by an Administration official on behalf of a village in the tailings area:

> I have been asked to explain to Bougainville Copper that the harvesting of the smaller streams was peculiarly the job of the women and that apart from the loss of variety in the local diet, loss of the streams has affected the social life of the women and the children in that small co-operative fishing parties are no longer possible. As well as bringing about social changes at village level, as a few examples will show. Prior to the coming of Bougainville Copper it was the custom to take a newborn child to a particular pool just downstream from the concentrator for a ritual wash; an especially important event if the child was the first born . . . Another pool not far from the confluence of the Kawerong and Barapinang creeks was used [magically] when a man wished to recover lost pigs or increase his fertility. [quoted in Bedford and Mamak 1977]

The special kinds of compensation just listed — for damages to subsistence items, for cash-crops, for 'bush', for fishing and for various other 'nuisances' and deprivations — were claimed and paid for in addition to the basic rents or occupation fees paid by the company for all land within the leases.

Under terms of its Prospecting Authority, the company began in 1969 to pay an annual occupation fee of $2.47 per hectare for all land under lease; the money was lodged with the Administration on the understanding that it would be paid to the landowners once their boundaries had been established. Later, when mining leases were granted, the occupation fees were increased to $4.94 per hectare pending final agreement on the unimproved value of such land. A first step in this direction was made by the Administration, who proposed that it be valued at $98.80 per hectare ($40 per acre). This figure was acceptable to the company, but it was not accepted by the Public Solicitor, who declared it to be low, nor by most landowners, who demanded $1000 per acre. The former was influenced by his conviction about what his prospective clients, the landowners, would accept; the latter by the prices paid for land in Arawa and Loloho—and by a penchant for round numbers.

To break the deadlock the company employed an Australian land valuer, who, understandably, found the job complex in many unfamiliar respects. As he said, 'We have only a money measure with which to express factors which have substantial value implications, but no established monetary expression. One cannot avoid the fleeting thought that it may be *our* system which is primitive rather than the Bougainvillians'.' (Quoted in Bedford and Mamak, 1977.) After a year-long field survey and consultation with all parties, the valuer proposed values based both on the location of the land (e.g., Special Mining Lease, lower Pinei River area, tailings area) and type of land (e.g. arable, village environment, river and river bed, hunting and forest, swamp). Per acre values ranged from $230 (for village environment) in the lower Pinei area, to $30 for swampy hunting and forest land in the tailings lease. Because of the daunting complexities of administering this schedule of values, the company proposed to simplify it with a twofold classification: $105 per acre for the best arable land and $45 per acre for all the rest. That, however, was not acceptable to the Public Solicitor, who proposed increasing the values to $150 and $100 respectively. Further discussion between company and Public Solicitor produced a compromise, whereby all land, regardless of location or type, would be valued at $130 per acre. When even this proved unacceptable to the landowners, the Public Solicitor withdrew from the negotiations. Finally, in 1973 a committee of landowners entered into direct negotiations with the company, which produced an accord whereby the company agreed to pay an occupation fee of $50 per acre ($123 per hectare) per annum for all land within the leases, and to have the figure reviewed in 1976. In their discussion of this bizarre arrangement, the chroniclers of this to-ing and fro-ing had the following to say:

> although some Bougainvillians are receiving large annual payments for land which, under Australian principles of valuation, would be classed as virtually worthless, the principle of paying the same rate for all classes of land probably accords more closely with a Bougainvillian's conception of

property values. To the inhabitant of a mountain village the steep forest-covered ridge and valley topography would, no doubt, be worth as much as any lowland plain; after all, it is home. [Bedford and Mamak 1977]

With this general background on compensation arrangements, we can now turn to the tailings area in particular and see how its residents and neighbours were affected by residing at the back-door of the mine.

Some communities in this area had to be resettled. The most afflicted of these was Dapera, whose people resided in several small hamlets just southwest of the mine, along the banks of the Kawerong River in places where most of the mine's rocky wastes were to be dumped. The Daperans were first offered a coastal plantation site near Kieta, but insisted on remaining in their mountains. A 'show-place' village was constructed for them on a rock-waste site near the mine, but was not occupied because of a change in mine engineering plans. Another community that had to be resettled numbered twenty-six households of Banoni, whose village near the mouth of the Jaba River would be inundated by the aggraded and widened river bed.

Most other Bougainvillians directly affected by the dumping or spreading of mine wastes were those who owned or used land within the lease area but who resided outside the area which the tailings were expected to cover. Such people were discommoded by the tailings in several ways: by losses of cash crops, by lost opportunities for fishing, and by the various kinds of 'nuisance' caused by the tailings. Examples of the curtailment of fishing and of the kinds of 'nuisances' experienced have already been given, along with the measures to compensate for them. Yet to be described is the ingenious way some owners of cash-crop land made the best of their situations.

It will be recalled that in order to avoid further disagreement over the valuations assigned to cash crops—disagreements that were very costly in terms of time and public relations—the company agreed to a quick-fix formula: $15 per coconut palm and $23.50 per cacao or coffee plant, regardless of age, for every plant within the lease. Many owners of land within the tailings area abandoned their cash crops when compensation was paid, assuming that the company now owned the crop. In several cases, however, the owners continued to plant and harvest crops on their former tracts until waste materials covered them.

By the end of 1974 most of the occupation fees had been set and most of the claims for compensation for damages, construction and leasing had been paid, either in lump sums or in standardized instalments. It may be useful, therefore, to provide some figures on the amounts and distributions of those payments. First, how much rent was paid and who received it? (As was the case for much of the information already reported in this chapter, these figures are taken from the 1977 study by Bedford and Mamak, a model of factual reporting and analysis that is enhanced by the authors' even-handed treatment.)

By the end of 1974 the company was paying about $820,000 a year in occupation fees for land within leases of the mine site, tailings and eastern roadway (not including the rents being paid for the port and town sites). Only one-third of the registered owners of the leased land were or had been residing within the leased areas themselves, most others having resided in villages just outside. In this connection, more than half of the $820,000 was paid to owners of land in the tailings area; not because of the economic value of their land — it was in large part swampy — but because of its relatively large size.

Among the registered owners receiving annually paid rents, the average amount received was $590 and the median $450; the range was very wide, from $103 to more than $60,000, with several receiving more than $10,000. It should be recalled however that very few of the 'registered owners' were sole owners. Figures are not available on how many individuals each registered owner represented (and with whom he or she, presumably, shared the rent), but during 1974 they numbered altogether at least 2000 men, women and children — and their numbers were increasing every year.

More certainty attaches to the distribution of rents among villages. If we base our calculations on the official residence of each registered owner, fifty-three villages shared in the rents, with amounts ranging from $33 to $183,592 (the latter having been the small village of Jaba, at the mouth of the river). However, more than one-half of the villages received less than $1000 a year.

Turning now to the distribution of compensation, most of the lump-sum payments had been made by the end of 1974. Over a period of seven years, more than $1.6 million was paid. Of this the range of lump-sum payments to individuals was from one or two dollars to more than $100,000; most were less than $1000. Again, in terms of the villages where those recipients resided, the range was very wide — from $55 to $201,217; most villages received less than $10,000.

The significance of these figures can best be judged by comparing them with the usual incomes earned by Bougainvillians from other sources.

During the period of 1968–74 many Bougainvillians earned cash incomes from the company by working on construction projects or by selling vegetables and fruit to feed the workers. In the absence of details about the amount of wages paid to Bougainvillian construction workers, and the number of Bougainvillians involved in producing food, those income figures cannot be used in our calculations. Since many of those vegetable sellers and construction workers shared in the receipts of compensation payments and rents, the absence of those details is less significant. Nor are figures from other wages very relevant to our comparison. Before the mine most wage-earning on Bougainville derived from unskilled labour on expatriate-owned plantations, and even before the mid-1960s only a few Bougainvillians were employed in what to them had become regarded as 'dogs' work'. Therefore in the comparison we are now seeking, the only figures to be used will be the incomes derived from cash-cropping.

Let us look first at incomes derived from the production of cacao, the only cash crop grown in most parts of the road-mine-tailings area. Separate figures are not available for cacao grown in those areas, so we shall have to be content with those recorded for the province as a whole.

The earliest figures on cacao production available to this writer cover the twelve-month cacao-growing period of 1977-8 after the kina had been adopted. In that period the province's (Bougainvillian) small-growers produced 8330 tonnes. Dividing the total value of this production (about K22.5 million) by the total population of Bougainvillians at that time (about 104,000), gives a figure of about K216 per capita annual income from this source — which is to be compared with an average per capita annual income of K410 received in occupation fees by the 2000 lessors of the road-mine-tailings area. Needless to say, this comparison is very tenuous: not all Bougainvillians shared equally in the proceeds from cacao production, and not all members of the landowning lessor groups share equally in the total proceeds from rent. Nevertheless, the exercise does provide an approximation of comparative magnitudes which is better than sheer guesswork.

Comparing cacao-derived incomes with compensation payments is too tenuous an exercise even to attempt, for a variety of reasons. For example, nearly all of the latter payments were in lump sums, and most of them were made over a period of seven years. Moreover, there is no credible way of estimating how many persons shared in those proceeds. About all that can be said, for purposes of comparison, is to repeat an earlier statement, that such payments during the dollar era ranged from one to two dollars per payee, to $100,000. (As a matter of fact, if such data were available, it would probably reveal a similar kind of distribution among Bougainvillians with respect to cacao-derived incomes — some persons received nothing, or next to nothing, while others received $10,000 or more per annum.)

The second major money-earning crop of Bougainvillians was copra. It was grown by relatively few people, including mainly those residing on Buka and on some of the larger island's coastal area. The available figures on production (actually, on export) do not distinguish between plantation and small-grower production, thereby rendering comparison tenuous.

From this comparison a striking contrast emerges between the monetary incomes of rent and compensation received by a few very highly paid recipients and the total non-wage incomes of all other Bougainvillians, including most other recipients of rents and compensation.

Reactions to these disparities varied. Company officials deplored some aspects of them because of their high costs, but accepted them in order to get on with mining. The Administration (not including the Public Solicitor) disliked them because of the unwelcome precedents they set for the rest of the Territory, but like the company agreed to them in order to hasten the receipt of revenues from an operating mine. As for Bougainvillians, there is much evidence to support the conclusion that many of them felt a growing dissatisfaction or jealousy about some of the larger receipts. While some

Panguna, 1988.

Resettled village, early 1970s.

Shovel and haul truck.

recipients appeared to be satisfied with the amounts received, many, perhaps most, wanted more.

This exercise in comparison is instructive, but only up to a point. While revealing that recipients of rents and compensation had on the average acquired higher monetary incomes than the average non-recipient Bougainvillians, it leaves out of consideration the non-monetary costs borne by many of those recipients, especially by those who resided on or near the mine site. These costs far exceeded any monetary gain.

The mine's first full years of operations, 1973 and 1974, were also high points in concentrate production (650,200 and 640,000 dry tonnes, respectively) and in net earnings (K158,400 million and K114,200 million). After that earnings dropped to an annual average of K49,000 million, in consequence not of variations in production (which remained fairly steady) but of major swings in market demand and price. During 1987–8 net earnings began to rise again, but were subsequently halted by closure of the mine. Over the years the copper and gold content of the mined ore decreased in grade, so that increasing quantities of ore had to be mined in order to keep concentrate production steady. That increase was accomplished through improvements in technology rather than through additional personnel.

Meanwhile, the pit became wider and deeper, having reached a rim-to-rim area of 400 hectares by 1989. During the same period waste materials and tailings were growing in volume year by year; as mentioned earlier, however, a 31-kilometre pipeline was being constructed to convey the tailings directly to Empress Augusta Bay.

By the end of 1988, the last full year of mining, a sum equivalent to K1000 million had been invested in the whole mining project. Part of that had already been discounted through depreciation; what value it will have after an extensive period of disuse and neglect, in Bougainville's tropical climate, is anybody's guess.

Another major casualty of the mine's forced closure was the termination and dispersal of its workforce. During construction, the workforce exceeded ten thousand. Thereafter it numbered on the average about 3840, and ranged (from year end to year end) between 3560 and 3565 (in 1988 and 1971, respectively) to 4222 (in 1974). At beginning of mining Niuginians, including Bougainvillians, numbered 72.7 per cent of the whole workforce; thereafter this preponderance of Niuginians rose steadily up to the end of mining, when their percentage of the total was 82.2 (40 per cent of which were Bougainvillians). This reflected the company's announced policy of 'localization', in terms both of overall numbers and of higher job levels. At the beginning and the end of the mine's operations Niuginians were distributed among the workforce as shown in the table.

Most of the advancement to more skilled jobs resulted from the company's training programmes, of which it sponsored two types: an on-site one, including a Mine Training College; and a scholarship fund that financed tertiary training for Niuginians in educational institutions elsewhere, includ-

Niuginian participation in the mine workforce in 1988 (last full working year)			
Category	National	Non-national	Localization (%)
Management and professional	180	250	42
Sub-management and sub-professional	573	6	99
Supervisory and skilled	787	350	69
Semi-skilled	1 320	4	99+
Unskilled	90	—	100
Total	2 950	610	80

ing Australian universities and, especially, the Papua New Guinea University of Technology at Lae. For on-site training the company employed up to ninety-three full-time teachers; half of these were attached to the Mine Training College at Panguna, the others to the working departments of the project. From 1970 to the end of 1987 nearly 15,000 Niuginians were enrolled in on-site training courses, which included heavy equipment operation, electronics, computer operation, management, medical and security. During that period another 400 Niuginians were given scholarships and maintenance funds for study elsewhere; and while these trainees were welcome to join the company's workforce upon completion of their studies, they were not obligated to do so. The same held true for on-site trainees: many of them chose to work for the company, but many others left to join or open businesses for themselves elsewhere.

A workers' union, headed by a Bougainvillian, was organized with company encouragement at an early stage. Grievances were voiced from time to time but they were usually settled quickly and satisfactorily, and occasioned very few work stoppages. Fortunately for the company, its employees proved to be more amenable to the logic of Western-style economics than were the owners of the land it occupied.

Even wages were not as fractious an issue as one might expect. The company's preference was to pay equal money for equal work regardless of ethnicity. But as its chairman, Frank Espie, said in a speech delivered in 1972:

We cannot pay indigenous employees expatriate level wages until this country as a whole reaches something approaching similar wage levels. The company would serve its own short-term interests by paying such wages: in this way we could attract, and hold, the cream of the country's talent. It would, by doing so, have destructively distorting effects on the whole manpower and economy of a developing country. It would particularly affect the balance between people who work for wages and those who live in agriculture, as do most of the people of this country. Such an imbalance has created grave problems, which are not yet solved, for a number of African countries.

The percipient reader may perhaps consider this a contradiction to the company's willingness to pay higher-than-average occupation fees for lease-area land in order to 'get on with the job'. However, the contradiction is less sharp when measured against the basic inequity, by any standard, of the Territory's official scale of land rents.

We turn now to measures taken by the company to adapt to the unfamiliar, exotic, and sometimes hostile social environment in which it had to operate. In the beginning of its operations Papua New Guinea was still a colonial dependency of Australia, and its administration was still in Australians' hands. However, company officials recognized early that that familiar (though not always like-minded) regime would not last very long. Even from the beginning the company was obliged to operate in a local social setting that was both exotic and suspicious. The company's preference, from the start, was to leave it up to the Administration to deal with the social situation, in order to do what its own employees were qualified to do—to operate a mine. Alas, the colonial Administration of that early period of exploration and mining was not empowered nor fully equipped to play that role. Several Administration officials recognized the inappropriateness of applying Australia's Mining Ordinance to the Territory, but were able to revise it to fit the Territory's conditions only fractionally and slowly. Moreover, even if the Territory's Administration had had the will and the manpower to mediate between the company and Bougainvillian landowners, most of the latter wished them not to do so—both because of their deep-seated mistrust of the Administration, and because of their expectations of better 'deals' from the company.

Therefore, from the very beginning of the mining project the company was obliged to treat directly with landowners and other Bougainvillians. It did so at first in an ad hoc, play-it-by-ear manner, then by means of an experimental Community (or Village) Relations staff. During its early years of operation that department had a fairly large staff, headed by high-level directors (at first expatriate, later Bougainvillian), and was kept very busy. Some members specialized in matters of compensation—hearing demands, visiting village and bush locations for corroboration, recommending settlements, etc. Others provided information and other assistance to Bougainvillians wishing to set up small business enterprises (mainly to service company needs). Trained specialists advised villagers about crops and livestock. Others registered and if possible resolved complaints concerning matters such as unwelcome visits to villages by company employees, the need for bridges over tailings-widened rivers, and even cases of illness not accessible to Administration medics. During one period of the mine's operation, a representative of the Administration was stationed at Panguna, to work partly with and partly instead of the company's Community Relations staff.

After several years of successful collaboration, however, that official was with-drawn because (it was said) of a shortage of provincial government funds.

In addition to maintaining its own staff for direct dealings with Bougain-villians on matters respecting compensation, etc., the company organized and supported, fully or partly, a number of separate organizations to help Bougainvillians adjust to or benefit from the massive presence of the mine. The first of those was the Bougainville Copper Foundation (BCF; initially named the Panguna Development Foundation), which was set up in 1971. Its broad aim was to improve the welfare and economic development of Bougainvillians by providing them with opportunities to purchase shares in promising enterprises, which over the years have included supermarkets, warehouses, poultry and pig farms, a medical clinic — all in central Bougainville — and several properties in Port Moresby. The company funded the BCF with a donation of two million BCL shares and an outright grant of three million kina. As the foundation became increasingly self-supporting, the company was able to decrease its donations. The foundation was managed by an executive board which included representatives of the company, the provincial government, and prominent Bougainvillians.

In 1975 the provincial government, in co-operation with the company, set up the Bougainville Development Corporation, to assist in the establish-ment of enterprises that would produce goods and services needed by the company. The first firm sponsored by the BDC was one to mine limestone. The company acquired a 12.5 per cent interest in this new project and con-tracted to purchase all of its output in hydrated lime. It joined with the BDC and the national government in a study to test the feasibility of using the Lalvai River, southeast of Panguna, as a source of hydro-electric power for the company, thereby reducing dependence upon the relatively expensive power plant at Loloho.

In addition the company, through its Business Advisory Service, operated for a number of years to encourage and assist individual Bougainvillians to establish enterprises that would supply various kinds of goods and services to the company, such as trucking and food. In 1988, for example, the company purchased over K30 million in goods and services from such firms.

A perennial problem faced by the company was its relations with Bougain-villians owning land in the various leases.

During the first fifteen years of its presence on the island the company had no option but to deal individually with the landowners of its several leases, except on the few occasions when some of the owners banded together in order, for example, to present claims through the Public Solicitor or the Mining Warden's Court. The company encouraged the landowners on several occasions to band together, believing that it would save effort and facilitate negotiations.

Eventually, in 1979 a former schoolteacher, Michael Pariu, persuaded some of his fellow landowners to organize a Panguna Landowners' Association (PLA), and extended membership to all other landowners within the company's lease areas—mine, tailings, and road. (Arawa town and village are outside the company's lease area, hence were not part of the PLA.) In its early years only a few of the PLA members played active roles; most of them were relatively well-educated village leaders (including some company employees) and over forty years of age. The PLA leaders commenced their activities by drawing up a list of festering grievances and presenting them to the company. For one reason or another the company was slow to respond, whereupon several members of the PLA invaded and looted the Panguna supermarket. At that the company *did* respond, thereby providing a precedent which, it turns out, some of the association's younger members filed away in their memories for future reference. From the negotiations 'stimulated' by that action there emerged a new agreement (1980), which among other things, introduced price-indexing of annual payments of occupation fees, and established a new entity, a Road–Mine–Tailings Lease Trust Fund (RMTL). Portions of certain kinds of compensation payments (payments for nuisances and for loss of 'bush') were paid into the fund in order to provide the landowner members with capital for income-generating businesses and other benefits. Under these new arrangements, for loss of bush K3.09 per hectare would be paid to the landowner, K4.32 to the fund; for nuisance (e.g., from dust, mental disturbance, damage to cemeteries, damage to village property by vagrants), K15 per hectare would be paid to the fund. Membership in the RMTL consisted of seventy-five representatives of landowners, headed by an eight-person executive committee.

For its first few years the RMTL operated to the satisfaction of most of its members. This was a remarkable record for any voluntary Bougainvillian association; but as we shall see, that utopian situation did not endure.

In the mid-1980s another old source of dissatisfaction began to reemerge: the relocated villagers. The passing years and the island's climate had inflicted their destructive effects on the modern houses, toilets, roads, and power and water supplies. Moreover, the children of the resettlement pioneers were growing up and getting married, and, in accord with age-old Nasioi tradition, were in need of separate dwellings of their own. The company agreed to do something about the water supplies, but resisted the other demands on grounds that may have been reasonable by Australian business standards, but were inappropriate and short-sighted on Bougainville.

Meanwhile, except for a few isolated acts of militancy, and the background noises of discontent, the company proceeded peacefully and successfully to mine and sell ore. Its major practical difficulty was the interdiction on its search for new ore bodies outside the current Prospecting Authorities. Its main political and social difficulty was the national government's refusal to give the provincial government a voice in the periodic, statutorily required reviews of the Bougainville Copper Agreement.

The first (and only) renegotiation of the initial (1967) BCA took place in 1974. From the company's viewpoint it could not have occurred at a worse time. Due to record high prices of copper and to fluctuations in exchange rates, the company's net earnings amounted to $158.4 million, and cash dividend payments to $80.2 million. During the twelve-year period prior to 1967, when the original BCA was negotiated, copper metal prices on the London Metal Exchange had averaged about 50 cents (US) per pound; during 1973 they reached a high of 148 cents and averaged 93 cents. Clearly (some critics argued) the company was earning too much for itself and its corporate parents and private shareholders, and was not paying the emerging PNG nation a large enough share of its earnings. The company's 'misfortune' in having been so unexpectedly 'successful' was aggravated by the prejudice of some, mainly academic, critics, that most multinational businesses were by their nature predatory and ruthless, especially those that 'preyed on' Third World societies. In such an atmosphere it is not surprising that little weight was given to the company's arguments that its profits had been earned only after a very costly risk investment, and that the 1973 earnings could not be guaranteed to continue. Accordingly the BCA was revised in full favour of the PNG partner. Among changes contained in it were the following:

- The company's tax-exempt period, originally scheduled to terminate on 1 April 1975, was terminated on 1 January 1974.
- The company lost the benefit of accumulated depreciation allowances and the 20 per cent tax exemption for income earned from production of copper.
- The company became liable to tax at a marginal rate of 70 per cent when its taxable income exceeded a certain level.
- In what was to become a major impediment to company planning, it was formally agreed that the company would not initiate new mineral exploration even within the existing Prospecting Authorities without prior government approval.

The new taxation measures formalized by the 1974 BCA served, of course, to reduce future dividends to the mine's shareholders (including the PNG government), but such reduction did not conspicuously hamper overall company operations. What did was the constraint against further exploration, which placed a dead hand of uncertainty over efforts to plan and provide for future operations.

The 1974 Bougainville Copper Agreement stipulated that review of the agreement in terms of mutual 'fairness', etc., would take place every seven years. For its part, the company appears to have welcomed the prospect of such review, and prepared for it by drawing up lists of proposals to be discussed, but the national government failed to participate. Changes of government and bureaucratic inertia may have been partly responsible for that default, but another impediment was the company's insistence that the provincial government also take part in the review, a proposal that the national government refused to accept.

There were doubtless several reasons, personal and institutional, for the stand-off between national and provincial governments; the weightiest was the demand by the latter for a larger share of the national government's revenues from the mine. Indeed, it appears to have been the national government's refusal of that demand that led the provincial authorities to withhold approval of the company's request to resume exploration. Evidently, company officials sympathized with the provincial government's demands but were unable to persuade the national government to follow suit.

Despite being helplessly wedged in the middle of the inter-government conflict, the company's top officials managed to retain good relations with both governments. Partly because of this, perhaps, the company's Community Relations activities decreased and its staff became reduced in status and influence within the company. Other factors may also have led to the decrease in the company's direct contacts with landowners, including a growing sense of frustration with the proliferation of type, number and price of the latters' demands. However, a hopeful feeling seems to have developed among the company's on-site managers, that congenial relations between themselves and the leaders of the provincial government had reduced the need for continual rapprochement at levels lower down. Such hopes were not to be fulfilled.

Before we end this chapter, some words should be added about the individuals who initiated and supervised the development and operation of the mine. Technology, manpower, financial resources, and market conditions provide the possibilities for such developments, but the vision, skills, energies and commitment of a few leaders are required to transmute those possibilities into an economically successful enterprise. This is especially so in a geographic setting as impedimental as central Bougainville's, and in a cultural and political setting as exotic and unstable as Papua New Guinea's.

The outstanding economic success of the Panguna mine cannot be disproved: in terms of dividends to its co-proprietors, revenues to the PNG government (and therefore relief to the Australian foreign-aid budget), and direct financial and other benefits to thousands of Niuginians, including its numerous employees and suppliers of goods and services. Many individuals, including Niuginians, worked together to develop and manage the company's economic success; regrettably only those in the most responsible managerial positions can be identified in this brief account.

Mention has already been made of Sir Maurice Mawby, whose entrepreneurial daring encouraged mineral exploration in Australia and elsewhere; and of Haddon King, born in British Guiana and educated in Canada, whose insights into geological environments and processes led him to develop CRA's search for copper in the southwest Pacific. Much credit must also be granted to K. M. Phillips, who was senior geologist of CRA Exploration and who led the field search for copper to Bougainville. Records of copper mineralization in Bougainville had existed since the late 1930s, but it was Phillips' imagination and general exploratory flair which recognized, and realized, the possibilities.

But the man responsible for establishing the whole enterprise and for guiding its operations during its first fifteen years was F. F. Espie.

Frank (later Sir Frank) Espie was — is — a big man: physically large and robust enough to create and shape what was for Australia a very large and new kind of mining project in a new and daunting geographic and political setting. With seemingly boundless energy he travelled almost continually for a number of years: to attend board meetings in London; to secure financing in Europe, America and Japan; to obtain and maintain co-operation from officials in Canberra and Port Moresby; to plan, review, and oversee field operations on Bougainville; and, not least, to visit Bougainvillians in their villages. Although he very much wanted the mine to 'go', it was just as evident (to myself and others) that he was sensitively aware that the mine would inevitably bring about radical changes in the lives of many Bougainvillians, and that he tried in every way known to him to protect them from the injurious aspects of those changes and to make it possible for them to share in the financial benefits produced by the mine.

Most importantly, from the start Espie broke entirely new ground in Australian mining. He personally visited both governments and mining operations, private and nationalized, in other developing countries, and ensured that his senior managers did likewise at the project stage and thereafter. These visits were frequently sobering, the more so when on-ground experience modified or contradicted official and academic accounts of events.

Not the least consequence was a huge commitment to training and education of local people years before production started, an investment then regarded by many white 'old hands' in the Territory and in other Third World mining areas as expensively quixotic and of problematic value. Another consequence was healthy awareness of the unfortunate long-term results elsewhere of appointing token or figurehead local 'senior managers'. A third consequence of Espie's travels was that the project management hired few and retained even fewer of the expatriates then in world oversupply because of decolonization and nationalizations — the type likely to be characterized as 'a sound chap with bush experience, knows how to keep a firm hand on the locals'.

Espie travelled intellectually as well, attempting to secure advice and experience from sociologists, anthropologists, geographers. In the world mining industry of the 1960s this was a decidedly unusual priority, common though such consultations have now become. Equally unusual was his willingness to enter the open forum, whether at the Waigani seminars at the infant University of Papua New Guinea or in local liaison.

The late Colin Bishop, the long-term site engineer who with his wife and young family developed a special talent for community relations, would retell with relish the blank disbelief of local villagers when his large sweating Chairman appeared among them in shorts and boots, and was introduced as 'nambawan bilong kampani' (top man of the company). Their reaction was 'no way'; with an upward jerk of the shoulder they would indicate that such superior visitors would be found only in the helicopters passing over the ridgeline, not on a muddy foot-track between village and gardens.

Espie was born to Australian parents in 1917 in Burma, where his father worked as a mining engineer. He studied mining engineering at the University of Adelaide and served in the Australian Army from 1940 to 1946, in North Africa and New Guinea. He began work in 1947 with what was to become one of the CRA group of companies. In 1965 he took charge of the investigations on Bougainville, later becoming managing director and chairman of BCL, and in 1974 deputy chairman of CRA. In 1979 he retired from executive duties but remained a director of BCL and CRA until 1985. Up to 1990 he also served as director of several other Australian corporations and of professional and charitable organizations. In 1971 he was awarded the OBE and in 1979 was created a Knight Bachelor. Altogether, he is a great man with a great heart, especially towards the many Bougainvillians whose interests he tried to safeguard and whose current predicament (which he could not have foreseen) he deeply deplores.

Alongside Espie, the individuals most directly responsible for the launching and economic success of the Panguna mine were D. C. Vernon, P. W. Quodling and R. W. Ballmer. The first two were closely associated with Espie since the project's beginning.

Ray Ballmer was general manager and then managing director of BCL, during its construction and early production phases, from 1969 to 1975. A United States citizen, he was born in 1926 and educated as a mining engineer, and then as an industrial manager at the Massachusetts Institute of Technology. He held various executive positions in the giant American Kennecott copper corporation before his association with BCL. Among the reasons for choosing him to head the new Bougainville project on site was his experience with large open-pit copper mines, a qualification possessed by no Australian of comparable executive credentials. An added qualification was his previous experience in working with American construction firms, in view of the circumstance that the huge American Bechtel Company had been engaged to construct the mine and its major facilities. Ballmer carried out his assignments with complete success and he did so in keeping with the humanitarian policies laid down by Espie. In 1975 with the mine well launched, he resigned from BCL and returned to the United States.

The next person to take over site management on Bougainville was D. C. Vernon. Born in 1928 and trained as a chemical engineer, he joined the CRA group in 1953 and was appointed manager of the Bougainville copper project in 1966, when he became Espie's principal aide. In 1970 he moved from Melbourne to Bougainville, where he eventually succeeded Ballmer as managing director. In 1979 he returned to Melbourne to assume a wider role in CRA, including the chairmanship of BCL upon Espie's retirement from that position. He retired from the chairmanship in 1986 but continues as a director of BCL, and of other mining companies, and as a member of several professional institutions.

Under Vernon's management, BCL continued to prosper and to grow; it also continued to remain sensitive to its local social setting and sympathetic to the concerns of its Bougainville neighbours. In my periodic visits to the islands during Vernon's management I was particularly struck by the *sincerity* of that sympathy—not only because it was good for the company, but because of his general liking for Bougainvillians and his concern for them as individuals. Under his management the company's Community Relations staff achieved a large measure of communication between villagers and company, and an influential voice in management.

When Vernon returned to Melbourne in 1979 he was succeeded by Paul Quodling, who served as general manager, then managing director, until his retirement in 1987. Quodling, who was born in 1926 and trained as an accountant, joined CRA in 1956 and was instrumental, with Espie and Vernon, in the establishment of BCL. During his long tenure on Bougainville the mine continued successfully to produce and market its products and to train and promote its Niugini employees to higher-level jobs. Also, except for the continuing ban on exploration, relations with the national government kept on a steady level. The company's relations with the provincial government were also co-operative and friendly during most of Quodling's tenure, but beginning in 1987 there began to surface the political forces, mainly external, that were to result eventually in closure of the mine.

Quodling's successor was R. J. Cornelius, who was born in 1932 and trained as a metallurgist. He joined BCL in 1982 after many years in other CRA group operations and served as general manager of concentrator operations at Panguna before becoming managing director in 1987. Like all his predecessors, he proved to be a highly competent manager of the mine; he was tireless in his efforts to maintain good working relations with both national and provincial governments, as well as with the mine's Bougainvillian neighbours. Unfortunately, the adverse external forces that had begun to surface during Quodling's term increased and erupted during Cornelius' term, leaving him with the unenviable task of securing what he could of the mine's property and evacuating its personnel. He accomplished this with remarkable composure and effectiveness.

Equally unenviable has been the task of D. S. Carruthers, who became BCL's chairman in 1986, a position he now holds along with many other assignments, including responsibility over all CRA interests in copper, lead-zinc and iron ore. After training as a geologist, Carruthers joined CRA in 1955, became general manager for the group's exploration in 1969, and in 1975 assumed supervision over the group's coal interest. Despite those numerous other jobs, he continues to devote a very large share of his seemingly endless working hours to the affairs of BCL, probing into the causes of the mine's shut-down, salvaging as much as possible of its material resources, and promoting and preparing for its eventual reopening.

The exigencies of publishing permit mention here of only the topmost executives who have been in charge of the establishment and operation of BCL's mine, but one exception needs to be made on behalf of a Niuginian who achieved a high position in the company's management hierarchy. Joseph

Lawrence Auna was general manager of the Personnel Services Division, and in charge of, among other functions, relations between the company and its Bougainvillian neighbours. Auna was born in 1939 in Buin. After attending secondary schools in PNG and the Royal Melbourne Institute of Technology in Victoria, he joined the PNG public service where he held a number of high-level jobs, including director of the Department of Forests and of the National Investment and Development Authority. He joined BCL in 1981 and rose rapidly to his present position. (Prior to that, from 1973 to 1981, he had served as the national government's nominated representative on BCL's board of directors.) Auna has proved to be an outstanding executive by any nation's standards, and one to be especially commended on account of his ability to serve as an understanding and even-handed arbiter between white and black cultures during a very stressful phase of the company's history.

9

The Other Bougainville

A S RECORDED IN CHAPTER 7 IN 1941 THERE WERE ABOUT 44,000 Bougainvillians residing in rural areas of Bougainville and Buka; another 2800 or so lived on European-owned plantations, on mission stations, and in the European-Chinese 'urban' centres of Kieta and Sohano (Buka Passage). At that time another 850 Bougainvillians were residing and working elsewhere in the Mandated Territory, mostly on plantations on New Britain. There were 200 or so indigenes from other parts of the Territory living on Bougainville: a few of them worked for the Administration, the rest on plantations. Added to all these were some 175 Europeans; about half of them were attached to the missions, the rest to plantations, retail stores, the Administration, and the small gold mine at Kupei. And finally, there were a few Chinese, the proprietors of most of the islands' retail stores.

As result of the war's toll in lives, born and unborn, a whole decade passed before the population of indigenous Bougainvillians regained its pre-war numbers. But thereafter their numbers increased at a rate much, much faster than before the war, more than doubling within the next twenty-five years. At the last published census, of 1980, the total number of Bougainvillians residing in the province was 108,726, including about 6200 living on the atolls. At that rate of increase, we can calculate the total population of Bougainvillians residing on the two main islands of the province as about 140,000 in 1990; except for a few non-Niuginian priests and a handful of Europeans and Chinese, that number constituted the total population of the two large islands. By mid-1990 nearly all expatriates had departed for safer places, and nearly all non-Bougainvillian Niuginians — 'redskins' — had gone as well. By early 1990 many Bougainvillians had also left their home islands, and many others remained elsewhere in Papua New Guinea, unable or unwilling to go home. The number of those exiles is unknown. Enough is known about several individual cases to report that fear of retribution and reprisals for past conduct, often combined with disapproval of the islands' 'revolution', were the principal reasons that led many to live elsewhere. The events that directly precipitated these recent changes in the ethnic composition of the province's population will be chronicled in later chapters. In this chapter will be described the more general occurrences that took place from the mid-1960s to the late 1980s and that led up to those events.

One of the most noteworthy occurrences revealed in the post-war censuses is that the natural rate of increase of Bougainvillians was about 3.5 per cent per annum, one of the biggest recorded in the world. Intensive research carried out in the 1970s on several small rural segments of Bougainville's population shows even higher rates of increase, in one case up to 4.7 per cent per annum, which would have resulted in a doubling of those numbers every fifteen to eighteen years. (In comparison, it would require seventy-seven years for the present population of Australia to double itself at present rates of immigration and low natural increase.) The reasons for that increase are manifold: better health care in general and better infant and maternity care in particular; fewer killing diseases (including malaria); and the relaxation of traditional rules regarding birth-spacing and childbirth. (Concerning the latter, several traditional Bougainville cultures had discouraged a woman from engaging in sex until her last child was weaned, and in some of them dietary and other restrictions imposed on mothers and new infants were, unintentionally, lethal.) The progress made in formal secular schooling during the last two decades suggests that some principles of family planning might have encouraged birth control, but countering this were other circumstances, including the teachings of the Catholic Church: most Bougainvillians are and have been Catholic. Also, many Bougainvillians have over the years voiced the sentiment, that admonitions to slow down their population increase are a *waitmans* ploy to keep their numbers small so they will remain politically weak. In any case, whatever the reasons for the spectacular natural increase in Bougainvillians' numbers, the phenomenon has affected, and will continue to affect, several other aspects of their lives.

Prior to World War II nearly all Bougainvillians satisfied their biological needs (for food, shelter, etc.) and their culturally-conditioned wants (for clothing, ornamentation, prestige symbols, rituals, etc.) by gardening, raising pigs, collecting wild food and, depending upon location, fishing and hunting. A few of them sold coconuts or copra to European and Chinese traders, in order to purchase objects not made by themselves, such as cloth, axes, machetes, lamps, kerosene, rice and tinned meat. In addition, most youths worked for a few years on European plantations, partly in order to purchase such items and partly to pay the Administration's annual ten shillings head tax.

After the war gardening continued to be the principal life-sustaining occupation of most Bougainvillians. Few of them chose to resume working on European plantations, and until the mid-1960s no other wage work was available to most of them. Instead, with the encouragement and assistance of the Administration, many of them turned to cash-cropping (oftimes with more hope than proficiency); after many false starts with different kinds of crops, they settled down to produce mostly copra and cacao. According to the census of 1966, out of 33,183 locally-born residents of Bougainville and Buka of working age (defined by the census as between twenty and sixty-five years) about 46 per cent were listed as 'wholly subsistent' (i.e., gardening, fishing) and another 35 per cent as 'mainly subsistent with some money raising'. In view of their antipathy to work on European plantations it is reasonable to assume that most of that 'money-raising' consisted of producing cacao and copra. This would leave about 19 per cent of the workforce, male and female, employed wholly or mainly in producing cacao and coconuts.

In other words, in the years just before mining commenced to transform their ways of life, Bougainvillians were still almost wholly a rural people, earning their living mainly by subsistence gardening, or gardening supplemented by some cash-cropping. These conclusions are also supported by census figures on 'urbanism': in 1966 only 2 per cent of these islands' total indigenous population, including non-Bougainvillian Niuginians, resided in one or the other of the 'urban' centres, Kieta or Sohano.

The next census, of 1971, is not very helpful in reporting Bougainvillians' occupational trends: this is partly because its scheme of classification does not permit precise comparison with the 1966 figures, and partly because a very large number of Bougainvillians were working at that time, but only temporarily, on mine construction projects. The only unambiguous occupational datum in the 1971 census is the number of persons listed in the 'mainly subsistence' category, who constituted about 18 per cent of those in the population of twenty years of age and older.

At the census of 1980 the labour-absorbing work of mine construction had long since ended and longer-range occupational trends were prevailing. Unfortunately (for our purposes) the data presented in this census are not precisely comparable to those of the census of 1966: different questions were asked and different kinds of analysis applied. By use of extrapolation (too tediously complex to describe here) it is possible to reach the conclusion that between 1966 and 1980 Bougainvillians had changed by large percentages from a 'pure' subsistence livelihood to one based partly or mainly on money-earning (mostly cash-cropping but also, in a large number of cases, wage labour or conducting businesses of their own). The only relatively unambiguous figures in this comparison of censuses concern the occupational category of subsistence: the 46 per cent of the workforce listed as 'wholly subsistent' in 1966 had by 1980 dropped to 19 per cent, two-thirds of whom were female. In other words, as time passed Bougainvillian men left it to their wives and unmarried daughters to do more and more of the food-growing, while they themselves turned more and more to money-earning, by cash-cropping, wage labour, or *bisnis* (business).

The direction of this trend is confirmed by figures on production and export. In the twelve-month period of 1963–4 production by 'indigenes' (i.e. Bougainvillian smallholders) totalled 174 tonnes of dry-bean cacao and 1935 tonnes of copra. By 1969–70 the figures on cacao had already risen to 1461; by 1970–80 to 10,151; and by 1988–9 to 13,841 tonnes, with a value of just under K1000 per tonne. (In 1977–8 Bougainvillian smallholders produced about 1.4 times as much cocoa as the province's plantations; in 1988–9 the relative amount had risen to three times.)

Recent figures on copra production show that in 1980 22,565 tonnes passed through the two receiving depots of Buka and Kieta. In 1989 that figure had risen to 26,756 tonnes. In both years it is estimated that about two-thirds of the total was produced by smallholders.

The trend from subsistence to money-earning was significant in itself, revealing, as it did, a growing dependence upon money and the 'foreign' things that only money can bring. But that trend is also noteworthy with respect to its effects upon the relations between Bougainvillians and their very limited quantity of usable land.

The two money-earning export crops that Bougainvillians have succeeded in producing (after unsuccessful experiments with several others) were copra and cacao. (Copra is the dried meat of the coconut, from which coconut oil is pressed. The cacao tree produces beans which are fermented, dried and ground to form cocoa powder.) Bougainville's climate and some of its soils are evidently very favourable for growing these two crops. With respect to climate there are some regional differences — some experts blame the relatively inferior coconuts grown in southern Bougainville on that area's greater cloudiness — but except in the higher altitudes climate is not the weightiest factor that influences cash-cropping's degree of success. Regional differences in soil and other surface features are weightier: there are several extensive areas described as being unsuitable for any commercial use. It must not be concluded from this, however, that all the rest of Bougainville's 7500 square kilometres are ideally suited to cash-cropping; factors of topography, etc. would reduce that figure considerably. But if there were no other inhibiting circumstances to limit cash-cropping in the province, there is no question but there is suitable land enough to accommodate much additional planting in years to come.

Unfortunately, those limiting factors are numerous. Bougainvillians have been selling copra for a century or more, first to visiting traders and then to locally-based trade stores. They became labourers on expatriate plantations located on Bougainville and elsewhere until World War II, since which time few of them have been willing to engage in that kind of onerous and generally low-wage work. Many Bougainvillians established plantations of their own, either as individual or co-operative ventures; and in some parts of the province, mainly on Buka and along the east and north coasts of the larger island, copra is the principal source of money for indigenous cash-croppers. Elsewhere copra production has not flourished as well, due either to local growing conditions or to transport difficulties, or both. But even where copra production is the principal indigenous money-earning activity, it seems to be impeded by natural or human obstacles — mainly the latter.

Throughout most of the province the quantity and quality of indigenously produced copra falls short of its potential. Generally speaking, pests and diseases are not as devastating in the province as they are in some other nearby islands. However, the poor ground maintenance that characterizes many indigenous plantations encourages pests and diseases that reduce yields, and also makes it more difficult and time-consuming to collect the fallen nuts.

In most places seed nuts are taken at random, without any selection for quality. Nor do many planters take the trouble to establish seedling nurseries. Few indigenous plantations have their trees properly spaced, and most would benefit from application of fertilizer or mulch. Fertilizer is costly, but where land is in short supply its costs might be offset by higher yields. Again, few if any indigenous planters trouble to plant leguminous cover crops, which would both improve the soil and retard excessive undergrowth.

In most places indigenous copra continues to be processed in small, family-owned smoke-dryers, which produce lower quality copra than hot-air dryers. The latter cost more to build and require larger co-operative groups to operate efficiently, but they have proved to be economically successful elsewhere in the Pacific islands.

Transporting copra from dryer to buyer is relatively easy and inexpensive for most coastal producers, and for those on all-weather roads, but because of its bulkiness, moving copra from isolated inland areas (such as Atamo or inland Teop) can be so costly that it discourages this form of money-earning altogether when copra prices are low.

Against all those liabilities serving to reduce the economic potential of the province's smallholder copra industry, one important asset must be noted, namely, the way in which the producer's own labour is put into it. Unlike most wage labour, which requires the labourers to devote many hours of many days continuously to a job, the small producer can earn money from his grove by working in spurts, at times more or less of his own choosing. Harvesting and drying require devotion of several days in succession, but this occurs only a few times a year. Undergrowth cutting should also be done with some regularity, but may be spread over time. Thus, the Bougainvillian copra producer can easily fit copra growing into his work regime, including subsistence gardening, other cash-cropping, and even some wage earning. (Some Bougainvillians managed to harvest and dry their coconuts during weekend or holiday leave from distant wage-earning jobs.)

Considering Bougainville–Buka as a whole there is probably enough suitable land to accommodate a fairly large expansion in coconut palm planting. However, the location and present ownership of that potential land are not well matched with the population that needs this source of income.

The errors and oversights that reduce indigenous copra profits find their parallels in smallholder cacao production. In most parts of Bougainville few growers exercise much care in selecting better seed, or go to the (profitable) trouble of establishing nurseries. Except in areas of heavy cloud-cover (e.g., in parts of southern Bougainville), cacao grows best under shade. A few indigenous growers follow the recommended practice, but it seems that most of these plant the cacao and intended shade tree simultaneously, which leaves the new cacao plant unprotected during its initial growth period.

Growing cacao requires more frequent weeding than coconuts, and in addition the plants must be kept free of vine; observation indicates that neither of these measures is carried out with enough regularity. High yield also depends upon regular and proper pruning of suckers, but according to one observer, 'fewer than one in a hundred' of the cacao farmers in this area understood the technique of pruning.

Like copra, cacao has the advantage of allowing growers to devote much of their time and labour to other activities, although it is not quite so undemanding as copra-making in this respect. Coconuts ripen and fall fairly steadily year-round, and may safely be left on the ground for days or weeks before processing. Cacao also ripens continuously, but unlike coconuts should be harvested at least once a month for best results. Moreover its ripening peaks in an annual flush, which requires additional labour within a brief period if all the crop is to be harvested and the plants kept relatively disease-free.

The difference in price between wet cacao beans and expertly processed dry beans is usually large enough to pay growers to process their own beans. Some smallholders do in fact ferment and dry their wet beans in small, makeshift village fementeries; this results in a dried bean which, though usually commercially acceptable, is often poor in quality and brings a lower price. Few individual growers, or family groups, produce enough beans to make it economically feasible for them to construct and operate the kind of costly fermenteries capable of turning out consistently high-quality dried beans.

As with coconuts, if one looks at Bougainville–Buka as a whole one could justifiably conclude that there is enough suitable land to accommodate a fairly large increase in cacao growing. But, as with coconuts, most of that 'extra' land will be increasingly required for other uses, including subsistence gardening. Also, the land is not in the form of a large undivided reserve that can be allocated to Bougainvillians according to individual and family needs: it is divided into innumerable estates, all of them firmly owned by particular individuals or kin-groups, and hence not available (under present laws and customs) for redistribution according to need.

Next to coconuts and cacao, the most important kind of cash-cropping engaged in by Bougainvillians in recent years has been garden vegetables and fruits grown for sale in three kinds of markets. The largest by far of these were the messes of BCL and its associated companies. Next in volume of sales were the weekly or bi-weekly markets of the urban centres. And last were the rural markets that have become weekly events in several rural centres, e.g., Buin Town, Konga, and Wakunai. (Some food bartering has been engaged in on these islands for decades and perhaps centuries, but institutionalized commercial food markets are a post-war development.)

The purchase of garden vegetables and fruits by BCL and its associated companies began in the late 1960s and developed into a highly organized enterprise, with most of the purchases being bought and transported from rural pick-up points by company trucks. The peak of the traffic was in 1971, during the mine construction phase, when recorded purchases reached $340,000 per annum. With the decrease in number of employees, the purchases decreased greatly, but continued to provide substantial incomes for many Bougainvillians.

Some communities have already reached the limits of locally-owned land suitable for cash-cropping, and, in view of their increasing populations must face some hard choices in order to meet their irreducible needs for food. Either they can cut back on cash-cropping and revert to gardening, thereby reducing their purchases of the imported foods, etc. which they now desire; or they can work for wages, if such jobs continue to be available; or they can move to some other part of their islands, if any such 'empty' and unowned land still exists; or emigrate to other places in the nation, if political conditions permit.

In this connection a question arises concerning the extent to which the mine and its adjuncts have reduced the island's actual and potential gardening and cash-cropping land.

An off-hand answer would be: not much. The middle and lower parts of the tailings leases were unsuitable for both gardening and cash-cropping, except for a small coconut plantation near the coast. Some Bougainvillian-owned coconut groves were lost to their owners at Loloho and Arawa, but most of the Arawa town site was expatriate-owned. Perhaps the most extensive gardening and native-owned cash-cropping land removed from production was in the area flanking the road from the mine to the port. To this must be added the, mainly garden, land lost to the villagers at or near the mine itself. But even including all those losses (which were of course highly deprivational for the scores of persons directly involved) it can be said that mining had reduced the whole island's quantity of arable land to only a small degree. But of course, for an indigenous population that is increasing as rapidly as Bougainville's, both in numbers and in their appetites for goods and services that only money can buy, any reduction in cash-cropping land becomes critical — especially if other opportunities for money-earning continue to diminish.

Another post-war change in Bougainville's economy was to indigenous-initiated *bisnis* (business). As more and more people turned to cash-cropping, they began to band together to process their products — to dry the coconuts and to ferment and dry the wet cacao beans. Undried coconut meat swiftly rots; wet cacao beans can be sold, but they fetch higher prices when fermented and dried. Some growers living nearby were able to sell their unprocessed coconuts and cacao beans to Chinese traders or to large expatriate-owned plantations, or to pay the latter to have them processed, but for most growers that was not feasible. Again, some growers smoke-dried their copra or fermented and dried their beans in small family or village dryers, but such work was irksome and the products inferior. In many places larger numbers of growers joined together to acquire better equipment and to share in the work of drying and fermenting. Many of these co-operatives purchased their own trucks and managed all their members' transport and marketing as well.

Some such co-operatives began soon after the war, but it was not until about 1964 that they began to proliferate. While there have been failures, due mainly to managerial inefficiency or dishonesty, several of them have persisted successfully and have added wholesale purchasing and lower-priced retailing to their activities. Cacao growing, especially, expanded after 1965, encouraged largely by rises in market prices but also by the operation of several highly successful co-operatives. However, some were dismal failures, and there were many problems of transport and marketing, as stated in the district *Annual Report* of 1961–2:

> The history of co-operatives in the district is a sad one. Societies have been torn apart by internal jealousies, misunderstandings of the concept of cooperation, under-capitalization, poor communications, inefficient and dishonest employees, high overheads, destructive opposition by some short-sighted individuals, lack of shipping points and lack of a [nearby] Copra Marketing Board Dept. Too often, because of irregular shipping, impassable roads and human frailties, copra is allowed to pile up and the [co-operative's] liquid resources are strained beyond reasonable limits.

In some places the earnings of growers increased quite spectacularly for a few years, and enthusiasm for the occupation reached the point of copra or cacao 'madness'. In Nagovisi, for example, sales of cacao beans by members of the local co-op, BANA, rose from $3952 to $99,842 between 1964 and 1969. Copra production also increased greatly between 1964 and 1970, including large numbers of new plantings by individuals.

In addition to the co-operatives established to process and market cocoa and copra which added purchasing and retailing to their activities, many individuals or small groups set up businesses devoted solely or mainly to retailing goods. Despite their knowledge about numerous store failures, many Bougainvillians persisted in viewing retailing as the shortest road to financial success. To preside over shelves holding readily selling staples and brightly labelled manufactures was seen by many to be far more attractive than planting, harvesting, and processing cash crops, or engaging in physically demanding wage work. Even the labour of acquiring goods for sale was viewed by many as more adventure than drudgery; a prestigious mystique attached to many retailing enterprises, however modest.

Some of the roots of this widespread attitude can be traced to the pre-war era, when virtually all retailing was in the hands of whites or Chinese, and

thus represented a mysterious and at that time wholly unattainable level of
material wealth. The war served to change that situation in two ways. First,
the war itself prevented access to all store-bought commodities for the
duration, thereby creating a universal and unsatisfiable demand. Later on,
the distribution of war reparations provided many persons with the cash to
satisfy that demand (and for a few, to enter the retailing business itself). As
a result trade stores sprang up everywhere; even in remote Nagovisi, where
before the war there had not been one indigenous store, by 1955 there were
twenty-six. Moreover, this proliferation was not a momentary fad; for
example, in the Buin and Kieta census divisions there were 322 such stores
in 1973, each having an average of only 108 steady customers.

To characterize such enterprises we can draw on a study by the anthropologist, Donald Mitchell, made in Nagovisi:

> The establishment of a village 'store' is a predictable Nagovisi pattern,
> which appears to be similar in many, if not most, respects to those in
> other parts of Bougainville. Village stores generally operate as small co-
> operatives, although the investors never conceive of them as such.
> Village households contribute share capital, usually in the order of $10
> per household, which is used to purchase an initial shipment of trade
> store goods from a merchant in one of the urban centers, or from a local
> wholesale co-operative (one began to function in 1969, in Nagovisi, but
> ceased operations in 1974). These goods are put in the store, marked up,
> then primarily purchased by the store's 'owners' who do not receive any
> discount. As stocks dwindle, the store's cash generally increases and a
> new shipment of goods is then brought in. Villagers may 'charge'
> goods, although nearly every store has a sign prohibiting *dinau* (debts).
> Profits, when there are any, are never distributed among the investors.
> Eventually, most stores collapse, usually through real or alleged fraud
> (more often, in fact, just mismanagement and poor accounting). Upon
> collapse, most investors get back their original investment and not much
> else. There are exceptions, of course; some stores are well run, appear to
> turn a profit (although even in the best-run stores, they are not
> distributed to the investors) and are usually well-stocked. More
> commonly, a store will have a lifetime of one or two years before
> collapse; after a short period, a new one will be organized, usually by a
> different kin group within the village, and will live out its own lifespan.
> [Mitchell 1976, p. 123]

Mitchell went on to state that 'a store run along standard individual
entrepreneurial capitalistic lines would not succeed in the [Nagovisi] village
context. The owner would be seen as personally profiting from the villagers
while they receive nothing.'

The account by another anthropologist, Eugene Ogan, about trade stores
among the Nasioi of Aropa Valley reveals a more complex situation. In

1962–4 (the time of his first field research there) many of the trade stores in the valley were associated with indigenous copra production, and some of them took on the aura of cargo cults.

> The pattern for these new 'bisnis' ventures was remarkably uniform: a man with some knowledge of European ways — who was at least fluent in Pidgin, at most the possessor of a mission secondary education — and more than average coconut holdings, solicited cash contributions from villagers in his own area and beyond. With this capital he attempted to act as agent for copra sales and manager of one or more trade-stores. In each case the goal of 'living like Europeans' proved unattainable with the limited material, educational and organizational resources available, at which time activities were noticed by Europeans which they identified as 'cargo cult'. [Ogan 1972, pp. 161–2]

In fact, wrote Ogan, '. . . "bisnis" and "kago" (i.e., material and spiritual riches obtained by supernatural means) were never differentiated in Nasioi minds, and . . . the emphasis on supernatural elements only increased as the, in European eyes, business-like aspects of the various operations failed.' Ogan provides an example of one store's pricing process:

> Jude [the store keeper] maintained that the regular mark-up over the price paid to Wo Fat [the Kieta retailer from whom Jude purchased his goods] for goods was 5 cents per 50 cents . . . However, on the one occasion when Sebastian delivered the order in person he set prices on a more haphazard basis, with mark-ups averaging 5 cents to 10 cents on every item over 40 cents, but never exceeding 50 cents on even the most expensive items . . . Special order items, on the other hand, such as beer for a party . . . apparently received no mark-up at all. [Ogan 1972, p. 168]

An examination of Jude's books revealed to Ogan 'a kind of bookkeeping in which no accounts balance'. For example, when Jude returned to the relatives of a recently deceased shareholder the latter's share in the store enterprise, he calculated the figure to be eighteen dollars. According to Jude's calculations, the deceased 'had made an initial contribution to the store of $10, and had purchased some $10 to $15 worth of goods before his death: therefore [sic] his "profit" came to $18.'

The anthropologist Michael Hamnett reported a similar situation in Atamo, a village on the border of Nasioi and Eivo. Several individuals had operated stores there since the 1950s, but few of them had kept track of profits and losses, and few had survived. During Hamnett's stay there in 1975–6 a store was operated by a young man with some business training, and evidently a better-than-average 'business sense'. However, his goods were priced higher than those of stores in the towns, reflecting his higher transport costs, etc., and his neighbours complained; one said, 'When we got our own stores *after* independence we were sure the prices would drop— but they haven't; prices here in the village are now higher than ever.' In another case reported by Hamnett, a store owner (then no longer in business) had deliberately priced his goods lower than cost, explaining, 'I don't need to make a profit; I have cash crops.'

The proliferation of village trade stores reached a peak in the early 1970s; after that their numbers steadily declined, on account of the increasing numbers and sizes of stores in the urban centres, and of their easier access (due to more and better roads).

While some villagers operated trade stores, others went into *bisnis* processing cacao beans raised not only by themselves but by some neighbours as well. Some of these fermenteries were owned by a single family group, others were organized along co-op lines. There were many financial failures for reasons similar to the failures of trade stores, but there were also many successes (at least, while cacao prices were high). Some of these enterprises were limited to bean processing; others included transport of the finished product as well; and some others served as retail outlets for their members, or as trade stores for the village in general.

Some rural *bisnis* were owned and operated by individuals (with the assistance of family members), others by groups of close kinsmen, and still others as co-operatives. Some of the latter were initiated and assisted by Administration officials; others depended on local initiative and resources. All types of enterprises have experienced successes and failures. It would make an interesting study to learn which of them had the most successes, and why. Such a study would reveal much about the nature of present-day Bougainvillian society and the distinctions between its more and less Westernized members.

By 1980 several Bougainvillians were operating businesses in the islands' urban centres; retail stores, transport companies, restaurants, etc. Regrettably, no comprehensive study of them has been published, but I gained a few impressions which I offer here.

A first impression is that the profits (and losses) were restricted to a very small number of individuals. A typical *bisnis* would be initiated by two or three venturesome and relatively sophisticated persons acting on their own or as representatives of a few clansmen and friends. Moreover, the numbers engaged in them were further limited through the circumstance that several of the principals were investors and 'directors' in two or more businesses, either separate companies or linked into 'holding companies'.

Another impression was that most of their founders occupied high-status positions, either in BCL, or politics, or both. Also, many if not most of the founders had been actively encouraged to start their enterprises with advice, finance, and technical assistance from BCL and the Administration. Any profits that such businesses earned went mainly to individuals (and their close associates) who were already earning relatively large incomes, or holding economically favoured positions.

A third impression was that the financing for the pre-1980 indigenous urban businesses came partly from public sources, such as development-bank loans, or the Bougainville Copper Foundation. Private investment derived mainly from BCL compensation payments and occupation fees, which at the time were almost the sole source of large amounts of cash readily available to indigenes. This was another reason why such enterprises were limited to the residents of areas within or near the mine complex.

The late 1970s and the early and middle 1980s witnessed a large proliferation of indigenous businesses on Bougainville–Buka, including many successes, in both rural and urban areas. Information on such enterprises is not available to this writer, but omission of description of them in this book should not be taken to mean that they did not exist, or that they did not form an important part of pre-Crisis Bougainvillian life.

We turn now to a post-1965 phenomenon that was entirely unprecedented on Bougainville, namely, the movement of large numbers of Bougainvillians from rural to urban areas. Even as recently as 1960 almost all Bougainvillians resided in small rural villages or on plantations or mission stations in rural areas. In the 1966 census the town of Kieta numbered only 755 residents, including 644 indigenes, and the 'urban' area of Buka Passage numbered 877 residents, including 800 indigenes. In contrast, there were at that time 70,847 persons residing in rural areas, 70,317 of them indigenes. (These figures do not distinguish between Bougainvillians and other Niuginians, but the latter did not number more than a few hundred at that time.)

By 1988 Buka Passage 'town' had been enlarged by addition of a market place, a few more offices and stores, and an airport for small planes, but on the whole it remained as sleepy as before. Not so in the case of Kieta. By 1988 Kieta itself had not greatly enlarged or changed but it had become flanked by a strip of urbanized settlements that extended along the coast for about twenty-five kilometres, from the airport in the south to Loloho in the north, and included the town of Arawa, the industrial sprawl of Toniva, and the large Marist complex of church, high school, and diocesan headquarters. Linked to this urban strip by a 25-kilometre highway is Panguna, the site of the company's mine and the residence of most of its employees.

The census of 1980 — the most recent comprehensive one available — records the total population of this entire urban complex (including Panguna) as numbering 17,523 citizens of Papua New Guinea (11,694 males, 5829 females). Of these, 7026 had been born in the North Solomon Province (and were therefore mostly Bougainvillians). In addition there were 2244 non-citizens aged ten years and older residing there at that time; their nationalities are not reported, but inasmuch as two-thirds of all non-citizens residing throughout the whole province (i.e., 3288) were Australian it can be inferred that Australians also constituted the majority of non-citizens living in the urban complex. Some other facts about the composition of this urban complex deserve mention:

• Of the 17,523 Niuginians residing there, 10,497 were from other provinces. In other words, three-fifths of the complex's citizen residents were non-Bougainvillian. Among those non-Bougainvillians aged fifteen and over (i.e., 6248) at least two-thirds were males without wives: there were only 2199 non-Bougainvillian adult females.

- Unlike the non-Bougainvillians living there, many of the older male Bougainvillian residents were living there with their wives.
- Of the 12,860 citizens aged ten and over, 7817 were working for salaries or wages, 2182 were engaged in 'housework', 1148 were full-time students, and 331 (279 males, 52 females) were 'looking for work'. Most of the males in this category were under thirty years of age.
- Residing in the complex were thirty-six citizens engaged in 'business', large or small.

Some commentary is called for to vivify these bare facts: first, to account for the presence there of so many non-Bougainvillian PNG citizens. At the end of 1980, the year of this census, BCL alone was employing 2200 of them, and many of them were living there with their wives and children. This was 66 per cent of the mine's national employees in 1980 (in 1988, of 2950 such employees a total of 1027 were from Bougainville). As for the complex's other 'redskins', (most of whom were single men), many had probably stayed on after the mine construction phase, and others were relative newcomers attracted by the promise of work. While most of these latter were employed (probably in wage work), some of them doubtless were unemployed, with no nearby relatives to depend upon, and no home village within their reach. In the view of many Bougainvillians, all male redskins were somewhat suspect; all single male redskins doubly so; and all unemployed single male redskins dangerously so.

With regard to the Bougainvillians, the question is: how did they come to be in this urban setting, which was so unlike their age-long rural ones; and how did they adjust to being there?

Their reason for being there is fairly certain: they had gone there to earn more money than they could earn by staying in their villages or by working on plantations, and under conditions they perceived to be less disagreeable. In addition, one cannot rule out the likelihood that a desire to see and experience a new kind of life had led many Bougainvillians to take up urban living, at least for a while.

Concerning Bougainvillians' adjustment to urban life, there are answers in another study by those indefatigable researchers, Richard Bedford and Alex Mamak. Their study was carried out in 1973, and while its statistics relate only to that year, the authors' general conclusions appear valid for a much longer period of time.

The population covered in this study consisted of 1965 Bougainvillians employed in the Kieta-Arawa-Panguna area during 1973. (There were 3850 non-Bougainvillian Niuginians residing there at the same time, but these were not a part of the study.) These Bougainvillians had come from all parts of Bougainville and Buka, and were employed as follows: 1121 by BCL, 147 by the Administration, and 697 by forty-four firms, many of which supplied goods and services to BCL. The 1121 Bougainvillians employed by BCL were the fortunate ones. Since the peak of the construction period, when the demand for labour was very high, BCL had always had more job applications than it could fill; for instance, in the period from January to April 1974 nearly half of the 209 Bougainvillian applicants interviewed for jobs had to be rejected.

The authors of this study identified three 'strategies' utilized by Bougainvillians towards urban wage employment:

> The first involved a spread of labour in any one year between activities in the village and wage employment in town. People who utilized this strategy worked in towns for relatively short periods, often at certain times of the year, before returning for a month or more to participate in cash cropping and other village-based economic and social activities. This group comprised a significant proportion of our unskilled and semi-skilled respondents, many of whom express the lowest degree of commitment to urban work and residence. The second strategy is associated with much stronger commitment to wage earning as the sole source of a cash income. Minimal direct involvement in the village is a feature of this approach, and almost no income is derived from other economic endeavour in town. The third approach demands participation in wage employment as well as other economic pursuits in town, especially business enterprises. Only a very small proportion of our respondents have chosen to spread their time, labour, and capital assets over a range of urban-based activities . . . We have termed the three strategies respectively, 'peasant', 'proletarian', and 'entrepreneur'. . . .
>
> A new class of Bougainville businessmen who are not in wage employment and derive their monetary incomes solely from their own town-based enterprises is also emerging. Policies favouring local participation in transport and service sectors of the urban economy have encouraged an, as yet, small number to leave wage employment altogether and run their own businesses. [Bedford and Mamak 1976: pp. 180, 185]

After 1973 many more Bougainvillians became almost wholly 'urbanized', either as 'proletarians' or as 'entrepreneurs'. I have no information concerning the living standards of the former, but have obtained some information about the latter, through interviews conducted by a third party with some of the towns' more successful Bougainvillian urbanites, who were asked: What do you do with your savings? Their responses were surprisingly alike:

> After obtaining the normal household possessions, including radio(s), TV and VCR's, the next highest priority was given to purchasing a 4WD vehicle in order to visit the [home] village. They were also keen to see their children were given every opportunity to receive a good

education . . . Most of them did not send money back to the village or
do anything to improve the lot of village people. They enjoyed visiting
the village which they did reasonably frequently, but otherwise tended
to stay in town . . . After satisfying their urban lifestyles many of them
built a house in the village for their use and sometimes that of their
immediate family. They then obtained additional refrigerators, radio's,
TV and video player, lighting plant and sometimes a boat . . . Beyond
this, some obtained land for cash cropping . . . About the only way in
which the village people benefitted from their urbanized counterparts
was to be put-up in town whenever they wished to visit. In this
circumstance they were usually given access to the car and made
welcome . . . Village people also benefitted to some degree as employees
working the smallholdings belonging to their urbanized counterparts.

What proportion of urbanized Bougainvillians lived in this way is not
known; but the fact that some of them did indicates that there are many
varieties of the Melanesian Way.

Turning next to the religious affiliations of this post, post-war era, the census
of 1966 provides the following enumeration for the province's 71,761
indigenes: Catholics, 51,635; Methodists, 11,283; Adventists, 6726; other
Christian, 1447; indigenous religions, 562. There were also 108 individuals
listed as having no religion, or who did not reply. At that time about 3676
of the indigenes enumerated in the census were from other districts of Papua
New Guinea; it is fairly safe to conclude that they made up most of the
number listed as 'other Christian' (Lutheran, Anglican, etc.).
    The census of 1980 enumerates religious affiliation only of residents of
the urban areas of the province. There is little doubt that had figures for the
whole province been given, they would have revealed little percentage change
in the declared religious affiliations of Bougainvillians, and in all likelihood
this surmise would apply to a census carried out in 1991. Bougainvillians
made giant strides in formal secular education during the last two decades,
but the social pressures to remain at least nominally a member of a church
did not lose their strength.
    In the years just before World War II nearly every village on Bougain-
ville had a school, where the local catechist (when he was not gardening,
etc.) assembled several of the village's older children (if *they* were not gardening, etc.). He drilled them in simple reading and writing — in the vernacular, if Catholic; in a kind of English, if Methodist or Adventist. Beyond that,
the only formal schooling was that given to older boys in mission schools — a
Catholic one near Buka Passage, a Methodist one in Buin. The main purpose
of these schools, whether avowed or not, was to train the pupils to become
village catechists, thereby promoting evangelization. Both Methodists and
Adventists emphasized English (the language of the missionaries themselves).
At first the Catholic missionaries (most of them German- or French-speakers)

resisted Anglicization. The fear of losing ambitious young adherents caused the switch to English; it was promoted by the deliberate appointment of English-speaking missionaries, including, eventually, an American as bishop. Along with those changes, the Catholic diocese raised its boarding school to secondary-school standard and initiated a programme to develop indigenous priests.

Because of budgetary limitations, as we have seen, the pre-war Territorial Administration left schooling entirely in the hands of the missions. The ethos of trusteeship that emerged after World War II led to a change in policy; the aim now was to diffuse education 'evenly throughout the indigenous community so that the whole population might share the benefits of [economic] development and participate in the political process [i.e., towards self-government and eventual independence]'. (Laracy 1976, p. 146)

For the first decade or so after the war the Administration's educational role on Bougainville was limited by lack of money and teachers. It granted subsidies to mission schools in return for some control over teachers' qualifications and over curriculum (including an emphasis on English). Gradually, however, Administration schools were established, partly in response to requests from the Bougainvillians themselves. For example, some leaders in Buin met with the Territory's Director of Education and requested him to establish more Administration schools in their area. When he replied that they already had mission schools, they countered with: '*tok bilong God, tasol*' ('those schools teach only religion'). This was not entirely true, but it represented a widespread opinion among Bougainvillians.

By 1970 there were still only seventeen Administration primary schools on Bougainville, with an enrolment of 1862; this is to be compared with eighty Catholic schools (enrolment, 9965), eighteen Methodist and Uniting schools (1646), and ten Adventist schools (617), according to the *Encyclopaedia of Papua and New Guinea*. During the next decade, however, the numbers of Administration primary schools and enrolments increased rapidly, as did secondary schools of both Administration and missions.

As a result of these developments, by 1980 of all citizens, male and female, in the 20–24 age bracket, 5291 had completed six grades of schooling; another 1814 had completed ten grades, and another 142 twelve grades. The census of 1980 does not record the number of Bougainvillians who had gone on to tertiary institutions, but information from other sources indicates that there were several who did. As early as 1971 the indigenous population of the district included three university graduates, and thirty-seven with 'other tertiary qualifications'.

More recent figures on comparable stages of schooling are not available to this writer, but there is evidence, in figures on their higher education, that Bougainvillians have continued, even quickened, their quest for Western-style education. First, the numbers enrolled in various PNG institutions of higher learning, year by year, was 1983, 396; 1984, 445; 1985, 326; 1986, 448; 1987, more than 372; 1988, 354; 1989, 410. Enrolment of course does not always culminate in graduation. For information on the latter, the records of the Papua New Guinea Commission for Higher Education reveal that as of February 1989 forty-six students from the North Solomons Province had graduated from programmes lasting one or more years, ranging from general arts and science to applied physics and accounting, from teacher training to communications engineering. Exactly how many Bougainvillians had attended or graduated from institutions of higher learning in Australia and elsewhere is not known to this writer, but the number cannot have been small. (BCL alone had sponsored 350 nationals for undergraduate and post-graduate studies in PNG and overseas. Through the 1980s, an average 35 were outside PNG as tertiary students each year. These were in addition to the students at the company's own Mine Training College, which awarded more than 17,000 certificates between 1972 and 1988.)

Before turning to the political aspirations and behaviour of Bougainvillians during 1965–88, it will be instructive to add a few more paragraphs about certain aspects of their economic behaviour at that time: first, to describe their material wants.

In pre-colonial times adult male Bougainvillians fell into two somewhat indistinct categories with respect to their desires for material goods. (For present purposes I assume that most women shared or at least supported the wishes of their husbands in this regard.) Most were content to have only enough vegetable food to satisfy hunger, plus special dishes on special occasions; enough pigs and shell money for domestic feasts and for prescribed social responsibilities; enough housing for weather-proof shelter; enough tools and (in some cases) water craft for subsistence. A second and smaller category, usually labelled Big Men, wanted to obtain or control the dis-tribution of some of these items — mainly food, pigs, and shell money — in order to maintain or acquire more-than-average social influence or power. (It may be that all or most Bougainvillian men had some yearnings towards those political goals; however, in the societies that this writer is familiar with, through field research or reading, only a few men *acted* as if they believed themselves qualified to achieve those goals.)

In terms of their material wants Bougainvillian men of the 1970s and 1980s presented a more complex picture; instead of two categories they appear to have differentiated into five.

First, there were those who acquired goods, traditional and modern, mainly for the use of themselves and their families, including additional kinds and amounts needed for domestic celebrations and for carrying out social obligations (e.g., bride-price, mortuary donations, hospitality). Those individuals correspond to the first of the two pre-colonial types described above, except that they must now be differentiated into sub-types which approximate to older and younger. The older men had spent all or most of their lives in their villages and lived fairly 'traditional' lives. Some of them earned modest cash incomes, principally through cash-cropping, with which they purchased small amounts of modern goods: tinned meat and fish, rice, bottled beverages, tobacco, sugar, clothing, lamps, kerosene, etc., plus perhaps radios, bicycles and bedding. Somewhat younger, on the average, were the men who lived in urban or peri-urban places, or who circulated between village residence and urban employment, or who earned fairly large incomes from cash-cropping. They were similar to the older men in being strictly 'consumers', but they acquired and used more modern goods than the latter: store-bought food, up-to-date houses and furnishings, vehicles, radios, etc.

There were also to be found a number of men who corresponded to the pre-colonial Big Man. These were men who wanted goods not only for consumption by themselves and their families, but also for distribution in the form of feasts and other displays of hospitality, for the purpose of acquiring social and political influence. These also were of two types. Some of them continued to use only traditional goods in their distributions. Others included many modern goods and services in their distributions, such as kina, store-bought food and beverages, vehicular transport of goods, information about and intervention with the Administration.

In addition, there were Bougainvillians whose material wants differed so widely from those of their compatriots that they constituted a category apart. These urbanized Bougainvillians had chosen to live European lifestyles as completely as their incomes permitted, in housing, food, clothing, recreation, etc. A significant minority of these men who had achieved higher education became even more thoroughly urbanized, reaching very senior positions in the police, the law and other professions, and in public service away from their home island.

10

The Political Dimension

C HAPTER 6, WHICH CHRONICLED THE POST-WAR HISTORY OF Bougainville to 1964, ended with the election of Paul Lapun to represent the whole district in the Territory's first House of Assembly. From 1964 to 1967 Lapun was Under-Secretary for Forests; during 1967–8 he was a leader in the Pangu Pati, whose motto is 'One name, one country, one people'. Lapun's resounding victory in 1964 was the first sign of anything approaching a sentiment of unity, or at least majority, among Bougainvillians' numerous cultural, regional and religious factions, excepting their self-identification as blackskins vis-à-vis the rest of the Territory's preponderant 'redskins' (who are in fact various shades of brown). Segments of Bougainvillians had for decades expressed feelings of antipathy towards whites in general, or colonial institutions in particular, but those had not led to any island-wide unified organization or concerted action.

The Territorial Administration sought in 1949 to create larger-than-village-size unities by setting up regional elected local govenrment councils. As mentioned in Chapter 6, by 1965 eight of these councils had become established in the district, but some entire language and cultural regions (e.g. Keriaka, Baitsi) had rejected them, as had several communities within regions in which they were established (e.g., the Hahalis Welfare Society communities of Buka, and several communities within the Kieta region).

Mention was also made of the practice, inaugurated in 1963, of having representatives of each local government council meet together annually to discuss common local problems, such as appeals for Administration assistance for roads, medical aid centres, and schools. Beginning with the 1970 meeting of the councils, however, topics of district-wide import were discussed and embodied in resolutions to the Administration — demands for a district teachers' college, for the return of expatriate-owned plantations to their 'traditional' owners, for a ship to provide more frequent service to the outlying atolls, etc. Several demands were also made directly to BCL — for money to finance district roads, for larger representation on BCL's board of directors, and, ironically, to construct a copper smelter on Bougainville. (This proposal was put forward in 1972 by MHA Father John Momis, despite warnings by a BCL spokesman about the destruction of vegetation by smelting.) Then, in the early 1970s, the large and fateful issue of secession began to be discussed at conferences of local government council; it was heatedly debated, but not resolved.

Meanwhile, some other institutions were formed to discuss and in some cases to agitate for district-wide concerns. Many of the leaders of the Administration-sponsored local government councils were to play important parts in those non- (or anti-) Administration organizations. The local government council system itself slipped into such a steep decline that it died, officially, in 1975 and was superseded by new forms.

One of these institutions came to be known as Village Government. It began in 1967 when a Siwai man, disillusioned with his local government council for not doing enough for his own village, was encouraged by his MHA, Paul Lapun, to 'follow his own local customs' by getting together with fellow villagers to decide what they wanted to do about running their own lives. What they first wanted to do was to record their own 'traditions' to conduct their own courts, according to local, not Western, standards of right and wrong. For their 'chairman' they agreed—without a Western-style vote—on an elderly Big Man, a 'traditional leader'. At first they appeared to have considered their organization more complementary than in opposition to the Siwai Local Government Council—for example, they continued to pay taxes to it—but after a while they shifted all of their 'public' interests to their own village government. Their reasons for starting their association were summarized (in Pidgin) by one of their university-trained fellow villagers. Translated, it reads:

> The members can meet in one place to discuss their customs. They can meet to revive their traditional customs. They can meet to get their own ideas moving ahead so that they will be ready for self-government and independence. They can consider together their own customs and those of Europeans. They can select new village leaders who can carry out the kind of role that was done by village leaders in the past. They can establish good life styles for all the people in their area. This is *unity*.

It was not long before several nearby villages followed their lead and set up village governments of their own, and in time the concept of local village governments, in contrast to multi-village local government councils, became so widespread that the provincial government terminated the latter and officially recognized the former. By that time the (self-defined) functions of village government had become extremely varied, including many of the things that the local government councils had done, or had tried to do. Some features were common to all of the village governments, including the perception that they were doing 'traditional' things in 'traditional' ways, including:

- having their 'traditional' leaders lead, without subjecting them to the humiliation of a Western-style election;
- setting their own moral and ethical standards regarding marriage, adultery, theft, slander, lying, etc.;
- conducting their own courts and defining their own penalties for wrong-doing.

The official statement by the North Solomons Provincial Government regarding village government appeared in the pamphlet, *North Solomons Now*, which was published in November 1976 to commemorate the transition from District to Provincial government:

> The Village Government concept . . . is a most basic political structure of the North Solomons Government. [It will eventually become] the seat of the North Solomons [Provincial] Government. [Its structures are] political and administrative [and developmental]. In so far as it is political, Village Government allows village leaders to take part in decision-making regarding what type of development they want. The philosophy behind [it] is that it would promote economic, political and social development at the village level, where in fact the majority of the people live. The system owes its success to the popular support of Village leaders. . . . It is an essential element of the North Solomons philosophy of humanism that the people are the best judges of their own needs. We believe that the government exists to meet the needs of the people, therefore, the Village Government is established to serve the people, and not the people serving the institution.

Noble sentiments, with a familiar sound! It was also somewhat ironic, inasmuch as the Province's 'villages' had been in most cases a creation of whites. As mentioned earlier, the most usual traditional settlement pattern was one of small hamlets, which the Germans, then the Australians, coercively combined into line-villages in order to simplify census-making and taxation. The measure met with much opposition for decades, and when Administration sanctions were removed the line-villages were often abandoned.

We turn now to a Bougainvillian 'government' that eschews all other forms of government, namely, that made up of the Fifty Toea Movement, whose members resided in several Nasioi-speaking villages in the mountains southeast of Panguna. It was founded in the mid-1960s by Damien Damen, who continued to be its leader. Its original name was the Bauring Society (*bauring*, the/his child, in Nasioi), but became widely known as the Fifty Toea Government, from the circumstance that 50 toea was the monthly membership fee. The tenets of its 4000 or so members were ultra-conservative: they maintained traditional ways, and opposed formal schooling for their children. They were against development, advocating that minerals should be left in the ground, and were against the government, having refused to vote in national or provincial elections. The tenets also included beliefs in *kago* (cargo), but of what kinds and how obtained I have not been able to discover.

And then there was (and maybe still is) the Pontoko Onoring Private Government of Sir James Tobu, as described in his letter of 4 April 1988 to BCL's managing director:

> Dear Sir,
>
> On a hill east of World Nations that Sir James Tobu of Pankaa Village in Bougainville to bring Forth Empire nearly Thousand Years ago in a far corner of the earth claim to be Forth the King of Solomons islands by time of Century 2000 years at the end.
>
> Well in the time to the end of Century Traditional Laws of Engova office have the informal controls according to ruses in order of the state Solomons Islands by the fact itself of Traditional Power on ste [sic] in the year of our Lord born unto us.
>
> It is truly Sir James Tobu to be honest the State of being a Royal Personage by his Law of World Nations the end of Century. Explaining itself further his Holiness set forth through the coming of the Holy Ghost in the day of Judgment is made on Earth.
>
> Royalty Money of Bougainville Copper Ltd, For my right of a citizen from the Law of Solomons Islands, Please the end crowns work at the point of death above the nobleman thoughts Sir James Tobu one who supports a King Hope for better now share paid to Engova Office.
>
> Sir James Tobu being judge in his opinion the act of creating which tells us now that danger for the other Nation in here Bougainville so since I say everyone for the other Land must be out.
>
> To bring into life to set going our Freedom to do as one Pleases, Permission, An Impudent Act happening once in a Hundred Years and to hold a special Church service in Memory of some event, for the best way to our Freedom of Expression.
>
> Next of King for my Brother   THANK YOU
> Lay Iruina Naki   Yours sincerely
> Died 22nd 8-68   Sir James Tobu
> Picking a Diamond in the rough
> takes a special kind of skill,
> in Kupe Gold field.

Returning now to events of broader political and governmental dimensions, we consider first the Territory's House of Assembly elections.

In the 1964 elections, Bougainville District constituted a single open electorate; for the 1968 elections the district was divided into two, north and south. In the 1964 elections the district was only part of the New Guinea

Islands special electorate (along with New Ireland and Manus), which had been reserved for European candidates; in the 1968 elections it was constituted as a separate 'regional' electorate and was open to any local resident, regardless of ethnicity, who possessed some secondary education. The returns in this election offer some intriguing insights into the thoughts and feelings of Bougainvillians in this period of rapid change. Above all, they provide additional proof of the inaccuracy of attributing to them, in 1968, a uniform political stand.

The open electorate in the south was contested by only two candidates: the incumbent, Paul Lapun, and Andrew Komoro. Komoro, one of the losers in the 1964 elections, entered the 1968 campaign very late, reportedly in response to some 'hints' from Administration officials.

Lapun, though a resident of Banoni, had many social ties throughout southern Bougainville. Although at the time not uncompromisingly opposed to the Administration, he had been on many occasions critical of official actions and omissions. His performance in the House of Assembly had been positive and effective; he served for a time as Under-Secretary for Forests, but gave up that office in order to play an active role in the comparatively radical Pangu Pati. Perhaps his most important action in terms of his constituents was his successful sponsorship of the amendment to the Mining Ordinance, which, in opposition to the Administration, allocated 5 per cent of the expected mine royalties—i.e., 5 per cent of 1.25 per cent of total copper revenue—to the Panguna landowners themselves.

Komoro, a native of Buin, was, like Lapun, literate and fluent in English, and a product of Catholic schooling, which included a year at the Marist Brothers' College in Sydney. Komoro, however, was more closely identified with the Administration because of his position as clerk of the Buin Local Government Council; and in contrast to Lapun's somewhat equivocal attitude towards the Catholic mission, Komoro was avowedly critical (although he had not severed all his ties with the church).

Neither candidate campaigned extensively, Lapun even less than Komoro. In fact for the three months preceding voting, Lapun was in Port Moresby, or in Canberra, where he attended a seminar. Komoro gave a few speeches. Lapun evidently considered his own views well enough known to require no further statement (although there was anything but consensus among voters on what those views actually were). In any case, Lapun won handily; of 11,827 formal votes cast (about 65 per cent of all registered voters in the electorate) he polled 9735.

Three candidates stood in the district's northern open electorate contest: Sampson Purupuru, a Methodist; Francis Hagai, a leader in the Hahalis Welfare Society; and Donatus Mola, a resident of northern Buka. Both Hagai and Mola had served as Catholic mission teachers, but whereas Hagai had become anti-mission and anti-Administration, Mola had remained closely connected with the mission and was a prominent member of the Buka Local Government Council. Mola won with 58 per cent of the vote, indicating that a pro-Administration stand was not a handicap in the north. Moreover, mission association seems to have been less of a handicap — or perhaps more of an advantage — than opposition to missions.

The district's regional electorate was contested by three candidates: Joseph Lue, a Siwai; John Lee, an Australian planter resident on Buka; and Gayne Cook, another Australian planter and resident of Tinputz. Both of the latter campaigned actively; Lue hardly at all. In the end Lue won, having led Lee by a narrow margin; his winning attributes appear to have been his skin colour and his backing by Paul Lapun.

Before the 1968 election there occurred two other events of district-wide political significance. The first was a meeting held in Kieta on 13 July 1968. It began with discord between members and non-members of the Kieta Local Government Council, but ended with a common plea for unity — not only within the Kieta sub-district itself but among all of the Territory's districts, and against the colonial Administration. Both sentiments were provoked by the Administration's 1967 Bougainville Copper Agreement with BCL, which in the view of many Bougainvillians was unfair to landowners in the mine lease area. The most vehement anti-Administration speaker was Damien Damen, the founder and leader of the Fifty Toea Movement, who later was to provide sanctuary for the new-born Bougainville Revolutionary Army.

The second of the events of political significance came to have much more important and durable consequences than the impassioned but fruitless Kieta meeting. This one took place in Port Moresby on 8 September 1968, when twenty-four Bougainvillians attending tertiary institutions on the mainland met with the MHA Paul Lapun to discuss grievances towards the colonial Administration. They produced a statement asking the Administration to

> allow Bougainville to go it alone, or else arrange for a plebiscite or referendum to find out conclusively the Bougainvillians' views on the following political alternatives:
> - whether Bougainville should form an independent nation on its own; or
> - leave Papua New Guinea and unite to form one nation with the British Solomon Islands south of Bougainville; or
> - remain with Papua New Guinea.

One of the framers of this proposal, university student Leo Hannett, listed the reasons why he favoured the alternative of secession. He cited differences between Bougainvillians and other Niuginians in terms of culture, race, and colonial history:

> Culturally, Bougainvillians feel and accept that they differ essentially from PNG cultures on our main types of dances like kuma and panpipe or bamboo dances. These are not found anywhere in Papua New Guinea. Most important of all, Bougainvillians firmly believe that we

have a distinctly marked different cultural and political ethos — on such
fundamental principles as natural and social justice and respect and
reverence for life. We believe in fair play at all times and that everyone
has his or her rights that must be respected at all times. We do not
believe in mob fighting or in the law of jungle that might is right, and
that individuals or groups of individuals can be a law unto themselves
... Numerous examples of wanton killing of human life in Papua New
Guinea by Papua New Guineans have led many Bougainvillians to doubt
whether Papua New Guineans share with us this conviction that human
life is sacred and must be upheld with utmost reverence.

Brave sentiments — but not very credible anthropology! The proposal was
not persuasive enough to induce the Administration to conduct a referendum on secession. However, the Port Moresby gathering did eventually
crystallize into an association that came to play an important part in Bougainville politics. Its founders gave it the name Mungkas, from the Buin word
for *black*.

Another organization with a considerable role was Napidakoe Navitu.
(*Napidakoe* is an anagram of the names of ethnic units in the Kieta sub-
district); *navitu* means 'grouping' in the Nasioi language.) It is reported to
have originated at a public meeting held at Kieta on 27 April 1969 at which
participants 'drew up a manifesto which asked the Administration to desist
from resuming Arawa land and asked the Administration to sit down to a
fully representative conference on land problems in the CRA mining area'.
After a few more meetings the organization's aims, according to their official
organ, *Bougainville News*, had gone beyond the consideration of land issues,
although this was the principal immediate problem. 'We realized', wrote
the organization's secretary, 'that the Copper Project was going to radically
transform the island, that the Administration was determined to adhere to
its policy of small-scale compensation and that it had no coherent plans to
safeguard the economic rights or the social integrity of the local people.' That
secretary was Barry Middlemiss, an Australian and at the time foreman of
Arawa plantation. Subsequently he left plantation work and devoted all of
his considerable energies to promoting the new organization, and to bringing
out its journal, in Pidgin and in English. In the first issue Middlemiss defined
the aims of Napidakoe Navitu, as follows:

- To further the social, political and economic development of
  Bougainville. This had to be stressed. We looked to the development
  of Bougainville, not just the Nasioi or the Kieta area of South
  Bougainville as a whole.
- To promote political autonomy.
- Better education for the people of Bougainville through Church and
  Administration schools.

In the same issue the 'rules' of the organization were listed:

(a) To encourage and foster economic, social and political development on the island of Bougainville and to unite the Bougainville people as one economic, social and political community.

(b) To restore, maintain, foster and encourage among the people of Bougainville an understanding and appreciation of their traditional culture and customs.

(c) To maintain a respect and appreciation for marriage and family ties and to encourage respect for and obedience to traditional customary marriage laws and to safeguard the stability of family groups.

(d) To create and endow scholarships for school children to assist in their advancement and education and to provide financial assistance for university students seeking to study such courses as the association may from time to time decide upon.

(e) To nominate candidates for election to the House of Assembly of the Territory of Papua and New Guinea or the parliament for the time being under whatever name.

(f) To unite all radical groups and political and religious bodies on the island of Bougainville.

(g) To endeavour to encourage and seek early domestic internal self-government for the Territory of Papua and New Guinea while depending upon Australia for financial assistance and guidance and control for [*sic*] external affairs.

It is not clear from this statement whether 'island of Bougainville' was intended to refer only to the one island, or to Buka also, or to the whole district.

By mid-1972 officials of Napidakoe Navitu were claiming some 8000 members, consisting largely of Nasioi-speakers, but including many others from Nagovisi and several from Buin and Siwai. The association had also branched out into some small business enterprises including a taxi service, a bus service and the mail contract to Panguna. Meanwhile, throughout 1970–7 *Bougainville News* (edited and largely written by Middlemiss) took stands on a number of politically important issues, such as land resumption and the related matters of Bougainville unity, secession, and the 1972 House of Assembly elections. In addition, editorials and letters published in the journal kept up a drumbeat of attacks against the Administration— particularly the incumbent (Australian) District Commissioner, and the official Radio Bougainville— and against anyone opposed to Paul Lapun, the organization's chairman, and its secretary, Middlemiss. As for the organization's attitude towards the copper company, if *Bougainville News* was representative, it could only be described as pragmatic and slightly ambivalent. No sentiment was voiced about prohibiting mining or getting rid of the company. In fact one could discern from the journal's editorials, news stories and letters a growing pride in the copper project ('the biggest industry of its kind in New Guinea') and acknowledgement of the economic benefits it would eventually bring to Bougainville. On the other hand, editorial complaints continued concerning distribution of the company's earnings. At first it was proposed in the journal that PNG should receive a larger share than the Bougainville Copper Agreement provided, but as time went by a larger share was demanded for Bougainville itself. ('Why should *our* riches go to pave more roads in Port Moresby and Lae?') Editorial complaints against the mining enterprise and its many ramifications were directed more at the Administration than at the company.

Napidakoe Navitu began as a local Kieta-area association, but with the passage of time its journal put more and more stress on the unity of Bougainville–Buka (presumably including Nissan as well, but with no mention of the atoll islands). The organization's original rules spoke of 'early domestic internal self-government for the Territory of PNG', but it was not long before its journal began to advocate new kinds of territorial alignments, and even secession from PNG.

On the occasion of a brief visit to Kieta by the Australian Prime Minister, John Gorton, Paul Lapun (speaking as the representative of Napidakoe Navitu) petitioned him for a referendum, saying in part:

> Sir, we ask you to grant to us an Official Referendum to determine the future of Bougainville. We earnestly entreat you to grant our wish as this is most important to us. At this stage we do not wish to remain with Papua New Guinea . . . We have no common links with Papua New Guinea, and we have no desires to remain with them. To us they are alien. We have absolutely nothing in common with them. If we are not ready for Independence when they are we would much prefer to stay with Australia until such time as we are ready.

Gorton is reported by *Bougainville New* to have turned down the request, asserting that he did not think that Bougainville should secede from the Territory, there being economic and social benefits that could come from being part of a large nation. He concluded by saying: 'Think and talk a lot more before asking for a referendum.'

Bougainville News took matters into its own hands and published a ballot for an 'unofficial' referendum (*referendum bilong yumi*), which readers were asked to mark and return to the journal's secretary. The options listed were the same as those listed by the Port Moresby group, along with a fourth one, union with the New Guinea islands (New Britain, New Ireland, Manus). The results of the referendum were mysterious to say the least. Rumours abounded that whole packets of pro-secession votes were contributed by the referendum's sponsors, and that ballots distributed for voting were impounded by its indigenous opponents or by individuals opposed to the sponsors themselves. In any case, that first, unofficial, attempt to sample public opinion on this important political question ended so inconclusively that its sponsors returned to their demand for an Administration-sponsored referendum. (There was no evident awareness of the irony that a referendum held by the Administration, whatever its outcome, would be more credible to Bougainvillians than one carried out by themselves.)

The referendum issue dominated the next conference of the district's local government councils, which took place in 1970. However, from it there emerged not one but four different, regional, points of view about the issue. As reported in Mamak and Bedford (1974B):

(1) [Representatives from] north Bougainville argued against the holding of a referendum on secession from Papua New Guinea because it would lead to a shortage of technical and unskilled mainland labour necessary to cope with north Bougainville's increased agricultural production.
(2) [Representatives from] Wakanai and the northeast coast favoured the formation of an official committee headed by Bougainville's four MHAs to present villagers with arguments for and against secession.
(3) [Representatives from] central Bougainville [were] in favour of holding a referendum but against Navitu carrying it out.
(4) [Representatives from] south Bougainville felt the time was ripe for secession.

In the end the second proposal was unanimously adopted, but nothing was done to implement it because of disagreement over the composition of the proposed committee.

In February 1972 Bougainvillians went to the polls again to elect, or re-elect, representatives to the Territory's third House of Assembly. In terms of the district's political future this election was crucial, as the new House would largely determine the timing of PNG's national birth and the structure of its nationhood—or so historians would conclude. Locally, in contrast, the issues resembled the mixture as before, with some changes in scope or emphasis.

The established order came under increasing attack during the election campaign, including any institution or individual identified too sympathetically with it. 'Secession' continued to be used as a rallying cry during the campaign, but by this time it had become broadened to express opposition to a variety of disliked institutions and situations—including the Administration, expatriate enterprise in general, redskins, and land alienation. During this campaign the more sophisticated Bougainvillians increasingly distinguished between 'referendum' (a hoped-for opportunity to voice popular choices) and go-it-alone 'secession'. In place of the latter, words such as 'separatism', 'regionalism', and 'local autonomy' seemed to be more audible.

Land alienation, past and future, remained a major issue, and candidates were often weighed in terms of stands they had taken on such episodes as the Rorovana land resumption. 'Custom' versus 'progress' also entered into some contests, but apart from such generalities as 'family stability', 'respect for elders', 'female chastity', and the like, it was difficult to discover what was included within 'custom', or how much and what kinds of innovations were implied by 'progress'. Cargo-logic, on the other hand, seems to have diminished somewhat as a way of thinking and obtaining desired ends, and consequently as a basis for favouring one candidate over another. Finally, the universal issue of the known versus the unknown entered consciously or unconsciously into the choices of many voters.

By the time of this election, national political parties were well established on the PNG political scene, and some of the Bougainville candidates bore party labels. From the information available to this writer, however, party per se was not a decisive factor in the victory or defeat of any aspirant on Bougainville.

For the 1972 election Bougainville District was allotted four seats in the House of Assembly. As in 1968, the district as a whole selected one regional member, but for the open electorates the district was divided into three parts: south (the Buin sub-district), a newly created central one (comprising the Kieta sub-district), and a northern one (comprising the rest of Bougainville Island, Buka, Nissan and the other atolls). The victors in the open electorates were: south, Paul Lapun; north, Donatus Mola; central, Raphael Bele. Both Lapun and Mola were incumbents; Bele, a Rorovanan, was a prominent member of Napidakoe Navitu.

In the regional electorate two candidates stood: Joseph Lue and John Momis. Lue, it will be recalled, was a Siwai and a former Marist brother; his political association was with the PNG People's Progress Party, and he was the incumbent Bougainville regional member in the second House. Momis is the son of a prominent Buin (Terei-speaking) father and a mixed-blood New Ireland mother, and is a Catholic diocesan priest attached to, but not a member of, the Marist order. As the church's principal education officer on Bougainville, he was already well known throughout the district, and his position as a priest undoubtedly lent weight to his political appeal. He was not, however, the official mission candidate, although he had been given permission by his bishop to stand for office and thus was considered by many to be so. (Indeed, it was reported that many of the older expatriate priests disapproved of his candidacy as inappropriate to his priestly office.) Prior to and during the election, Father Momis had no declared party tie, but his stated views were close to those of the Pangu Pati.

The results were overwhelmingly in favour of Momis: 20,772 to 3749. A number of factors contributed to this large disparity. Lue was generally characterized as a Port Moresby man, whereas Momis spoke out repeatedly for Bougainvillian interests. Although Momis did not campaign explicitly for secession, he did declare himself in favour of it when asked to describe his stand, and he vocally supported referendum. Lue was widely reputed to be pro-establishment; Momis was generally anti-Administration, and was specifically opposed to BCL on several occasions, such as the Rorovana land resumption. And he was assisted in his campaign by the positive support given him by Paul Lapun.

After the election Bougainville's representatives were allotted important positions in the third House of Assembly. Lapun was named Minister for Mines, and Mola Minister for Business Development. Father Momis, by then a declared member of Pangu Pati, was made deputy chairman of a newly established all-party committee charged with preparing a draft constitution for an independent Papua New Guinea; this was a key position, since the chairman was the Chief Minister Michael Somare. Some observers interpreted those appointments as a calculated move on the part of the coalition's pro-unity leadership to bind the Bougainville members to their cause — but if this was the strategy, it proved to have some opposite effects.

The next event of district-wide significance occurred at Kieta in December 1972, when members of the Mungkas, home from Port Moresby on Christmas vacation, held a 'seminar' about the district's political, economic and social problems. The meeting was attended by many older Bougainvillians, including councillors and some 'traditional village leaders'. Although it did not result in complete consensus on every issue, nor lead to any specific programme of action, it did serve to bridge what had become a widening gap in communications between older leaders and the more educated younger students — at least, for a while.

Much more galvanizing, and unifying, was the tragedy that befell two prominent Bougainvillians on Christmas Eve of that year. Two senior civil servants, Peter Moini and Dr Luke Rovin, fatally injured a girl in a driving accident in the Eastern Highlands; they were seized and brutally murdered by the girl's relatives and neighbours. (It was subsequently reported that the two men had been partying and were intoxicated, but that information was not circulated on Bougainville at the time.) The news of the murders was received on Bougainville with angry public demonstrations against the 'savage' Highlanders in particular, and mainland Niuginians in general. Several Bougainvillians working in the Highlands returned home; demands for repatriation of redskins increased; and calls for secession became more vocal. Those and other grievances and demands were aired shortly thereafter at a public meeting at Kieta between the Territory's Chief Minister Somare, who was then on tour, and local people, including local government councillors and officers of Napidakoe Navitu. Demands for secession were expressed in other meetings around Bougainville, and tempers became even more inflamed when in March 1973 a Highlands court acquitted the murderer of one of the slain Bougainvillians.

The response of the Chief Minister, Somare, to all this was to set up a committee to explore ways to improve relations between the central government and the Bougainville District. A Bougainvillian student, Leo Hannett, was named to head the committee, which he set about to do with characteristic purpose and energy. He began in Port Moresby by assembling the district's four MHAs (Momis, Lapun, Mola and Bele), several Bougainvillian public servants, and the members of Mungkas. The proposals to emerge from the meeting were:

- to transform the district from a mere administrative unit of the Territory
  to a distinct district *government* empowered to collect and distribute tax
  revenues and to define its own relations with central government, including
  the right to secede;
- to have the district government's top offices held by Bougainvillians;
- to have at least two Bougainvillians appointed to BCL's board of directors.

With those proposals in hand Hannett returned to Bougainville, where in
the meantime still another political movement was taking shape.

Secession had by then become a rallying cry for many Bougainvillians,
but by no means for all. It was more popular with young people, especially
the tertiary-educated ones, than with older and more experienced political
leaders who held that the district was not yet prepared for it in terms of unity.
It was far more popular in the south than in the north. Moreover, the issue
was complicated by the sentiment, shared by many of the local government
councillors, that political initiative was being arrogated by the Mungkas,
in league with the mainland-habituated MHAs. The latter, it was claimed,
rarely if ever returned to their electorates, and were more concerned with
their careers in the national government than with matters at home. However, even the anti-secessionists desired that 'something' be done to redress
their various grievances against the central government, BCL, redskins, etc.

The first opportunity for doing that 'something' occurred at the conference of local government councils in late February 1973, which was attended
by Hannett, Lapun and Momis. Although disagreements persisted on some
issues, especially on secession, consensus was reached to set up a separate body,
a Bougainville Special Political Committee (BSPC), which would replace
the Chief Minister's committee and would seek to arrive at district-wide
agreements on the district's place, if any, in Papua New Guinea. The new
committee was to be made up of one representative from each local government council, along with one from each of the following: Napidakoe Navitu,
Hahalis Welfare Society and the Port Moresby and Lae branches of Mungkas.
(Later on the committee was enlarged by representation from other aggre-
gations of Bougainvillians.) Hannett was elected to head the committee, and
the MHAs were named advisers to it. Soon after the initial conference
Hannett travelled around the district in order to sample opinion at the grass
(sweet potato?) roots. He found repeated the four kinds of factions that had
surfaced at the conference; as reported in Mamak and Bedford (1974B), they
were:

> The uncompromising secessionists: Kieta, Buin, and Siwai. The people
> are unsure about their specific goals or the nature of their problems, yet
> leaders are very committed to the search for a legitimate government
> separate from Papua New Guinea and based on a combination of
> traditional and modern socio-economic values. Politics is very much an
> emotional matter.
>
> The autonomists: Buka and Teop-Tinputz. Leaders express a strong

preference for some measure of autonomy for the District. (At the
conference the President of Buka Council said: 'If we were to take a
vote today on whether Bougainville should break away from Papua New
Guinea, north Bougainville would vote against it. However, should
Bougainville decide to go ahead with secession, then north Bougainville
would break away from the rest of Bougainville.')

Animistic secessionists: members of the Eivo Council and people
living on the south central coast, who advocated secession on
unconditional terms. They have a strong desire for ancestral spirits to be
involved in the affairs of the people.

The uncommitted: the rest of the people: Wakunai, Torokina,
Hahon, Nissan, the outer Islands. There are no outstanding leaders who
can voice the opinions of these people. They are adopting a 'wait and
see' attitude. The central government's impact on these people has been
minimal. [pp. 24–5]

The tour of this remarkable and highly influential young man (he was
about twenty-six years old) seems to have moderated his own political
expression somewhat, because thereafter the agenda of the BSPC changed
from secession to consideration of ways by which the district could obtain
and operate its own, more autonomous government within Papua New
Guinea. To supplement the operating funds allocated to it by the Chief
Minister, the BSPC canvassed all adult Bougainvillians, male and female,
for small contributions, with some success. Also, it is worth noting that
two of the members of the committee's six-member executive held high
positions in BCL, and a third, Henry Moses, was leader of the official union
of the company's employees.

In April of the same year the Bougainvillian employees of BCL got
together to form their own Panguna Mungkas Association (PMA), to
safeguard their distinctive interests. The PMA maintained a low profile until
the end of May 1973, when a Bougainvillian visitor, suspected of shoplift-
ing at the Panguna supermarket, was roughed up by company security
guards. Offended by this action, the officers of the PMA opened its ranks
to all Bougainvillians living nearby and demanded that the company Bougainvillianize the workforce, punish all non-Bougainvillians involved in fights
with Bougainvillians, and repatriate them at the end of their contracts, etc.

In due course the PMA established links with the BSPC (some of its
members held offices in both groups), but the PMA itself did not become
as effective as its founders had hoped. This was partly because of the divided
loyalties of most of the leaders, who also occupied high positions in the
company, and partly because of the understandable reluctance of the company
to deal separately and distinctively with such a large part (over 30 per cent)
of their workforce.

In the meantime, continuing agitation by several organized bodies of
Bougainvillians persuaded Chief Minister Somare to replace the incumbent

District Commissioner, an Australian, with a Bougainvillian. The new appointee, Alexis Sarei, D.Phil., a native of Buka and a former Vatican-trained priest, had left the priesthood and become one of the Chief Minister's principal aides. Sarei's arrival on Bougainville coincided, most auspiciously, with the BSPC's first district-wide conference at Arawa in May 1973. The conference's active participants included the district's four MHAs, the presidents of all of the district's local government councils, representatives of the Hahalis Welfare Society, Napidakoe Navitu, Mungkas, the outer islands, and of Bougainville's non-council areas. In addition, 'village elders' and urban workers were there as invited guests. The purpose of the conference was to obtain widespread agreement on the establishment of a district government, one which would continue to be a part of the Papua New Guinea nation but on terms defined by Bougainvillians themselves.

The conference lasted three days. Characteristically, its official participants and guests raised many of the same kinds of issues that had consumed the proceedings of the local government councils, including such parochial ones as specific stretches of road and more frequent shipping to one or another of the outer islands, along with somewhat broader ones such as referendum, secession, and BCL's failure to do this or that. Once again it brought to the surface some deep-rooted differences, such as: Napidakoe Navitu's, including Lapun's, jealousy of the BSPC; the stay-at-home councillors' wariness of the more sophisticated members of Mungkas; the gulf between north and south. There were also, of course, the perdurable differences between pragmatists and radical activists, between the barely and the highly educated, and between the cargoists and the 'realists'. The conference also revealed the growing dissatisfaction of several participants with what they perceived to be the failure of their MHAs to fight for Bougainville causes. Despite those differences and digressions, the conference convenors managed to frame and enact a resolution calling for the formation of a district government, within Papua New Guinea, while reserving the option to hold a referendum on secession if later on required. The resolution was adopted without dissent.

Before concluding this account of that important event, one other feature warrants mention: this was the participants' frequent avowal of almost reverential respect for 'traditional leaders', 'village elders', Big Men. A few of the latter attended the conference; some even spoke, although not very effectively, and often not to the point. Nevertheless, their 'wisdom' was usually rewarded with applause, and it was generally agreed that they should be included in the leadership of the envisaged district government. Towards them there was an ambivalent mixture of condescension and nostalgic respect.

After the conference the BSPC leaders prepared a blueprint for their desired district government, which they presented to the national government's Constitutional Planning Committee when that body visited Bougainville in July of that year. This called for structuring of the district government by Bougainvillians; the election of many district officials, including the head

one, and the empowering of those officials to do the following unprecedented things:
- to initiate district development plans;
- to make laws specific to the district (e.g., regarding vagrancy and education);
- to exercise control over exploitation of the district's natural resources;
- to settle all local land-ownership problems;
- to permit village courts to settle local disputes by custom; and, a huge concession;
- to prescribe and collect all major taxes, licence and custom fees, court fees, etc. some of which would be remitted to the national government (a complete reversal of the existing procedure). More specifically, the proposal asked that mine royalties be paid directly to the district.

Not surprisingly, the national government promptly turned down the submission and answered with a plan of its own. It would permit the district government to make rules and decisions about certain matters ('within the framework of national policies'), but retained all financial powers, including taxing, for itself. While the national government would retain the authority to collect mining royalties, it would as a consolation remit an equivalent amount to the district.

Bougainvillian leaders reacted to this rebuff with anger and consternation. Some of them renewed their threats of secession, but others stated either that they did not want secession, or that they were not yet prepared for it. The possibility of secession moved the Chief Minister to proffer negotiations — whether with sincerity or as a delaying tactic is not clear. In any case, negotiations did take place, and on 5 July 1974 Bougainvillians got their district government. Called the Provincial Government, it was composed of elected officials and had a few administrative duties and perquisites, but no substantial taxing or legislative powers.

As a postscript to this series of events, Chief Minister Somare sacked Hannett from his position as adviser on Bougainville affairs, for having publicly called for the dismissal of Lapun and Mola from the Cabinet. As mentioned earlier, those two MHAs had been for some time unpopular with many of their constituents; but the thing that moved Hannett to make his attack was information that they had not even attempted to support in Cabinet the Bougainvillians' proposals for district government. Back home, however, Hannett's sacking did not diminish his popularity, for he was called on to play the principal role in framing the structure and the functions of the new 'Provincial' government — which, as reported earlier, singled out village government as its most basic and important element.

Planning for the new provincial government took a long time and included much bargaining with the national government over the division of functions and powers. The national government compromised on some issues, but its adamant stand on others, especially on financing and taxing powers, moved many Bougainvillian leaders to declare and celebrate 'independence' from Papua New Guinea on 1 September 1975, just two weeks before Papua New Guinea celebrated its own independence from Australian control. Despite the existence of this hypothetical republic, the North Solomons Provincial Government continued to operate.

In 1976 the appointed Administrator (subsequently District Commissioner), Alexis Sarei, was elected premier of the new province, and presided over the organization and administration of the authorities and powers that had been wrested from the national government. These were still too little in the eyes of many Bougainvillians, but enough to replace *secession* with the new catchword, *development*.

Development, however, proved to be as elusive as secession and, as happens in societies everywhere, Bougainville's voters expressed their frustrations at the next election, in 1980, by replacing the incumbent with a new premier. This was Leo Hannett, whose campaign slogan was 'Let's get moving'. Slogans can win elections, but Hannett quickly recognized that the funds doled out to the province were insufficient to build the new roads, etc. needed for 'development'. To Hannett the most obvious source of additional funds was a larger slice of the money being remitted to the national government by BCL. The latter was receptive to Hannett's idea, and proposed to discuss it at the next scheduled review of the Bougainville Copper Agreement, due to take place in 1981. However, the national government was unwilling to include representation from the North Solomons Province, and because of that (and perhaps for other reasons not publicized) the review did not take place. Despite that rebuff, Hannett made effective use of the provincial government's 'business arm', the Bougainville Development Corporation to initiate several new enterprises which supplied goods or services to the company and in doing so earned money for its Bougainvillian managers and shareholders, including the provincial government. In retrospect the period of Hannett's premiership was marked by growth in some parts of the province's economy, and by stable and generally harmonious relations between provincial government and BCL—at the top, at least.

That harmony was weakened by the elections of 1984, when for the first time national party politics became a factor in provincial elections. That came about through the efforts of Father Momis, who had been opposed to the company from the beginning and, for that and perhaps other more personal reasons, wished to supplant Hannett. As co-founder of the locally popular national Melanesian Alliance Party in 1982, and with considerable local influence of his own, he succeeded in bringing about the election of a majority of Alliance-endorsed candidates to the provincial Assembly, who in turn chose Alexis Sarei as premier — not because of his party connection, which was in fact weak, but because of his wide personal popularity. After two years in office, however, Sarei resigned, his relationship with Momis having cooled, and his control over the Melanesian Alliance partisans in his government undermined.

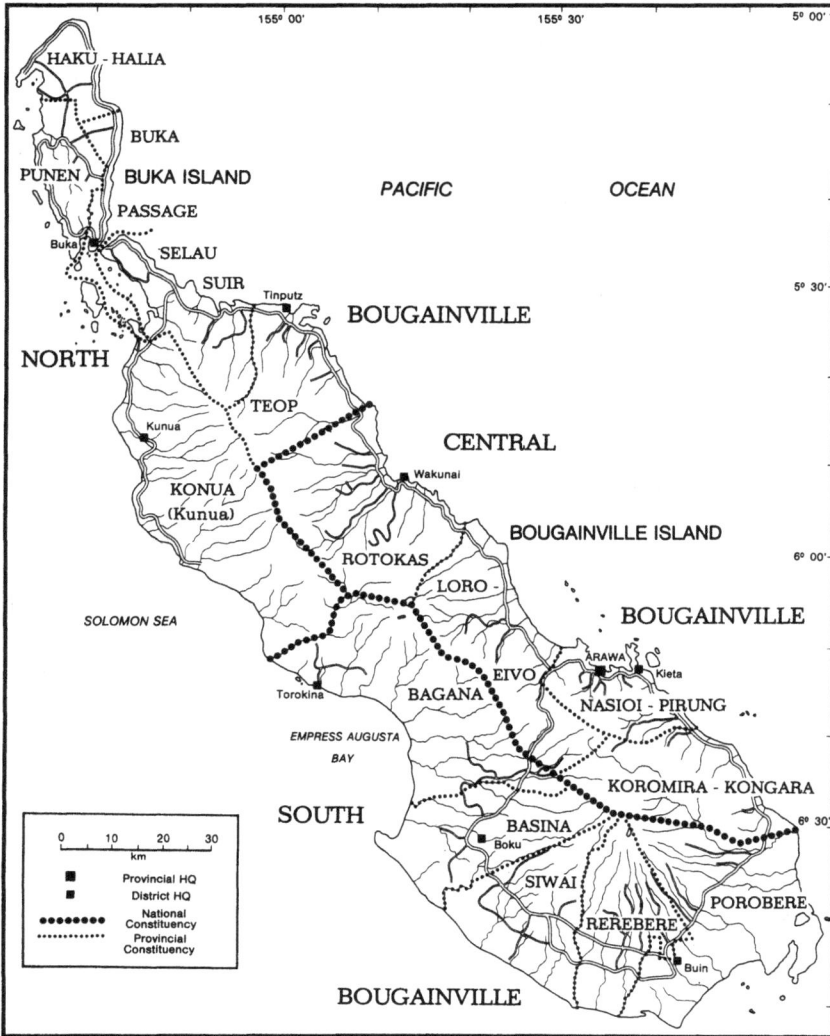

Figure 15 Administrative subdivisions 1987

In Sarei's place the Assembly chose one Joseph Kabui, then still under thirty years of age, whose electorate included the Special Mining Lease area and whose home village was one of those most severely discommoded by the mining and tailings disposal. Kabui was also a member of the Panguna Land-owners' Association, although his land rights in the lease area were secondary, deriving from paternal ties. Prior to his election in 1984 (with Melanesian Alliance endorsement) he had been an industrial relations officer with the Bougainville Mining Workers' Union. With that background it is not surprising that the provincial government's attitude towards the company became distant and suspicious. At the same time, a rift developed between that government and the Bougainville Development Corporation, mainly because of Hannett's chairmanship of the corporation but also because of BCL's involvement in some of its enterprises.

Two years later Kabui was re-elected, this time without specific Melanesian Alliance endorsement. In view of the increasing prominence of younger voters, Kabui's youth may have had something to do with his success (although he may have come to wish that he had lost!). In any event, Kabui was at the head of the government that was in office — if not in control — when one of his Nasioi neighbours, one Francis Ona, began dynamiting company property and went on to proclaim a 'government' of his own. But before chronicling those events, we will examine a set of demands which Father John Momis put forward during the 1987 election campaign for the national House of Assembly.

With apparently unbeatable electoral support, it is puzzling (to this outsider) why Momis felt it necessary to add a new element to his campaign. Or perhaps it was because of that support that he felt confident enough to put forth the audacious proposal known as the Bougainville Initiative. In any case, although that proposal received short shrift from the national government, it doubtless served as a model for Bougainvillians bent on increasing, to cargoistic magnitude, their demands against BCL. The substance of the Bougainville Initiative was contained in a letter, dated 4 May 1987, written and hand-delivered by Momis to Paul Quodling, who was on the point of retiring from his post of managing director of BCL.

The letter, five pages of closely spaced typing, began with a friendly 'Dear Paul' and an offer to him 'to crown your long career with the achievement of gaining the friendship of the people'; it ended with an implicit threat to close the mine unless its proposals were heeded. These quotations convey its flavour, although the tone and content may owe more to Momis' Australian speechwriter than to Momis himself.

> The fundamental truth is that BCL has colonized our people, it has taken their land, it has reduced them to passive dependence. Our people — who to you are just Bougainvillians, or even more anonymously 'nationals' — are now servants in a land where once they were masters. They keenly feel it, and deeply resent it.
>
> I do not come to you to cause you pain. As you will see, I have come to offer a fresh initiative, based deeply in our Melanesian culture, which can do much to resolve the impasse which frustrates us all. Yet the starting point must be an honest acknowledgement of the truth, even if it is painful.
>
> . . . The BCL mine has forever changed the perceptions, the hopes and fears of the people of Bougainville. You are the invaders. You have invaded the soil and the places of our ancestors, but above all, your mind has invaded our minds . . .

Leo Hannett.

Father John Momis.

Dr Alexis Sarei.

Bishop Gregory Singkai.

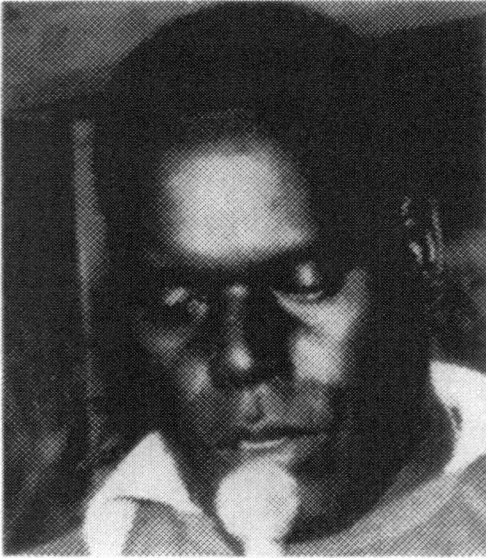

Francis Ona.

Sam Kauona.

Pylon toppled by explosive charge, 15 April 1989.

You say you are good corporate citizens. You pay taxes to the national government, you pay royalties, you trumpet your public relations stunts such as the Bougainville Copper Foundation, which doles out small sums to local sports clubs and the like. You build infrastructure, you compensate landowners. Yet those very monies are also very much the problem.

You have been so determined to take our earth and send it to Europe and Japan as quickly as possible, that you have created an operation on a scale which makes it overwhelming. Because of that massive scale you pour fifty million tons of our earth into the Jaba River every year. Because of that massive scale, you have made such massive profits that our economy has been reduced to colonial dependence.

You are not ruthless exploiters. You are the modern faceless corporation which takes care to observe local laws, and incorporates local people, locking them into minority shareholdings in trading operations you so generously call a foundation. The vast sums of money you make by selling our gold, silver and copper has created amounts of cash undreamt of before. And that cash has tragically seduced some of the best leaders of our people. It has robbed us of some of our finest men, who have ceased serving their people, and instead line their pockets with the money of BCL.

. . . Not only does BCL have almost nothing to offer that is relevant to assisting the self-reliance and dignity of our people, it actually saps the confidence, reducing our people to dependent wage slaves, cogs in a wheel. You have parked your bulldozers in our living rooms, yet you pretend that you can simply go about your business, and we go about ours, and never the twain meet. I tell you this is not so. . . .

The modern corporation does not obey the natural rise and fall of life and death, as does our Melanesian tradition, which distributes a leader's wealth when he dies. The modern corporation obeys the ideology of the cancer cell, to ever grow and grow, without ceasing.

The principal demand of the letter was succinctly put:

that BCL set aside each year three per cent of its gross income from selling our minerals, to be given directly to the North Solomons Provincial Government [as] untied aid, without strings. It is up to the government of the province . . . to determine how that money may best be spent to foster the application of the Melanesian entrepreneurial tradition to creating small-scale ventures which meet the needs of the people, and which return their benefits to the village people. . . . If we are to recover our confidence in ourselves, this is essential.

BCL's response to the Bougainville Initiative was explicit:

In responding to your letter of 4th May, I must first assure you that it has long been my objective to see a larger proportion of the revenue generated by Bougainville Copper Limited remain within the province. However, there are certain legal and statutory constraints under which I am bound to administer the financial affairs of this Company. These place your 'Bougainville Initiative' outside my authority.

That response was clearly not consonant with Momis' characterization of the Melanesian Way, but it was in accord with the principles, still in force, by which the company was required to operate.

(The Melanesian Way is a fluid concept, and has even less semantic substance than the American Way (e.g., equality, free enterprise, apple pie). There is even more cultural diversity among the thousand or so peoples inhabiting 'Melanesia' than the self-styled ethnic units (and social classes) inhabiting the USA. Morever, even the most ardent Bougainvillian devotees of the Melanesian Way differentiate sharply between themselves and, say, the New Guinea Highlanders. Nevertheless, the phrase continues to serve as a clarion call for Melanesian politicians engaged in advocating this or that local principle or practice against those of their former colonial governors.)

On 25 May Momis wrote to both Paul Quodling and Robert Cornelius, the retiring and incoming managing directors of BCL. The core of this message was the same as the first one, but it also contained charges that BCL was dong less than other PNG mines such as Ok Tedi and Misima in being 'sensitive to the local cultures', in localizing employment in the higher grades, in disposing wastes cleanly, etc. And this time the writer was not 'alone':

I have with me representatives of the local villages, Ministers in the Provincial Government with portfolio responsibility for the mine's impact on our culture. I also have with me the leader of the trade unions. You will also be aware that the [Catholic] bishop of Bougainville is keenly awaiting news of the outcome of these talks, to discover what progress has been made. Indeed, all people of Bougainville now expect concrete results, and your detailed response to our original Bougainville Initiative.

To cap this phase of the campaign a letter, dated 10 June 1987, was sent to Quodling by the province's Premier, Joseph Kabui. Although supporting the Bougainville Initiative it was neither condemnatory nor threatening. Instead, it called attention to the surge of people, including those from other provinces, who, 'attracted by the glittering Panguna jobs', had placed an intolerable financial burden upon the province in its efforts to provide schooling, health and other public services. The letter also charged that the national government was doing nothing to help the province to cope financially with its supernumerary costs.

Evidently, BCL officials were well aware of the inequities, as Quodling had acknowledged in his 11 May reply to Momis. Unfortunately for both the province and the company, the latter was legally bound by terms of its current Bougainville Copper Agreement to address such issues only with the national government during the periodic reviews of the agreement, which were prescribed to take place every seven years. (The reviews, however, had been indefinitely postponed, largely because of the national government's refusal to include representatives of the province in them.)

The grievances which gave rise to the Bougainville Initiative were real and substantial, as events were to prove. But before we move on to examine those events, it may be useful to review the general situation on Bougainville in mid-1988.

At that time Bougainvillian society as a whole was relatively quiet and stable, considering the periods of discord and political ferment it had recently experienced. As a whole, it was economically prosperous compared with most other provinces in the nation, and viewed with envy by some of them. However, there was restlessness among semi-educated youths unable to find work they were willing to do. There were also pockets of perennial disaffection among people whose living had been disturbed by mining. More diffusely, there was growing disgruntlement from the have-nots towards the haves: people who, for reasons of remoteness or poor education, etc., had not shared in the mine-generated affluence resented those who had. Enveloping all those discontents was the universally held sentiment that the province was not receiving its fair share of the revenues generated by the mine.

Most of those dissatisfactions had to do with money, but there were also resentments of other kinds, especially against the many mainland Niuginians in the province. Although the anger roused by the murder of Rovin and Moini had abated, Bougainvillians in general expressed anxieties concerning the many unemployed and semi-employed redskins who inhabited the urban centres, and envy concerning the highly paid non-Bougainvillian Niuginians in company employ. (There was, however, no resentment of redskins employed in plantation jobs, which Bougainvillians disdained!) In addition, there was growing apprehension by people living within or near the mine-lease areas regarding the visible or imagined effects of the mine upon *health* — of themselves, of the flying-foxes they sometimes hunted, and of the marine life they sometimes fished. Some of the province's indigenes felt strongly about one or another of these issues, but for the population as a whole the principal issue was money — specifically, how to obtain more of it. While most Bougainvillians continued to cherish some aspects of their 'traditional' life — i.e., what they selectively recalled, or imagined, about that life — nearly all of them desired some Western-type objects or conditions which only money could obtain.

Some Bougainvillians continued to believe that money and other kinds of 'cargo' would arrive, either through supernatural intervention, or in fulfilment of promises made by campaigning politicians. Others wished, or expected, to obtain money in the form of compensation from a company which was believed to be almost supernaturally wealthy; this compensation for losses or deprivations was susceptible to boundless definition and monetary value. Most Bougainvillians, however, had learned by mid-1988 that the money they wanted had to be earned by work of some kind.

Fortunately, there were in mid-1988 enough money-earning opportunities in the province to go around. These were not enough to permit all would-be money-earners to do the kind of work they preferred or to earn as much as they wanted, but there were enough ways to earn money. On the other hand, short of some change unforeseeable at the time, that favourable situation would not last. Even with the continuation and expansion of mining, the number of paid jobs in that capital-intensive industry would remain small relative to the province's current population; even smaller relative to its relentless rate of natural increase.

As for the province's other major money-earning activity, cash-cropping, in mid-1988 there was still enough arable land available to accommodate even a growing population for a decade or two (provided that would-be producers of some land-poor kin groups were permitted to grow cash-crops elsewhere). At the rate the population was then increasing, the time would soon arrive when a choice would have to be made between using the limited amount of arable land for subsistence gardening or for cash-cropping: either to produce all one's own food, or to purchase it with the earnings from cash crops. The latter alternative would remain viable only if the market for cash crops remained profitable and accessible, and the price of imported food not too high. Few Bougainvillians appeared to be aware of this predicament. And even those few seemed unwilling or unable to broadcast their concerns in terms other than the demand for 'more money' — from the national government or from BCL or both.

In summary, if the circumstances that prevailed in mid-1988 continued, Bougainvillians' per capita holdings in money were destined to decrease within the foreseeable future. Meanwhile the desire for more money per capita would likely increase, due mainly to the rising expectations for imported goods that usually accompany increases in level of Western-type schooling.

But of course, all such assumptions about the province's future were nullified, for at least several years, by a revolution whose non-revolutionary commencement took place in late 1988.

11

The Mine Shuts Down

ON 17 MAY 1990, LEADERS OF THE BOUGAINVILLE REVOLUTION-ary Army held a ceremony to mark the unilateral declaration of independence (UDI) of the Republic of Bougainville. This event is described in the following chapter, but we pause here to examine in detail the general conditions which led to the UDI.

As of this writing (May 1991) the Republic of Bougainville still exists — at least in the minds of its founders; but it is still very temporary and isolated. That it materialized should come as no surprise to anyone who has read thus far. The situation which made it possible had been present for a long time (for example, cultural ideologies, demographic trends, educational advances, ethnic differences, institutional jealousies); but of course even Melanesian 'revolutions' do not eventuate out of general conditions. Accordingly, the first task of this chapter will be to describe how particular manifestations of those general conditions came to lead some Bougainvillians to celebrate their exclusive Bougainvillianness with a non-Bougainvillian ceremony in a very non-Bougainvillian setting.

Perhaps the most fearful of those general conditions was the province's fast rate of natural population increase, at which the population is predicted to double every fifteen to eighteen years. Even the departure of most non-Bougainvillians (which was one of the revolution's objectives and which had been largely accomplished by June 1990) can have only slight effect on this inauspicious trend. There is no evidence that the founders of the Republic of Bougainville were aware of this issue — or if they were, that they would be in favour of addressing it. In this they did not differ from other PNG leaders, who have yet to adopt a population policy and who, according to one critic, 'wanted big families and nothing to do with any policy that limited the number of children'. (Oseah Philemon, *Post-Courier*, 17 May 1990) The long-range economic consequences of this population increase on Bougainville have already been mentioned. Its relevance to the events of 1988–90 is more particularistic: it resulted in the existence of a large cohort of youths who were educated enough to alienate them from their former village life and occupations but not enough for employment in urban jobs. In any case, such jobs were less and less available.

A second general condition underlying the UDI was the perception, held by many if not most Bougainvillians, that they differ significantly from other Niuginians—not only in skin colour (which is true) but also in customs and temperament (which would be impossible to demonstrate). With respect to customs, no two native societies in Papua New Guinea are alike in all details; the same is true of the twenty or so distinctive societies in Bougainville–Buka. In some of their customs some Bougainvillian societies resemble mainland New Guinea societies more than they do others in Bougainville. One example of this is the leadership institution of much of Buka, which is based on hereditary chieftainship; it possesses many more parallels with, say, the type of leadership found in the Trobriand Islands than with that found in central Bougainville. As for temperament, the distinction drawn by some Bougainvillians between themselves as 'peaceful' and 'progressive' contrasted with the 'savage', 'backward' mainland Niuginians would be impossible to sustain in view not only of episodes during the recent militancy but of recorded evidence from their pre-colonial past.

Another general condition (this one explicitly referred to in the independence declaration) was the grievance, only recently identified but loudly voiced, that Bougainville had been appropriated, first by colonial powers and then by the leadership of mainland Papua New Guinea, without the inhabitants' consent. Concerning the first appropriation, by Germany, the charge is entirely correct but largely irrelevant, there having been no 'Bougainvillians' as such a century ago. There were only members of small tribes, which were often at war with one another. One wonders how such a referendum on a proposed dependency would have been conducted. (*Yupela kanaka laik mipela masta lukautim yu?* (Roughly, 'Would you blackfellows like us masters to take care of you?') As for the inclusion of Bougainville within Papua New Guinea, there had been a token protest against that (see page 194), but it was not large or representative enough to matter, much less to prevail.

Other, more particular, antecedents for the UDI were the widespread complaints among Bougainvillians: (1) that Bougainville was the Territory's most neglected district in colonial times; (2) that Bougainvillians should have received much larger benefits from the nation's BCL mine revenues because the ore was in Bougainville, and because Bougainvillians had suffered excessively from the mining; (3) that Bougainvillians residing near the mine had profited much more from mining than those residing elsewhere; (4) that the elders of landholding matrilineages had received much larger portions of compensation payments than the younger ones; and so on. All this was true and legitimate, except perhaps for the second item; the province did receive a larger-than-average allotment of the nation's public services in road building, educational facilities, etc., but evidently it was not large enough to satisfy local demands. Whether or not this complaint had much substance, it was voiced loudly and frequently, not only by the provincial government but by BCL officials as well. As we saw in Chapter 8, the refusal of the central government, prior to 1980, to grant more concessions to the provincial government produced bitter resentment within the province and resulted in indefinite postponement of the long-overdue review of the Bougainville Copper Agreement.

Another particular circumstance behind the independence declaration was the sentiment of the province's Catholic clergy towards secular authorities in general and BCL in particular. This is not to imply that the bishop had officially advocated secession, but rather that he and some of his clergy were judged to have been less than neutral during the militancy that led up to the UDI. Moreover, someone described as 'representing' the Catholic Church was (is?) a government minister in the revolutionary republic. The role played by the bishop during the militancy will be described later on. The local clergy's attitudes towards BCL, had ranged from mild or passive neutrality to bitter outspoken hostility. The relevance of these sentiments to the UDI is that four out of five Bougainvillians were Catholic; that most of the province's recent political leaders were educated in Catholic schools, including some who were once priests; and that John Momis, the province's most influential politician and harshest BCL critic, still is a priest.

It will be recalled that in 1979 the 'elders' of the clans claiming ownership of land contained in the mine-lease areas organized themselves into a Panguna Landowners' Association (PLA), which thenceforth represented them in dealings with BCL. In the 1980 agreement worked out between the company and the PLA, compensation payments were fixed for each type of damage and part of such payments was required to be placed in a Road-Mine-Tailings Lease Trust Fund (RMTL). This fund is now to provide assistance to the land-owners in the form of loans for education, health care, the establishment of new businesses, etc., both during the lifetime of the mine and thereafter. Its trustees were also empowered to increase the fund through investment in profit-making enterprises not directly associated with its members.

During its first years the fund operated to the satisfaction of most of its members, making loans to them for building materials, transport, and new businesses; but after many of those loans remained unpaid, and uncollect-able (to a total exceeding K300,000), the fund's management tightened its lending policy by reducing loans to individual members and using some of its assets to purchase or buy shares in larger-scale and more promising enterprises, such as copra plantations and the Bougainville Development Corporation, which was sponsored by the provincial government. One-quarter of the fund's annual profits was used for reinvestment in such enterprises, three-quarters for donations to local (lease-area) activities, such as health centres, schools, clubs, road repair, but no loans were made to individual members of the trust. To anyone familiar with any known Melanesian Way, the reactions of the fund's rank-and-file members were predictable: they accused its managers of corruption ('eating the profits'), and worse. Younger members accused the older members, especially the fund's chairman, Severinus Ampaoi, who had been BCL's Village Relations officer for a number of years, and had since then become a successful businessman (which in the minds of many members made him doubly suspect).

Meanwhile, the PLA, within which the RMTL Trust Fund was established, was experiencing difficulties of its own. There was a revolt by some of its younger members against the older, more entrenched ones who had founded the association in 1979, and who at that time had been recognized as representatives of the groups—the matrilineages—which held full or residual titles to the various land estates within the mine-lease areas. At that time and for a number of years afterwards there had been no publicly evident opposition to such leadership: it was those founding elders who represented the association in dealings with BCL, and who were the recipients of compensation paid to their respective landowning groups. That harmony had, however, changed with time. Many of the young people became much more formally educated than their elders, and their social values and material expectations changed. No longer willing to accept their 'traditional' status as junior, and hence subordinate, members of their landowning groups, they voiced demands for larger shares of the compensation payments received by the older members of their groups. As result of their increased exposure to the cargoistic side of political dissent, they also increased their expectations about the size of the payments themselves.

The generation-based revolt exemplified in those changes in expectation was fuelled by the size of the compensation payments received in 1986, which were smaller than expected. Many members—especially the junior ones—were angered, and blamed the elder negotiators for the shortcoming. (In fact, the 1986 payments constituted an increase over those of previous years, but the reason why the increase was smaller than previously was due to the indexation built into the Bougainville Copper Agreement, and not because of bargaining failure on the part of the PLA's representatives.)

Dissatisfaction with the management of the RMTL Trust Fund was part of a wider discontent with the leadership of the PLA itself, which, as will be recalled, consisted of a small number of men who in 1979 were in their forties and who, because of their better education (for that era) occupied positions of leadership in their villages and landowning groups. Some were working for BCL, or had had experience in negotiations with BCL. And it was some of those leaders who had helped to establish and direct the operations of the fund.

During the PLA's early years its leadership acted vigorously and efficiently in dealings with BCL—including organization of a violent demonstration in 1980 which had prodded the company into fixing the more favourable compensation payments. Over time however, discontent with the leadership grew; the causes were manifold. One was the coming-of-age of a younger generation of members, many of whom had received more formal schooling than their elders. Another was a nationwide demand for larger shares of the revenues generated by foreign-owned enterprises, which in the case of the PLA was stimulated and legitimized by the Bougainville Initiative formulated by Father Momis. Linked with these was the perception, by junior members, that the current leadership was not pressing hard enough to obtain those larger shares – either because of indolence and incapability or because they had been corrupted by BCL. And finally, some of the discontent is traceable to the resentment felt by the junior members of some of the landowning groups towards their seniors because they had received smaller shares of compensation payments than they felt they deserved. The fact is that due to natural population increase the compensation payments had to be divided among more and more members. Moreover, as the younger members grew older their need for money increased, especially in the case of those unable to obtain wage jobs.

Money, however, was not the only source of the dissatisfaction with the PLA's leadership. Many if not most members of the landowning groups were becoming increasingly concerned with the visible or imagined hurtful effects of the mine: with the ever-increasing size of the pit (and hence the irretrievable loss of land); with the ever-increasing piles of waste rubble over once-arable land; with pollution known or believed to be emanating from the mine (causing, it was believed, poor crops, human illness, extinction of wildlife, etc.). The pit was indeed consuming more and more land day by day, and the company could not honestly give assurances that the land covered by waste material could be eventually restored to an arable state. As for pollution and its effects (some of which were aggravated by their invisibility), the dimensions and weightiness of the issue were inflated especially by younger members of the land-owning groups whose schooling had sensitized them to acid rain, rainforest depletion, and the like – concerns that their parents and grandparents had never heard of, much less shared.

Unfortunately for this chronicle there is to my knowledge no running account of how the various factors just mentioned interacted to bring about the partition of the PLA, but it is recorded that the breakaways held their first formal meeting on 21 August 1987, thereby becoming the *New* PLA. Twelve days later BCL received a handwritten letter from Mrs Perpetua Serero, chairman of the New PLA, demanding that all future dealings between company and PLA be conducted through herself and the association's new executive committee: 'Any slight move or involvement of your office with those out-going committees will result in a massive demonstration.' Mrs Serero was thirty-four years old at the time. (She died two years later.) She was a native of Guava village, and a first cousin of Francis Ona (who was secretary of the New PLA and about whom more will be heard). She had been among the first graduates of Goroka Teachers College; after teaching for a time she became the first female announcer with Radio North

Solomons. She was married to a Niuginian of Manam Island who worked for BCL.

On 29 September BCL received a letter from the original PLA, this one from Michael Pariu, who had founded the association seven years earlier and who was its secretary. He stated that he and his committee members were leaders in their respective' villages, had been 'traditionally' elected to head the PLA, and would continue to represent the association in all its dealings with BCL. As for the 'so-called' New PLA, it 'is manipulated by so many members with personal problems, some Union members and those expelled from the BCL' who, he added, do not represent the whole of the area included within the (true) PLA.

Between receipt of these contradictory letters, the company's managing director (Bob Cornelius) and its manager for personnel services (Joseph Auna, a native of south Bougainville) were summoned to a meeting with the province's premier, Joseph Kabui (then in his early thirties, and a member of a land-owning group in the upper tailings lease area). Kabui informed the company officials that he had attended the organizing meeting of the New PLA, along with the MHAs Momis and Bele, that his provincial government had given official recognition to the New PLA, and that BCL would be 'playing around with fire' if it did not do likewise.

Placed in a dilemma between the obligation set forth in its legally binding compensation agreement with the original PLA (which had been sanctioned by the national government) and the demands and veiled threats emanating from the New PLA and the provincial premier, the company, following the Western rather than the Melanesian Way, continued to deal with the old. One factor in the decision was lack of information about the membership and constitution of the New PLA. The company requested such information from the premier and from Mrs Serero, who did not respond.

During the next few months the company continued to be bombarded with demands and threats from the New PLA, culminating in a mine-closure ultimatum, which demanded:

- that the company pay the New PLA K10,000 million, for 'all the resources destroyed on [their] land, from 1963 to 1988' and thereafter pay 50 per cent of all its profits to the landowners and the provincial government;
- the the company consult with landowners before undertaking any new projects;
- that BCL become a local company belonging to the landowners and the people of Bougainville after five years;
- that the national government return 50 per cent of all money received from BCL from 1972 to 1988 to the provincial government; and that
- the provincial government build roads in all mining and tailings lease areas; fund all village government offices; establish services such as water supply to all villages; etc.

The times specified for response to or compliance with these demands ranged from twenty-four hours to 'as soon as possible'. Clearly, 'time' was becoming a new, unprecedented factor in this particular Melanesian Way.

In the event, the company did not pay the demanded K10,000 million so representatives of the New PLA set up road blocks on the mine access road on 17 May 1988. The barriers were promptly removed by police, and for the next few months, which were punctuated by further demands and mine-closure threats, overt acts of physical militancy subsided.

The PNG government, in the manner of governments, instituted an independent inquiry into the circumstances of the mine. In response to 'strong representations' from landowners in the mine area, it employed Applied Geology Associates Ltd, a New Zealand firm of independent consultants. Their brief was:

- to determine the overall impact that mining operations at Panguna have had on the social and environmental aspects of the area; and
- to determine the likely future impacts of continued mining operations at Panguna on the environment and the people in the area, taking into account the (nearly completed) tailings pipeline.

The areas covered in the study included those of the Special Mining Leases, the tailings lease, the road lease, the port, Empress Augusta Bay, and all relevant villages in surrounding areas. The field investigations took place in November 1988 and were carried out by a team of six highly qualified scientists, including a geographer, John Connell, who had spent many years engaged in sociological and economic research on Bougainville. The team's members travelled widely and conducted public hearings in many villages. The topics investigated ranged from soil and water analysis and terrestrial and marine biology, to health, economics and social conditions of the mine-affected populace. Special attention was devoted to grievances voiced by villagers during the survey, to those listed by village leaders in their previous communications to BCL, and to the positions of the provincial and national governments (including Momis and Bele). No attempt will be made here to summarize the AGA's findings and recommendations, which make up a densely worded document of two hundred pages. This writer is not qualified to evaluate the report's scientific findings about, for example, the amounts of chemical pollutants in rivers, or the possibility of revegetating waste dumps. But it does not require a qualification in, say, chemistry or medicine to detect that some of the authors did not write as they might have for professional journals subject to peer review. The report's bias against BCL which some readers might discern here and there is nothing to compare with the vituperative charges of the New PLA and its political supporters, though it is somewhat out of keeping in a scientific report. Thus, while statements by landowners tended to be accepted, or at least reported, at face value, some of those by company officials were queried or ignored. All in all, however, the report contains masses of information and a host of what could have become valuable recommendations. In some respects it is a useful blueprint for amelioration; unfortunately, it came to serve as the spark that ignited the militancy which turned into a revolution. It did so, not because it was somewhat critical of the company, but because, in the feelings of certain Bougainvillians, it was not nearly critical enough.

Upon completion of their field investigations, the members of the AGA team held a meeting in Arawa on 18 November 1988 to report their preliminary findings. During the meeting some of them stated that, although mining operations had resulted in extensive damage to the physical environment, they had found no significantly high levels of chemical pollution. They described as unlikely the opinion held by many Bougainvillians that BCL was responsible for the decrease in wildlife and the decline in soil fertility (except of course in the pit and waste-dump areas), or for certain illnesses then prevalent in the lease-area villages. Predictably, many of the landowners in the audience disagreed vehemently. Francis Ona stormed out of the meeting, after calling the survey a 'white-wash'. (Thereafter Ona did not return to his job with BCL, where he was then employed as a driver.)

That evening, or shortly thereafter, Ona was interviewed, in Pidgin, by the National Broadcasting Commission. (Radio receivers, either centrally powered or battery-charged, were to be found even in the most remote hamlets, and programming included news of local events.) Claiming to be acting as spokesman for the New PLA, he said that he and his associates had no confidence in BCL or in the national government; that the AGA scientists (whose investigations, he said, he and his associates had requested) had wrongly exonerated BCL; and that in the absence of other measures to protect the lives and welfare of Bougainvillians, the New PLA had decided to close the mine. Included in his remarks were many bitter charges against what he described as the affluence and racist attitudes of resident expatriates in general, and of Australians in particular. One alleged transcript of the interview also quotes Ona as having said that 'today my people have decided to break away from Papua New Guinea and be an independent state on our own', but a report by one listener asserts that Ona's remarks contained little secessionist sentiment, either expressed or implied. What Ona really said remains something of a mystery, inasmuch as the tape of the interview has 'disappeared'.

Three days later several landowners, including Ona, blocked a road leading to a company facility, but a representative of the provincial government persuaded the crowd to end it a few hours later. A more ominous action occurred on 22 November, when three hooded men armed with axes and a knife broke into a BCL explosives magazine and held its guard at knifepoint while they made off with a large quantity of explosives. A week later more explosives were stolen from the Bougainville Limestone Mining Company at Manetai. Then during the night of 26 November one of the pylons supporting the power lines from Loloho to Panguna was toppled with explosives, effectively interrupting operation of the mine's ore-processing plants. Other acts of sabotage had been occurring during the previous weeks, but this was the first one serious enough to carry out the threat to 'close down the mine'.

On the previous day BCL's managing director, Bob Cornelius, attended a meeting, convened at Port Moresby at his request, of the Minister for

Minerals and Energy, Patterson Lowa, the Minister for Provincial Affairs, John Momis, and the premier of the North Solomons Provincial Government, Joseph Kabui. According to the minutes of this meeting, the government officials were more interested in reiterating the usual grievances against the company than in the threat posed by the theft of explosives. However, when the meeting resumed the following day, after news of the pylon toppling, there was a noticeable change in the government mood; whatever the ministers' personal feelings may have been about the company, the mine's continued operation was essential to the financial welfare of the nation. Suggestions by the provincial premier to 'forgive and forget' if the rest of the explosives were returned, and by Lowa to offer inducements to the saboteurs if they would consent to cease and desist, were rejected. It was decided to send more police to Bougainville to end the militancy, and to have Momis 'talk to' the landowners. Shortly thereafter Momis counselled the landowners to avoid physical violence, but he also accused the company and the national government (of which he was and still is a minister) of disregarding the 'people's wishes'. (*Niugini Nius*, 30 November 1988)

On 4 December power was restored to the mine, but on the following day it was cut off again when a second pylon was toppled. That break was also promptly repaired but, in view of those and several other acts of threat and arson, company officials decided to cease operations, to avoid further risks to company personnel and property, until the national government gave assurances that it was safe to do so. That assurance was received from the Prime Minister, Rabbie Namaliu, on 10 December, and the mine resumed operations the following day. Meanwhile, the 120 police sent to Bougainville to assist the local police had made a few arrests of suspected saboteurs. The ringleader, Ona, had effectively 'gone bush'.

Francis Ona was born 15 February 1953. It has been impossible for me to establish facts about his pedigree except that he was a first cousin of Mrs Perpetuo Serero, the New PLA's chairman, and that the only land rights he had in his home village, Guava, were of a secondary nature, having been received through his father. Whether he had inherited any other land rights in the mine-lease area through his mother (which would have entitled him to primary rights) I have been unable to learn. In any case, it was his rights in the Guava land identified with his father's lineage which were the focus of his grievance. Ona charged that Matthew Kove, the brother of his father and 'traditional' leader of the father's matrilineage, had not given Ona (nor the other secondary rights holders) a large enough share of compensation and other payments received from BCL. Added to that was the complaint that his uncle Kove had not tried hard enough to obtain additional money and other benefits from BCL. It was such grievances that had led to the formation of the New PLA.

My information concerning Ona's early schooling is limited. He received professional training in surveying, and was employed as a mine-pit surveyor by BCL in November 1976, at the age of twenty-three. After ten years in that job he applied for transfer to pit operations as a haul-truck operator. (According to company records, he had not been performing well as a mine surveyor, and was 'counselled' by his supervisor after one of his survey assignments had led to 'problems' in the pit.) The transfer involved a slight downgrading in his company housing, which he accepted — 'reluctantly', it is recorded. He continued to work as a driver until he 'resigned' after the meeting with AGA in November 1988. Meanwhile, he had married a woman from Madang and had three children; his wife was in Madang expecting a fourth when Ona 'went bush'. For a postscript to this woefully scanty resumé: soon after the outbreak of sabotage Ona's uncle, Matthew Kove, one of the leaders of the original PLA, disappeared, and is reported as having been murdered in a 'traditional' way. One observer of PNG mining situations wrote that

> . . . [compensation] 'deals' done with one generation of landowners, or their leaders, will be repudiated by the next generation regardless of the manner in which deals are negotiated. . . . the authority of [the elder generation] leaders within their communities may well be undermined by the very fact of their having been party to some previous agreement.

The author of this sagacious judgment is the anthropologist, Colin Filer; its applicability to Kove's murder is poignantly on mark.

Ona and his fellow militants were described by Premier Kabui as being 'mostly young people aged between 18 and 24 years'. (*Post Courier*, 24 January 1989) They disappeared into the bush, reportedly to seek shelter (or join forces?) with Damien Damen in the heavily-forested mountainous Kongara region southeast of Panguna. Damien, it will be remembered, was leader of the cargoistic Fifty Toea Movement, which disavowed all official forms of government: council, village, provincial, and national. Damien was a magnetic and forceful leader, and it has been reported, but not confirmed or circumstantially described, that he exercised a strong influence over Ona and his followers. It may be assumed that he played an influential role in transforming the militancy from anti-BCL *sabotage* to anti-PNG *secession*.

During the rest of 1988 and the first few weeks of 1989 acts of sabotage ceased, due perhaps to the presence in and around Panguna of 400 police, and the imposition of a dusk-to-dawn curfew. However, the police were unable to track down Ona, and he did not respond to their threats of 'surrender or die'. Meanwhile, other prominent Bougainvillians began to take sides. David Sisito, MHA for Central Bougainville, described Ona as 'a man of dignity among his own people, fighting for his land for the benefits of the future generation'. To this he added a blast at BCL and the national government for 'failing to control pollution', which 'has now reached [an] uncontrollable stage and if BCL is allowed to do more mining the province will become a living hell'. (*Post-Courier*, 2 February 1989) A sharply contrasting view was expressed by Lawrence Daveona, a native of Guava village holding primary rights in some of the lands with which Ona was associated. Daveona, who holds a bachelor's degree in economics, grew up with the mine and was well versed in the legal and political side of its history. His view was expressed in an interview published in Port Moresby's *Post-Courier* on 3 February 1989. This writer does not know how representative were the views expressed in this interview, but they differ markedly from those expressed by Sisito and many other Bougainvillians. In brief summary, it defends BCL for recognizing the need to review the Bougainville Copper Agreement and for attempting to include the provincial government in the review process. It condemns the previous national government of Julius Chan (which included Momis and Bele) for refusing to include the provincial government in the review process. It also condemns the provincial government for misusing funds allocated to it by the national government, and criticizes 'a good number of prominent members of our community' for, among other things, damning BCL without having even read the agreement by which BCL is required to operate.

The charges and counter-charges were also exchanged at higher levels, where, for example, Julius (now Sir Julius) Chan asserted that Momis 'should shoulder the bulk of the blame for the Bougainville crisis'; this moved the latter to respond with an attack on Chan for his indifference to the plight of 'simple villagers', while enriching himself by shady deals. (*Post-Courier*, 24 January 1989) While these exchanges were taking place, and while Police Commissioner Paul Tohian, in charge of the police force ranged against the militants, was issuing challenges to the militants to 'surrender or die', Prime Minister Namaliu, and North Solomons Premier Kabui were appealing to Ona to come out of hiding—not to be shot or jailed, but to 'negotiate'. In response to these invitations, the one threatening, the other entreating, Ona remained in hiding and issued another set of demands:

- that Bougainville Copper Limited pay the K10,000 million compensation demanded by landowners, *and* immediately shut down its operations;
- that PNG adopt a new economic order (because, he stated, the country's economic base is controlled by a white mafia); and
- that North Solomons Province break away from PNG unless the new economic order is adopted. [*Post-Courier*, 24 February 1989]

In the meantime the Crisis was being complicated by the tactics used by the police to enforce the curfew and to press the search for Ona and his followers. Some of the curfew-breakers were arrested and, according to some reports, very brutally handled. Houses were burned—how many cannot now be established—and some of them were apparently looted. Charges were also made of rapes having occurred; whether or not the charges were true, talk about them and other alleged atrocities spread widely, and served to increase popular support for Ona and his men.

Police action failed to end the militancy. In response to mounting political pressure in the national parliament, the national government offered the unhappy landowners in particular, and Bougainvillians in general, special benefits which came to be known as the Bougainville Peace Package. After meetings with representatives of the North Solomons Provincial Government and of BCL, the offer — directed mainly to the dissident landowners — contained these elements:

- sale at cost price (half to the landowners, half to the provincial government) of 4.5 per cent equity in BCL, to be paid for from future dividends, plus another 5.1 per cent at market price (the shares to come from the 19.1 per cent equity owned by the national government);
- a higher annual compensation grant to the provincial government of K5 million; and
- a national government-financed development programme, including the construction of roads, schools, and health facilities. (*PNG Times*, 4–10 May 1989)

As its contribution to the package, BCL made the commitment to undertake a number of construction projects desired by landowners in the mine-lease areas, at a total cost of approximately K41.5 million.

During the months required to secure the agreement of all parties to the proffered package, a number of other, more decisive, events were occurring, including an intensification of the militancy, which resulted in the shut-down of mining in September 1989. But to follow the Peace Package further along its tortuous way (and to reveal the militants' contempt for its offerings): two days before its scheduled ceremonial signing, set for 12 September 1989, militants shot and killed John Bika, a minister in the provincial government who had played a leading role in formulating the final terms of the package, and who was to be one of its signers.

While the national and provincial governments were putting together their offer, the Catholic Church on Bougainville continued its efforts to end the militancy — which, according to some of its critics, certain members of the clergy had actually abetted. Bishop Gregory Singkai, a Bougainvillian, was criticized in some quarters for having publicly condemned acts of violence by the police, but not those of the militants. During the early stages of the militancy the bishop and other clergymen in his diocese had met Ona and talked with him in his bush hideout, but to no effect. Eventually those more or less 'personal' attempts at persuasion were superseded by the clergy's efforts to conduct government-authorized mediations, armed with three 'carrots': (1) a comprehensive programme aimed at meeting landowners' 'legitimate' demands; (2) removal of the police and Defence Force personnel from the province; and (3) amnesty and reconciliation for Ona and his followers. (*Niu-gini Nius*, 19 April 1989)

The first of these proposals was taking form as the Bougainville Peace Package; the second, the withdrawal of government forces, did in fact take place almost a year later. As for amnesty for the rebels, it was hotly discussed during 1989 and 1990. The Prime Minister offered it several times, but individual members of his government, and of the Opposition, argued for or against it. In the peace talks that were announced at intervals over the next twelve months, but not actually held until July 1990, Ona and his principal military commander, Sam Kauona, were offered safe passage but no unequivocal promise of permanent amnesty. In fact, with the intensification of the militancy (accompanied by mounting criticism of government inaction) the 'surrender or die' signals from some national government officials reached a crescendo at one point with the offer of a K200,000 reward for Ona's capture.

Very little has been published or broadcast about the inner workings of the militants, and about the course of their militancy. After 'going bush' towards the end of 1988, Ona and his followers — the numbers of whom are not known to the author — made for the Kongara region southeast of Panguna, which remained their base until March 1990. During the early months of the militancy, some trained military men joined its ranks. Reports reaching Port Moresby in 1989 stated that 'former PNG Defence Force soldiers' had joined Ona's saboteurs (*Niugini Nius*, 29 December 1989); their number is not specified. One of them, Samuel Mabiy Kauona, born circa 1962, was to become well known.

This writer's knowledge about the 'elusive', 'enigmatic' Kauona is limited to brief newspaper reports. According to one of them, he was 'from' (born in?) Marura village in the Kongara region; he had been a platoon commander in the PNG Defence Force; and in March 1989 he had gone on leave to Bougainville. There, it is reported, he told the premier, Joseph Kabui, that he wanted to 'help his people', which he then did by joining Ona at the latter's base in Kongara. (*Post-Courier*, 29 December 1989)

An article in the *Sydney Morning Herald* (26 January 1990) provides some details about Kauona's life in the PNG military:

> Sam Kauona first came to Australia in the mid-'80s, chosen by his military leaders to complete the army officer training course . . . He completed the course and graduated . . . Military sources have told *The Herald* that Kauona was a model soldier when he returned to PNG. He was already marked for leadership and was quickly selected to undergo further training in Australia. In 1988 he completed the other course which has served him so well in his jungle war against the mining company and the security forces. It was the 18-month Ammunition Technical Officers Course, run at Bandiana Army Camp near Wodonga in northern Victoria. That course deals mainly with the storage and handling of all types of explosives . . .
>
> Military sources in Australia say Kauona was different from the average trainee they deal with from PNG. Most are quiet, shy and

overawed by being here, but Kauona was an extrovert, always smiling
and talking to everyone, and more than willing to join his classmates at
the pubs, dances and discos . . .

He's been described as having a 'touch of the larrikin', but also as 'the
sort of guy who would organise and lead troops at the PNG military
level'. The same people also call him calculating. 'He would smile and
shake your hand but you could never tell what was going on behind his
eyes' . . . During the explosives handling course he did tell a colleague
that he wanted to learn all about demolition so that he could go to
work in the mines on Bougainville. This is one ambition he has achieved
with a vengeance.

And from the *Australian* (20 January 1990):

Sam Kauona was drunk the night he told an Australian Army colleague
he would desert the Papua New Guinea Defence Forces as soon as he
graduated from the course that taught him all about explosives. [That
same colleague added] . . . Kauona, though not the quietest of the 15
officers on the course, but a fairly innocuous figure: the pastor's son
attended church and read the Bible each night . . . The part of the course
that really interested Kauona involved training in setting small charges
to blow up ordinance that had deteriorated to a dangerous state:
Bougainville is still littered with unexploded bombs and shells from
World War II.

Clearly, Sam Kauona was just what the militancy needed. At one point
a rumour circulated that Ona had been murdered, or put under house arrest,
and Kauona had taken charge. In fact, there was one period when all communications coming from what had become known as the Bougainville
Revolutinary Army (BRA) were issued by its 'Commander', Sam Kauona;
in due course Francis Ona's name began to reappear, this time as 'Supreme
Commander', and Kauona is reported to have stated, 'Francis is in charge;
I just work for him'.

In addition to Kauona (and an unknown number of other former soldiers),
numerous other male Bougainvillians began to join the militancy in early
1989. It is generally agreed that most of them were young, and that many
of them had joined in reaction to the rough and destructive behaviour of
the national forces, including the police. What is not known is their number
at any one time, nor the degree to which all of them had 'joined'. There
were numerous reports of small groups of BRA, or of those calling themselves BRA, operating far away from the Kongara stronghold and the central
Bougainville 'battle zone' — stealing vehicles, ransacking dwellings, threatening and strong-arming civilians. They were behaving in a way commonly
attributed in PNG to *raskals*. Such behaviour commenced even during the
presence of the national forces — but well out of their reach. It increased so

much after the national forces had left in March 1990 that Ona, evidently concerned about the loss of popular support, announced on radio that he would surrender himself if they did not cease. I cannot discover what effects this appeal had.

The militants continued their attacks against BCL facilities and personnel throughout April and May of 1989, despite the presence of nearly six hundred national police and soldiers. In April Ona proclaimed a 'Republic of Bougainville' and demanded withdrawal of the PNG forces from what was to them a 'foreign land'; he added, 'We are not part of your country anymore. We have told our Premier [Kabui] but he failed to let your government know'. (*Nuigini Nius*, 12 April 1989)

During the first half of May 1989, militants made several attacks on BCL personnel and property around the periphery of the mine, including a bow-and-arrow attack on grass-cutters (which severely wounded two of them). As a result, on 15 May company management ordered all work in the perimeter area to stop. On the same day, in concern for their own safety, members of the two mine unions met and resolved to cease work for five days, and then to resume work only during daylight hours. This posed a double difficulty for company management: it could not continue to operate profitably on a daylight-only schedule (i.e. part-time); even if it were able to do so, it could not assure the safety of workers, essential to operations, who were posted outside Panguna itself.

All this was of course anguishing to the national government, not least because of the consequent loss of anticipated revenue. After receiving messages from the Prime Minister urging speedy resumption of operations, and promising the deployment of additional protective forces, the company management proposed to its unions to reopen on 21 May. On that and the following day two more pylons were toppled, and a number of company vehicles were fired upon, resulting in injuries to eight employees, the wounding of a visiting CRA employee, the shooting at employees at the Panguna school, an armed hold-up nearby and so on. This was weighty evidence that the militants' intelligence network was efficient and that the militants themselves did not wish mining to resume.

It did not. A day or so later the company made an attempt to reconnect the power transmission lines, but that also was abandoned as result of an attack on a pylon nearby.

Undeterred, the national government continued to formulate and promote its Bougainville Peace Package, evidently hoping to win over the non-militant landowners and thereby move them to repudiate the militancy. As an added inducement, on 25 May the PNG Executive Council declared a fifteen-day truce, during which the militants were guaranteed immunity from harassment and arrest. This would enable them to meet representatives of the national and provincial governments, church leaders, and BCL, and to negotiate a settlement 'in a spirit of reconciliation, compromise and peace'. (*Post-Courier*, 25 May 1989) According to the government's announcement,

the representatives of the militants were to include Ona and the cult leader
Damien Damen!

Predictably, the parliamentary Opposition leader, Paias Wingti, objected,
calling it 'the worst decision ever made by any Government both before and
after independence in PNG'. He went on to condemn the government's
'unethical decision to give immunity to those criminals in Bougainville'.
(*PNG Weekend Nius*, 27 May 1989)

Unpredictably, the elusive Ona was agreeable to the truce offer and the
opportunity to negotiate — or so said Bougainville's Catholic bishop, the Most
Reverend Gregory Singkai. However, the bishop added, Ona's acceptance
was conditional on a guarantee of his own safety, and the withdrawal of
the government's forces from Bougainville. (*Post-Courier*, 29 May 1989)
Shortly thereafter Ona announced, through Bishop Singkai, three more
conditions: that the mine remain closed; that a referendum be held to decide
whether Bougainvillians wished to secede from PNG; and that Ona be
granted, not only safety during the negotiations, but immunity from perse-
cution later on. (*Post-Courier* 1 June 1989) The good Bishop relayed Ona's
message to Prime Minister Namaliu, who characterized Ona's 'willingness
to talk' as a 'major breakthrough'. (*Post-Courier*, 5 June 1989)

Alas, the 'break' did not come 'through'. In his reply to Ona, Namaliu
agreed to grant the requested immunity but refused to withdraw the security
forces and rejected the demand for a referendum. Ona remained in his
mountain stronghold. Namaliu with new and uncharacteristic firmness,
announced:

> I am extremely disappointed that Mr Ona was not willing to come out
> and discuss his grievances in a peaceful and amicable manner as my
> government has given him ample opportunity. I regret that Mr Ona has
> displayed an attitude of not being genuine in his desire to talk to us and
> resolve this longstanding dispute. The national government therefore has
> no option but to bring down the full force of the law to deal with those
> who are responsible for acts of terrorism and sabotage. [*Niugini Nius*,
> 8 June 1989]

These decisions were announced in 'an Address to the Nation', which also
emphasized the government's wish 'to enable the mine to resume as soon
as possible' in order to 'maintain the strength of our economy, our capacity
to be a reliable exporter, and to enable all levels of business and industry to
function *no matter who owns them* [italics added].' Namaliu went on: 'Bougainville Copper remains our largest individual contribution to revenue. A cut
in revenue means a cut in revenue to you, and cut in jobs in the public service,
and even in essential areas such as health and education.'

Despite this show of verbal firmness, the Opposition continued to criticize
the government for its handling — or non-handling — of the Crisis, and added
the specific charge that Cabinet Minister John Momis and his Melanesian

Alliance Party were playing 'double standards' in efforts to resolve the crisis. (*Nuigini Nius*, 9 June 1989) The same newspaper report stated that 'many Opposition Members believe Father Momis is behind moves with the Panguna landowners to close down the mine for good.' In reply, Momis is quoted as saying: 'I do not want to see the Bougainville Copper Mine closed permanently,' [which is not the same as 'I want to see BCL reopen the mine as soon as possible.'] And, he added, 'I do not support secession.' (*Niugini Nius*, 14 June 1989)

The Prime Minister's Address to the Nation contained some harsh words for the militants — for 'killing innocent people', for committing 'acts of sabotage and violence', etc. Less noticeable, perhaps, were his words addressed to the security forces:

> I want to appeal to our security forces in North Solomons Province to protect the safety of innocent people and to use their powers in accordance with the law. We look to them to honour the very best traditions of our disciplinary forces in discharging their duties, and in bringing about a return of peace and harmony, and the rule of law.

From all accounts that appeal was overdue, and remained ineffectual. Mention has already been made of the harsh measures used to enforce the curfew, and of the burning of whole villages containing individuals suspected of harbouring or assisting militants. There is fairly credible evidence that several persons were arrested on suspicion of assisting the militancy, and then were killed outright, or fatally wounded while in custody (for example, a report in the *Sunday Morning Herald*, 29 January 1990). Other reports, of varying credibility, told of rapes and the killing of women. The most prominent among the victims were the provincial premier, Joseph Kabui (who was 'violently' assaulted by security forces) and a provincial minister, Michael Laimo (who was battered so badly that he lost an eye). There was also the unnecessarily rough handling suffered by many of the thousands of villagers evacuated from the battle zone around Panguna; the evacuation was intended to isolate Ona's forces and to protect the evacuees from cross-fire. There were doubtless many reasons (some of them understandable, but none of them good) for the behaviour of the security forces towards non-combatant Bougainvillians — not least was the ever-widening, mutual gulf between blacks and redskins (who were preponderant in the national forces).

The security forces were unable to accomplish their primary objective on Bougainville, for reasons that were hotly debated. In fact, their presence on the island appears to have had the effect of strengthening the militants' cause among the populace at large.

We can gain some perspective on the speed of the transformation of the militancy, from verbal protest to bloody insurrection, by taking a last look at the courageous young woman who helped start it all. On 15 June 1989 Perpetua Serero at age thirty-six died of cardio-respiratory failure and chronic respiratory disorder in the Arawa General Hospital. It had been less than two years since she had attended the organization meeting of the New Panguna Landowners' Association and had been elected its chair. During the next fourteen months she had written several letters to BCL (protesting against compensation payments, pollution, etc.) and had participated in one or two peaceful demonstrations; but she had never 'gone bush' with her cousin, Francis Ona. Two days before her death she was visited in the hospital by two members of the PLA, who quoted (probably paraphrased) her as having said:

> I am only a lady and worse still a sick woman. I will silently support whatever course the two of you take to achieve something tangible for our people [from the Peace Package]. We must get something for our people, otherwise all this fighting will be worthless . . . There is no more communication with Francis Ona and we do not know what Damien Damen has done to him. I fear secession and what it might bring. I do not want the whole province to blame us the landowners when and if we do [secede] and the secession does not deliver the goods to our province. The whole crisis . . . has developed into something totally different from the original demands of the landowners. . . . *Mi poret* [I am afraid] because what we have started has developed into something else . . . Our uncle Matthew Kove, nobody tells us what has happened to him. If he is dead, then his death must not be in vain . . . [*PNG Times*, 16 June 1989]

(As mentioned earlier, it appears that Kove had been in fact murdered, possibly at the behest of his nephew, Ona.)

We turn now to examine some of the strains that existed in the national government, which was formulating policy to deal with the Bougainville Crisis.

Prime Minister Rabbie Namaliu generally took a middle way in his responses to events on Bougainville, but he issued stern rebukes to the militants at times. Michael Somare, Foreign Minister and former prime minister, was also regarded as even-handed, and was trusted sufficiently by both sides to be nominated as a negotiator for the eventual peace talks.

The Deputy Prime Minister and Minister of State, E. R. 'Ted' Diro, is regarded as a hardliner, and his actions seem to confirm the view that he was one of the chief hawks in the Cabinet. He was reported to be angered at the vacillation of his colleagues in dealing with the Crisis. Others who took a hard line against the rebels were Benais Sabumei, Minister for Defence, and Matthew Ijape, the Police Minister.

Father John Momis, Minister for Provincial Affairs, had ambivalent loyalties. He supported national unity *and* Bougainville autonomy; he was opposed to BCL, but he also rejected secession. Julius Chan and Leo Hannett accused him of duplicity, and the BRA militants rejected him as a mediator.

Bernard Narakobi, the Minister of Justice and Attorney-General is a votary (and indefatigable re-definer) of the Melanesian Way. His frequent reverses in political opinion and mood cannot have simplified the Prime Minister's attempts to make and execute official policy for Bougainville. In one breath he attacked the militants, and Bishop Singkai for championing them; in another, he called for expelling BCL, nationalizing the mine, and taking over all its shares — 49 per cent to go to the national government, 51 per cent to the landowners and provincial government. The Prime Minister's reaction was to say: 'I will be gently reminding the Minister that such comments should be confined to the Cabinet and Party rooms at this time.' (*Niugini Nius*, 14 February 1990) As that journal commented: 'The statement by Mr Narakobi has already caused differences within the Cabinet ranks, which are expected to widen.'

The loyal Opposition, under Paias Wingti, kept up a drum-roll of criticism of the government for its handling, and non-handling, of the Crisis.

Notwithstanding the tensions inherent in the Westminster system, and the shouting evidence of its failure to resolve the Crisis, the government continued to survive and even increase its voting strength. To outsiders such as this writer, this is difficult to understand, much less explain.

Three of the most hawkish Ministers — Sabumei, Ijape, and Diro — made a visit to Bougainville on 27–28 June, and satisfied themselves that 'everything' was satisfactory and proceeding according to plan.

In July the national security forces occupied Guava village, where they found a network of defensive trenches, a cache of weapons, and a number of Defence Force training manuals. Up to this time, direct encounters with militants had been infrequent, and had resulted in few casualties on either side. More than six hundred people had been evacuated from the battle zone, and most of them accommodated in 'safe' villages or in the urban areas. (*Post-Courier*, 10 July 1989) At this time the security forces consisted of 500 soldiers, 200 riot police, and an unspecified number of Correctional Services officers. They were under orders to end the militancy as quickly as possible.

In August BCL's manager, Bob Cornelius, met with the Prime Minister, who assured him that a strong security force would be maintained where needed to permit repair and restoration work to be carried out, in order to reopen and maintain operation of the mine. The Prime Minister is quoted as saying, 'Additional security forces will be sent to the province for the restoration period.' (Memorandum from office of the PM, 1 August 1989) On the same day, the Minister for State, Ted Diro, announced to the press that the government had already reduced the Defence Force on Bougainville to 400, the police to 120, and the Correctional Services officers to 175. The reason, he said, was to let the militant landowners 'retain some honour and glory'. He added, 'If you defeat them in detail they are likely to come back after 12 months or two years . . . The situation lies in the hands of leaders of North Solomons.' (*Post-Courier*, 2 August 1989) To which the newspaper's reporter added, 'Mr Diro appeared frustrated with his fellow politicians and even Cabinet colleagues.'

Meanwhile the number of evacuees from the battle zone had climbed to nearly 4000. Most of them could be accommodated in tents or housing left unoccupied by fleeing non-Bougainvillians — but the line had to be drawn *somewhere*! As reported in the *Arawa Bulletin* of 8 September 1989:

> a member of the Emergency Evacuation reported that the Arawa Country Club will be used to accommodate about 300 villagers from the Jaba area for a minimum of one week. The committee member said that displaced villagers will not be allowed to go on the golf course and the tennis courts. The appropriation of the Arawa Country Club has brought a wave of protest from the Club Committee and members. A letter expressing extreme concern and regret has gone to the Prime Minister. . . . The letter says, 'We feel strongly that all the alternatives have not been considered. The closure of the ACC is seen as a step backwards as it has always been said that the community should remain as normal as possible. This cannot be with the club used as a care centre.'

Without doubt, war is hell!

Meanwhile Prime Minister Namaliu stepped back from the uncompromising position towards the militancy expressed in his 9 June Address to the Nation, and, on 26 August offered safe passage to Ona and his key men (including Sam Kauona and Damien Damen) to a meeting with government representatives, in order to discuss the Bougainville Peace Package. A vocal reply was given to a reporter, not by Ona but by Kauona, who stipulated that any such 'dialogue' must be concerned with a referendum on secession. (*Arawa Bulletin*, 15 September 1989) A more direct reply was the murder on 10 September of John Bika, Minister for Commerce in the provincial government. As reported earlier, Bika had played a leading role in formulating the peace package. His crime, according to a statement attributed to Ona, was to 'attempt to destroy support for Ona and for secession'. (*Post-Courier*, 13 September 1989)

A two-way dialogue between the Prime Minister and Ona is reported to have taken place at the beginning of November 1989, when the former announced that he had carried on a 'personal intimate contact' with the latter by telephone. Namaliu would not say whether Ona had agreed to come out from hiding to negotiate the Bougainville Peace Package, but said that the end of the crisis was 'in sight'. (*Post-Courier*, 3 November 1989) A few days later in an interview broadcast by the National Broadcasting Corporation, Ona rejected the package out of hand: 'I find it just rubbish and dust from BCL.' He attacked both the national and provincial governments, especially Father Momis and Joseph Kabui: the leaders of both governments were 'in the hands of the White Mafia', etc. (*Post-Courier*, 6 November 1989) Despite that diatribe, the Prime Minister insisted that 'the initial contacts made last week are now being followed up'. (*Niugini Nius*, 9 November 1989) When pressed by the Opposition to reveal the contents of his conversation with Ona, he refused, leading to the charge that he was playing 'cheap politics' in anticipation of an impending no-confidence vote. In any case, this particular 'breakthrough' eventually came to naught: in a letter to the *PNG Times* on 10 November 1989 Ona denied that the telephone conversation had taken place.

After operation of the mine ceased during May, non-essential employees of the company had been given early leave, and management, along with essential employees, had settled down to await the promised deployment of enough security forces to allow safe operation to resume. Despite assurances of safety from Port Moresby, acts of violence and threatening signs of it continued against the company throughout June and the first half of July, including the killing of three civilian employees of a BCL-related firm. After mid-July the violence slackened enough to encourage company management to make another attempt to repair the damaged transmission lines. Repair work commenced on 4 August and was completed, without incident, on 22 August. Two weeks more were required to restore the ore-processing facilities; finally, on 5 September, operations resumed—but very briefly, another pylon having been toppled the following day. That break in the power line was quickly repaired, despite an attack, but mining was not again resumed, because of the risk to company employees. The murder of John Bika on 10 September showed that the risk was real. Even though the mine remained closed—which was one of Ona's original demands—he and his followers issued a reminder of their proximity and intent by toppling yet another pylon on 12 November.

An ironical twist was added to this state of siege warfare on 27 October, when the non-militant landowners, both 'old' and 'new' PLA, held a *karekara*, a 'traditional' Nasioi peace ceremony, in Arawa, calling on all parties to cease hostilities and to restore law and order to Bougainville. Present at the ceremony and signing the peace agreement were the Prime Minister and several of his ministers, the premier and ministers of the provincial government, the chairman and general manager of BCL (who, although sceptical, could not have done otherwise), along with members of the PLA. Unfortunately, although urged to attend, there were no representatives of Ona and his pylon-topplers, and no let-up in their sabotage.

As a result of the continuing attacks on its employees and facilities, the BCL directors decided to retrench operations and to remove all personnel. This was finally completed in February 1990, when all employees were sent home to other areas of Bougainville and Papua New Guinea, or to Australia or elsewhere. A few of them remained in Port Moresby in order to facilitate the resumption of operations if or when it became safe to do so.

Thus ended, at least for the foreseeable future, an enterprise which had generated a total of about K1600 million, of which about K1000 million had gone to the government of Papua New Guinea in the form of taxes,

royalties and dividends. In addition, the enterprise had provided employment for PNG nationals for a total of about 100,000 man-years, and specialized training for 12,000 of its employees (1000 had completed full trade
apprenticeships and some 400 graduate and postgraduate studies under
company sponsorship). Not least among the nationals to be deprived of
income by closure of the mine were the company's 1200 or so Bougainvillian employees and their families. Like most developing countries, PNG has
no social security system.

12

Isolation

BCL SUSPENDED MINING IN SEPTEMBER 1989, AND HAD SHUT DOWN its operation entirely and withdrawn its personnel by February 1990. The following month a ceasefire was declared between the PNG government forces and the militants, and all national troops and police were withdrawn from Bougainville–Buka. Then at the beginning of May the national government announced a total blockade of goods and services, and a fortnight later came the unilateral declaration of independence (UDI) and the formation of the Repubic of Bougainville.

Before we proceed with an account of these events, there is an important issue underlying much of the Crisis which we must examine. This concerns the claim to an identity as 'Bougainvillian', which was defined largely in contrast to 'redskin'.

To many Bougainvillians 'redskin' means New Guinea Highlanders; to others it means mainland Niuginians; to still others it means all Niuginians except for natives of the North Solomons Province. In this writer's personal experience — which is not documentable — the trend in recent years has been for the reference to become wider. As Bougainvillians have come to widen their perception of their own ethnic category as Bougainvillians, their common, and sharply distinctive blackness has led them to lump other Niuginians into an equally distinctive category of less black, hence redskins. In the following account 'Niuginian' is used instead of 'redskin' to refer to non-Bougainvillian indigenes of PNG.

Mention has been made of Bougainvillians' special anxieties about adult male Niuginians who were single or unemployed; of the pressure on BCL to 'Bougainvillianize' its workforce; and of the rage against Highlanders aroused by their killing of Moini and Rovin. Those occurrences are recent; the attitudes reflected in them were conceived and nurtured over several decades. This development has been traced in a valuable study, by Jill Nash and Eugene Ogan (March 1990).

Before World War II few Niuginians resided on Bougainville, but Bougainvillians employed in New Britain and other parts of the Mandated Territory saw and worked with Tolai, Sepiks, Madangs, and so on. In the course of those encounters many of the Bougainvillians, whom whites indiscriminately labelled 'Bukas', doubtless accepted the stereotype held by many whites that Bukas were superior to other Niuginians in several respects, such as industriousness, loyalty, and placidity. After World War II those Bougainvillians who had been trapped in Rabaul during the war returned home to their relatively puritanical communities with lurid accounts of the New Britain and New Ireland women who had served in Japanese brothels there.

Mention has also been made of the disinclination of Bougainvillians to resume work in the islands' white-owned plantations after the war (which had led to the importation of hundreds of Niuginians, mainly Highlanders and Sepiks). The few Bougainvillians who were employed on the plantations usually worked in non-menial jobs, such as foreman or domestic, and this occupational distinction served to reinforce a widespread attitude that other Niuginians were more 'primitive' and 'backward' than themselves. Highlanders especially (most of whom came to be called 'Chimbus') were seen to be savage and repulsive. (Nash and Ogan provide colourful examples of Bougainvillians' scandalized and contemptuous reactions upon seeing photographs of Highland women breast-feeding pigs, and of Highland men either naked or wearing only a phallocrypt.) In the meantime stay-at-home Bougainvillians were also encountering more and more Niuginians from New Britain and Papua, who were assigned to their islands in higher-level Administration jobs as clerks and medical assistants. Although Bougainvillians usually resented submitting to orders from the latter, they could not escape concluding that those particular varieties of Niuginians were at least as 'progressive' as themselves.

That was the inter-ethnic situation before the mine. During the exploration phase most of the non-technical work done in connection with it (portering, clearing, rod-holding) was performed by Bougainvillians. However, during the next phase, construction, the company required more unskilled labour than the island could supply and was obliged to import labour from elsewhere in the Territory. At its peak in 1968–70 more than ten thousand labourers were employed, including six thousand Niuginians. The resulting scene is described by Nash and Ogan: 'The influx of men, almost all without families, to the relatively small Kieta-Panguna-Arawa area . . . created a drunken, brawling social scene more like a frontier town of the American Old West than the quiet, colonial backwater of the early 1960s.' (1990, p. 8)

Skin colour was not the only factor that led more and more Bougainvillians to distinguish themselves from other Niuginians, but it was easily the most visible one. As such it became a potent symbol, exemplified by its use as the name Mungkas, adopted by Bougainvillian students at the University of Papua New Guinea, and by the union-like association formed among BCL's Bougainvillian employees. Nash and Ogan (p. 9) provide another illuminating example of this symbolism:

> The potency of the colour symbolism can be seen in connection with the Papua New Guinea flag. As soon as the design was publicized, [Bougainvillians] noted that the upper half of flag was to be red, the lower half, black. Discussion in the villages maintained that this design was meant to announce the continued domination of 'redskins' over Bougainvillians in an independent Papua New Guinea.

In March 1989 a Kieta woman was fatally attacked by an axe-wielding man described as a 'Highlander'. In 'traditional' pay-back response, her people killed five Highlander labourers working at Aropa plantation, and later burnt down a settlement of other Niuginians near the Aropa airport. These overt acts added to the growing xenophobic, specifically anti-redskin, attitude of Bougainvillians living in or near the urban area. Niuginians in general made plans to depart, and an existing organization, the Highlands Welfare Group, co-ordinated the exodus. Five Highlands provincial groups contributed K5000 each towards the cost of the evacuation, and several of their premiers went to Bougainville to help supervise it. (*PNG Times*, 8–14 June 1989) In May 1989 the exodus began; soon afterwards 450 more Niuginians left by chartered ship, and a year later there were few of them left. Even those in the public service joined the retreat.

The militancy that had erupted in late November 1988 was at that time directly solely against BCL. Throughout the following months the publicly stated goals of its leaders proliferated, as did their targets. The first 'official' order aimed at non-Bougainvillians, presumably including Niuginians, was allegedly issued on 15 March 1990 by Sam Kauona (who by that time had become military commander of the Bougainville Revolutionary Army). The order directed 'all unemployed non-Bougainvillians' to leave the island by 17 March, which resulted, according to the news report, 'in a scramble for airline seats and places on ships'. (*Post-Courier*, 15 May 1990) Shortly after the news report, the self-same (or a different?) Sam Kauona, in an interview on Radio North Solomons, 'branded the report as false and a tactic used by journalists to add fuel to the present crisis'. In the same interview Kauona is reported to have said that Niuginian public servants working in the province should remain there in order 'to provide training to Bougainvillians in a bid to take over these positions'. But, he added, 'the presence of squatter settlers in the urban areas have made it difficult for the BRA to operate and maintain law and order.' BRA made no 'official' pronouncements about Niuginians who were employed, and presumably untroublesome, but other Bougainvillians took advantage of the anarchic situation created by the militancy to vent their spleen on Niuginians in general and Highlanders in particular.

The anti-Niuginian sentiments that provoked the exodus were fuelled by the presence of the national government's security forces of police and army, which consisted mainly of Niuginians. Many of the actions of those outsiders on and off duty, aroused the anger of even the anti-militant Bougainvillians, and hardened their ill-will towards redskins.

In some cases the ill-will included even the Bougainvillian spouses of Niuginians. In March 1990 the Bougainvillian wife of a Niuginian was raped, by men claiming to be BRA, because she was married to a redskin. In several other cases, the Bougainvillian wives of Niuginians were prohibited from accompanying their husbands when the latter left.

Strangely, it occurred to no one, not even the most militant xenophobe, to call public attention to the fact that the revolution's leader, Francis Ona, was married to a Niuginian, and that his cousin the chairman of the militant New PLA, was (or had been) likewise. Strange, also, that despite their stereotypic reputation for pay-back savagery, Niuginians outside Bougainville seem not to have molested the Bougainvillians living among them in widespread and equally savage reprisal. There were several isolated incidents of 'harassment', and reports about Bougainville students in Port Moresby and Lae feeling apprehensive enough to leave school, but there was nothing comparable to the anti-redskin mood that prevailed on Bougainville.

The anti-redskin attitudes, not surprisingly, had diffused very widely among the island's children. This had occurred years before the militancy began. Evidence for this is given in studies by T. K. Moulik (1977) and by Mamak and Bedford (1974). An example is given in the latter (pp. 5–6) of a comment, described as typical, of a sixteen-year-old: 'I think all Niuginians are bad because they want to make trouble among themselves . . . Bougainvillians are like brothers and sisters'. Self-deception is not exclusively a sickness of Western civilization.

Meanwhile, a curious series of events was unfolding which seemed to indicate that a 'world-renowned group in the mining industry' was preparing to take over where BCL had been forced to leave off.

In February 1989 Michael Ogio, MHA for North Bougainville, had issued a statement saying that he had information in hand that

'reveals a possible take-over of Bougainville Copper by a company called Bougainville Resources Development' which proposes to allocate shares to the North Solomons Provincial government (16 percent), Panguna landowners (16 percent) and the Melanesian Alliance Party (16 percent) . . . [and that] the parties involved in the present troubles at the Bougainville Copper mine continue so as to force the collapse of BCL, allowing them to bring in new developers. [*Post-Courier*, 6 February 1989]

In a statement published two days later, Father Momis and the North Solomons premier announced the establishment of a new joint venture between the provincial government and a 'foreign partner', to prospect and mine in the province. (*Post-Courier*, 8 February 1989) The premier, Joseph Kabui, went on to say that

> his government would be a 50 percent partner while the foreign partner, which he could only say was a world-renowned group in the mining industry, would own the other 50 percent, [and that] the provincial government intended sharing its stage in the venture with landowners as well as the national government, if 'it is interested'.

The newspaper article characterized the new venture as the 'brainchild' of Father Momis and other 'senior' provincial leaders (obviously not including Ogio!). It paraphrased Momis as having 'vehemently denied that his political party, the Melanesian Alliance, nor he personally, would hold any shares in the joint venture company'; nor, he added, would the proposed mining company take over BCL, saying 'it was an impossible task'.

The next the ordinary public heard of this venture was a year later. The most succinct account of it appeared in the *Australian Financial Review* of 20 March 1990, under the by-line of Rowan Callick:

> Several Australian businessmen, including Mr Martin Dougherty and Mr Don Willesee, are involved in moves inspired by the PNG Member for Bougainville, Father John Momis, to package a resurrection of the island to business audiences in Australia and elsewhere . . .
>
> Documents tabled in the PNG Parliament by the Leader of the Opposition, Mr Paias Wingti, indicate that Mr [Benedict] Chan has been involved with Father Momis and Mr Joe Kabui, the premier of the North Solomons province, which includes Bougainville, in preparing to take over the Island's prospecting authorities — despite a 19-year prospecting moratorium following problems associated with CRA Ltd's establishment of the Bougainville copper mine at Panguna. The mine has been closed since May 15 last year.
>
> Mr Chan — who is not related to the former PNG Prime Minister, Sir Julius Chan — is also a director of Mosaic Oil Niugini Ltd, a wholly owned subsidiary of Mosaic Oil NL. Among its top 20 shareholders is the Melanesian Awareness Foundation, with 461,250 25 cent shares trading on low turnovers at about 14 cents. The foundation is understood to be connected with Father Momis' Melanesian Alliance Party. The leader of the Bougainville Revolutionary Army, Mr Francis Ona, reappeared at the weekend . . . to make a radio broadcast . . . He did not rule out reopening of the copper mine, but seemed to oppose allowing CRA to resume management control. Father Momis recently recommended a company involving Mr Dougherty and Mr Willesee to the PNG Cabinet to help it to re-establish investment on Bougainville.

The verbatim record of the tabling of the documents in the Papua New Guinea Parliament, and of the flurry of charges and counter-charges accompanying it, sharply reveals the principals' sentiments and political rivalry. A brief summary of it appeared in a news report broadcast by Radio Australia:

> The Papua New Guinea Prime Minister, Mr Namaliu, has promised an investigation into allegations made in the country's parliament that Australian-based businessmen sought to prolong the Bougainville conflict. The allegations were made by the Opposition Leader, Mr Wingti, who produced a letter which he said showed the businessmen conspired with the highest levels of government to prolong the conflict so as to buy cheaply the closed Bougainville Copper Limited mine at Panguna. The Provincial Affairs Minister and MP for Bougainville, Father John Momis, who was alleged to have been involved in the conspiracy, told the House that the letter produced by Mr Wingti was a 'concoction of lies'. A statement issued to the media in Port Moresby by one of the Australian businessmen named, Mr Martin Doherty [sic], also denied the allegations, calling the letter purporting to be from himself and another businessman, a lie and a fraud. In his statement Mr Doherty denied he and other parties were involved in domestic political activities, saying that the letter produced in parliament had been concocted to give that impression.

To any observer of the PNG political scene there was nothing unusual about the Wingti–Momis exchange in parliament, but one of the tabled documents in particular was greeted with either rage or glee, depending upon the member's current political stance. Despite its length it is reproduced here—partly because of the insight it provides into the wider political context of the Bougainville Crisis, and partly in order to lighten somewhat the grim drama of the Crisis with some less tragic by-play, whether authentic or farcical. (The editorial accompanying the letter characterized it as the latter.) The letter purports to be from Martin Dougherty; the name of the addressee had been blocked out before the letter was tabled:

> Dear Sirs,
> Please find enclosed draft agreement which indicates what is expected of the investor group who wish to invest in the Bougainville Joint Venture.
> As to your investor's concern over the political situation in Papua New Guinea, I refer to our previous discussion, my partner Benedict Chan and I have the Papua New Guinea government in the 'palm of our hands'. Namaliu, the Prime Minister, and Father John Momis are totally under our 'control' (thence that A$4 million we required).
> As regard [sic] the opponents in Papua New Guinea Parliament, my partner Benedict Chan can give you the following assurances:

(a) Mr Chan is presently working towards removing Sir Julius Chan permanently from Papua New Guinea Parliament. He will no longer be effective. Please read enclosed newspaper reports and Report of investigating Authority into Placer Pacific Limited (engineered by Mr Benedict Chan).

(b) Steps to neutralize Paias Wingti, former Prime Minister will be put into motion to ensure Namaliu stay as Prime Minister for the next term. (By this way [sic], Benedict Chan can assure your group that Michael Somare is a has-been Prime Minister, he is 'history').

(c) Karl Stock [sic] will be removed from his present ministry (see annexed documents regarding corruption Stack/Sir Hugo hearing) and someone agreeable will be put in his place. All future dealing in timber concessions will be handled by Mr Chan's brother, Mr Albert Chan. Your group's timber request will have to go through him.

(d) T. Diro will be removed from his present position (enclosed are newspaper clippings of his corrupt activities).

Mr Benedict Chan is instrumental in assisting Mr Charles Copeman and Mr Howard Brady of Mosaic Oil NL in securing Papuan Basin leases. He has Patterson Lowa Minister for Minerals and Energy under his control and the Minister had been well 'looked after'.

Our strategy in Bougainville is:

(1) remove Leo Hannett from Bougainville Development Corporation and secure total control of the Corporation.

(2) engineer the Bougainville crisis to pick up Panguna Mine with minimum of costs.

My past achievements in Australia (in demoralizing Australia [sic] Politicians) will demonstrate my ability in publicly destroying Father Momis's political opponents in Papua New Guinea.

To demonstrate our influence over the Papua New Guinea government, I enclose copies of letters from Joseph Kabui for your information. If your group should require any further proof, I will have no trouble obtaining the necessary letters from the government.
Cheers
Martin J. Dougherty
Chairman

When the letter was published in the *Post-Courier*, the editor added in explanation, 'The word "Cheers" and the following three lines appear to have been typed on an electronic typewriter or computer. The typeface is quite different from that of the standard typewriter used in the body of the letter.'

The political supporters of Wingti might say that support for his charges surfaced three days later, when another link between Dougherty and Momis was revealed in another article by Rowan Callick in the *Australian Financial Review*, 23 March 1990:

Asia Pacific Communications, a Sydney-based company chaired by media consultant Mr Martin Dougherty, is being paid $330,000 by the Papua New Guinea government to 'help project a better image overseas and with PNG', the Provincial Affairs Minister, Father John Momis, revealed yesterday. Father Momis, who had recommended the company to the government, admitted that its principals knew little about PNG 'but would be provided with the raw material by the PNG government' . . . There had been no tender, but alternative companies had been considered, said Father Momis, who is the MP for Bougainville. . . .

Mr Dougherty is also associated with a company whose principal is another Sydney-based businessman, Mr Benedict Chan. The company entered into an agreement last year with the North Solomons provincial government—responsible for Bougainville Island—to seek prospecting authorities through the Bougainville Resources Joint Venture.

. . . Father Momis hit out in Parliament at the 'smearing' of Mr Chan, who was born in Rabaul, PNG, by the leader of the Opposition, Mr Wingti. Father Momis said: 'He [Wingti] now questions if the joint venturers and other partners are going to buy out Bougainville Copper Ltd and give the people of Bougainville a say in the control of their wealth. I am not saying that is the plan . . . but what is wrong with that if they do?

'Since when has BCL been any different from any other publicly listed company and not had to face the pressures of the open, democratic market-place?'

One wonders what meanings the directors, the shareholders and the out-of-work former employees of BCL, would infer from this rendition of 'the open, democratic market-place'?

While the government of Papua New Guinea was attempting to deal with the Bougainville Crisis by those and other methods, two other national governments became involved.

Australia has long been vitally concerned with PNG in general ways too obvious to require listing. With regard to the Bougainville Crisis, its interests seemed to be threefold: to protect the thousand or so Australians residing on the island prior to the Crisis; to discourage secession, in order to keep PNG unified; and to maintain BCL's operation of the mine, not because of its Australian connection but because of its contribution to PNG's economic welfare. Any sizeable decrease in PNG's revenues would induce Australia to increase its donations because of the historic, and continuing, special relationship between the two nations. Contingency plans were made to evacuate the Australians in January 1990 following the wounding of two of them by militants. The plans involved use of an RAAF Hercules aircraft and deployment of up to a battalion of troops. (*Australian*, 20 January 1990) This project, however, did not eventuate.

An effort to assist the PNG security forces on Bougainville proved less popular. Although the four helicopters lent by Australia to the forces were intended solely for transport, the BRA and its supporters objected to their potential, and allegedly actual, use as platforms from which military weapons are fired (misdescribed as 'gunships'). The charge was echoed by non-interventionists in Australia, including members of the Australia-wide Public Sector Union. (*Post-Courier*, 23 February 1990) The issue was highlighted when rumours spread that the security forces had murdered six to eight Bougainvillian civilians on 14 February and used one of the helicopters to dump their bodies into the sea. The usual 'investigation' was duly announced, but there the matter stuck for more than a year, to the point at which this narrative concludes. (*Sydney Morning Herald*, 9 March 1990)

As for Australia's interest in keeping the mine going, there were no direct measures that could be taken. BCL's parent, CRA, was registered in Australia, but BCL itself was registered in PNG. In fact, the only public action taken by the Australian government in this matter consisted of a statement by the Foreign Affairs Minister, Gareth Evans. On the eve of a visit to PNG he said that 'Australia warmly welcomed the impending talks between the PNG Government and the leadership of the Bougainville militants . . .'. Innocent-seeming as they were, those remarks provoked heated responses from PNG's Foreign Minister, Michael Somare, and Justice Minister, Bernard Narokobi, who said, in effect, 'Shut up, and do not comment on the proposed peace talks unless asked to do so.' To which Narokobi added that 'amongst the underlying causes [of the Bougainville Crisis] was the insensitive role of the Australian management of BCL.' (ABC Radio report, 27 February 1990, by Sean Dorney) From this and other utterances by Narokobi during the Crisis, the reader will understand why he came to be the BRA's choice of a PNG official with whom to 'negotiate'.

The second foreign nation to become involved in the Crisis was the Solomon Islands, the independent Commonwealth nation whose Shortland Island is only ten kilometres distant from southern Bougainville. Before and during the colonial era there was fairly frequent travelling between the two, and that continued after independence: residents of Shortland Island often visited Buin to shop. During late 1989 and early 1990 it was widely rumored that BRA militants were slipping across to Shortland, and beyond, to obtain firearms and ammunition. Acting on those and other suspicions, in October 1989 Paul Tohian, Police Commissioner and State of Emergency Controller on Bougainville, placed a ban on the movement of people between Bougainville and Solomon Islands. The ban was subsequently extended to include the searching of all cargo arriving from the Solomons by ship or air.

During this period the provincial government of Bougainville was under extreme pressure from both sides, and its members faced physical danger and even death. The BRA had assassinated John Bika, the Minister for Commerce and Liquor Licensing, in September 1989. The premier and one of his ministers had been bashed by national police two months previously, and the premier had been verbally attacked by leaders of the non-militant PLA, as well as by the chairman of the Bougainville Development Corporation (which was 'officially' the economic arm of the provincial government). With the national forces in visible control of the urban areas, and the BRA in invisible control everywhere else — except for Buka — the premier and his colleagues had little left to govern. Nevertheless, they played an active role in revising and promoting the national government's Bougainville Peace Package. And while Premier Kabui was one of the first non-militants to call for withdrawal of the security forces, he did not at that point publicly espouse the cause of secession. He did however continue his attacks on BCL — not surprising in view of his earlier co-authorship of the proposal for the new mining venture.

Contrary to the almost universal trend to division and discord, the organization that had started it all, the Panguna Landowners' Association, settled its differences during this period. The reconciliation between members of the original association and the non-militant members of the new was formally announced in a resolution dated 24 November 1989.

No such reconciliation took place between the Bougainville Catholic clergy and its critics. One sequence of exchanges between them began when Bishop Singkai charged PNG security forces with killing innocent people on Bougainville, and with abuses of human rights. Regarding the latter, the bishop stated:

> In recent months the Catholic clergy has been subjected to repeated harassment by the security forces . . . the diocese of Bougainville fears for the lives of its personnel . . . assaults on a Catholic priest, a brother, a seminarian and an elderly church accountant . . . all Marist nuns had to be evacuated from the island after being subjected to continual threats and vile language . . . [*Age*, 31 January 1990]

A response was quickly made by the Defence Minister, Benais Sabumei, as reported in the *Post-Courier* of 30 January 1990:

> The Catholic Church in the North Solomons has — for the first time — come in for serious condemnation over its handling of the Bougainville crisis. And sources said last night that two of its clergymen may have their residential status revoked . . . Mr Sabumei said it was hypocritical of Bishop Singkai to condemn the security forces for killing innocent people on Bougainville while at the same time condoning the actions of the outlawed terrorists who are also killing innocent people . . . Bishop Singkai has openly and publicly supported secession for Bougainville and the government is aware of this. As the head of a church which at least 90 per cent of the people follow, Bishop Singkai should now publicly tell the people of PNG whether he supports the killing of lives and destruction of property by the terrorists.
> . . . Mr Sabumei also accused foreign clergymen on Bougainville of

encouraging the people of Bougainville to fight for economic injustice done to them by Bougainville Copper Limited, and added, 'Those clergymen must accept some responsibility for the destruction of human lives and properties.'

This was countered by the bishop:

It is with deep regret and sorrow that the Diocese of Bougainville is compelled to respond to the hysterical and outrageous accusations attributed to the Defense Minister, Mr Sabumei . . . [We take] this opportunity to state categorically that the statements are totally false . . . [*PNG Times*, 1 February 1990]

Not only false, but very regrettable, announced Singkai's colleagues:

The Catholic Bishops of Papua New Guinea regret the controversy currently surrounding our brother bishop, Gregory Singkai, of Bougainville. We have every confidence in Bishop Singkai . . . [*PNG Times*, 1 February 1990]

Encouraged, evidently, by that show of collegial support, the bishop made another entry into the fray by recommending the 'official' closing of the mine:

BCL has been the bone of contention especially for landowners. Now that the BCL is legally non-existent, its running lease having expired nine years ago, [*sic*] the Government should now declare it officially closed. It is of no use clamouring that the mine is going to re-open as soon as the crisis is settled. This manner of speaking only goes to upset the fighting landowners who have been intent on closing the mine and also adds fuel to the conflict. Once the BCL is officially closed, there is some chance of settling the present dispute and coming up with a new mining lease agreeable to all parties concerned. [*Post-Courier*, 16 February 1990]

The most trenchant criticism of the role of Bougainville's clergy during the crisis is that of historian Professor James Griffin. One sample:

In the middle of such atrocities one might have expected moral leadership from the diocesan Catholic Church which has held such influence in the province. However, at the obsequies of John Bika, Bishop Gregory Singkai was more remarkably non-committal than even his previously even-handed attitude to the fifth commandment ('Thou shalt not kill') would have suggested. If there is one achievement for which the Christian missionaries have been generally commended, it is their contribution to peace. There is, in fact, a Catholic Commission on Peace and Justice in the North Solomons. However, one spokesman appears to be an American priest who last year mendaciously

pronounced that BCL had actually killed hundreds of people. BCL
challenged the statement; there was no reply. Another American priest,
noted for his opposition to BCL's presence in the '60s, assured a
journalist that the BRA were not really shooting at people but at the
mine! The Justice and Peace spokesman asked recently whether the
government would 'Take responsibility for the destruction of the K900
million worth of mining equipment and assets of BCL' if fuel shipments
are halted, because it would then be impossible for the BRA to protect
it! Singkai has always been a secessionist. Something epical was
suggested in June when his portly elderly frame struggled into the
mountainous Kongara allegedly to preach peace to Ona. One account
has it that there his life was threatened. His report to the Prime Minister
recommended an amnesty for Ona and, it is said, statehood in 18
months and independence in five years. After that, his main plea was the
withdrawal of the security forces. Singkai was offered the Education
portfolio; in Ona's Interim Government it has been deputed to a
Catholic Church nominee. Understandably, in view of the feudal
structure of the Church, the Conference of Papua New Guinea bishops
has only fraternal influence. But until recently it has seemed
unnecessarily reticent in remarking on the North Solomons violence.
The metropolitan, Sir Peter Korongku is, however, himself North
Solomonese. The Bougainville Diocese is now protesting that a total
blockade is 'inhuman' and contrary to charters of human rights. 'Peace-
loving, peace-seeking persons who want a negotiated settlement . . . can
only despair' it says, but while conjuring spectres of starving and
deformed children, there was no mention of the suffering of the
unemployed or the actions of the BRA in expelling 'redskin' parents and
'mixed race' children, while refusing to allow 'Buka' spouses to leave the
province. [Griffin 1990, pp. 77–8]

In the midst of all the verbal warfare, some physical fighting took place
between security forces and BRA. It was on a small scale, but enough to
result in casualties on both sides. News from the front line was occasionally spiced with reports about Ona himself: he had been killed by security
forces, or by his Army Commander, Sam Kauona, or put under house arrest
by the latter. Such reports were periodically spread, and periodically proved
untrue.

Meanwhile, on Buka, far from the fray, a visiting delegation from the
security forces claimed to have been told, 'The conflict is none of [our]
business and [we] do not want to be a part of it.' Here is a reporter's
paraphrase of the delegation's paraphrase:

The people condemned the actions of the Panguna militants and
endorsed the presence of security force personnel on their island. The
people are demanding the immediate reinstatement of government

services, banking facilities and the reopening of the Copra Marketing Board's buying depot on Buka. The people also want schools on the island to be reopened. They said they were not part of the dissident Panguna Landowners' Association and 'We are not part of the Bougainville problem, we have nothing to do with the Panguna mine,' they said. They were angry at reports that militant leaders Francis Ona and Sam Kauona were carrying out illegal activities on behalf of the people of the province. [*Post-Courier*, 27 February 1990]

To which the reporter added: 'Buka was the scene of a bloody battle between troops and militants, the latter of whom had infiltrated the island about two months ago on a recruiting drive. Four soldiers and an unaccounted number of militants died in the battle on the night of February 12.'

In about the middle of February 1990, another dialogue seems to have been established between national government and BRA; it led first to a ceasefire and finally to the withdrawal of all government forces from Bougainville. Just how this particular breakthrough came about is unclear, to say the least. One guess is that it was the government's impatience, fuelled by frustration in the field and political opposition at home, that initiated the process; there were deep disagreements on objective and method within the Cabinet itself. The militants on Bougainville had no cause for impatience — they had nothing to lose by waiting; nor was there any viable local opposition to themselves to threaten their 'official' presence. In any case, in late February both sides approved of a cooling-off period as prologue to a ceasefire.

It all started with an article in a book, titled *Waging Peace in the Philippines*, by Professor Peter Wallenstein, who was head of the Department of Peace and Conflict Research at Sweden's University of Uppsala. In his article Wallenstein analysed eighty-eight recent conflicts around the world and the ways they were, or were not, resolved. The book came into the hands of Graeme Kemelfield, who was director of the University of Papua New Guinea Extension Centre in the North Solomons. He concluded that Wallenstein was the man to resolve the conflict on Bougainville.

With some members of the provincial government, Kemelfield prepared a four-page 'confidential' document, which was evidently approved by both Ona and Kauona. The nature of Kemelfield's connection with the BRA leaders is not clear, but apparently they authorized him to present the plan to the national government.

There were at the time two other breakthrough peace initiatives already in the works (including the traditional Nasioi peace ceremony already referred to). When those failed to achieve the desired results, the eminent Swedish conflict-resolver, Professor Wallenstein, was brought in. To make a long, and exceedingly convoluted, story short: all the parties got together and agreed to a ceasefire and the withdrawal of all but seventy members of the government's forces then on Bougainville. (Clearly, one sure way to resolve

a conflict is to remove one of the parties to it.) Another time-tried part of
the negotiated agreement was to have the 'laying-down-of-arms' witnessed
and certified by a team of 'international observers'.

Events did not proceed exactly as intended. The BRA did indeed lay down
their arms under the eye of the international observers, but they promptly
picked them up again. As Kauona is reported to have said: 'We agreed to
lay down our arms, but not to *surrender* them.

Meanwhile the ceasefire took effect on 2 March, and the government
security forces left the island on the 16th. But in this case also, things did
not go exactly as planned.

The national government had wished to leave a number of police on
Bougainville in order to preserve an official presence there, but in the event
all forces — police, army and Correctional Services — were withdrawn. The
decision to do so was made by Police Commissioner Paul Tohian, who was
in charge of the government's campaign on the island. Reports circulated
that Tohian's decison was due to pique — frustration over the constraints
imposed by the government upon the actions of the forces, including their
final withdrawal, and the resulting humiliation of himself. Whatever the
motivation, he quickly made an effort to restore his esteem. On the night
of 16 March he attended a party of welcome in Port Moresby for the returned
security forces, from which, in the company of several of his mates (and,
according to a news report, well in his cups), he proceeded to the home of
Prime Minister Namaliu and attempted to arrest him, hoping thereby to
topple his government. Alas for Tohian, Namaliu was 'rescued' — according
to an AAP–Reuters report, by Australian defence personnel; according to
Tohian's 'acting' successor, by his own police. Tohian was immediately
sacked, but allowed to return to his Tabar Island home. Later on, as a result
of an investigation into the affair, he was arrested and charged with treason —
a charge that was subsequently dropped.

When questioned in parliament, the Prime Minister declared that, despite
the total withdrawal of the police from Bougainville, the nation continued
to maintain its authority in the province by means of the presence there of
the provincial government. But that presence was ineffective, to say the least.
On 16 March members of the now un-reined BRA arrested and assaulted
two of the province's principal officers, including Michael Laimo, who had
lost an eye from his previous bashing by security force personnel. After the
national forces had been withdrawn from Bougainville, the BRA was in
seeming control of the civilian population of most of Bougainville Island.

The leadership of the BRA moved its headquarters to Panguna and offered
in a letter to BCL to guard the company's property from 'angry civilians' — if
paid to do so by the company. Ona was quoted as saying, 'Just for your
information, Bougainville people don't want your company on the island
in relation to destructions that your company brought to the island. They
are ready to walk in and destroy everything if I pull the BRA boys out.'
(*PNG Times*, 3 May 1990) To which BCL's management replied that it was

authorized to deal only with the national government. At this writing I do not know whether the BRA's 'protection' has been maintained or withdrawn, and if the latter, what has ensued. In any case, that continued BRA 'protection' could not save the company's buildings and machinery from the ravages of Bougainville's climate.

Meanwhile the BRA's troops (including many new and teenage recruits) roamed freely around the island; their presence was less in the north and on Buka. They were charged with maintaining law and order, but according to several reports, they frequently accomplished that by bullying and terrorizing civilians. Under most conditions of warfare many such charges can be dismissed as normal propaganda, but these were given substance by Ona himself. Here are some reported (and most likely translated) excerpts from a radio broadcast he made on or about 18 March 1990:

> I have received a lot of complaints which are giving us a bad name . . . I want the good name to be resolved—that is my strong wish . . . If you don't follow my wish and you continue misbehaving, I can surrender now and everything will fall apart . . . Those you fought to protect now regard us as worse than the security forces. I have received too many complaints and am not too happy . . . A lot of you I don't consider as BRA . . . I want all vehicles that were taken from people to be returned . . . If you took it from BCL at Panguna, you must leave it there . . . I want all our BRA commanders to remember that cars are expensive to buy . . . Our name has not gone well with business people because a lot of you guys, the young ones go into shops and just grab goods . . it's against BRA law. [*Post-Courier*, 20 March 1990]

Further evidence that Ona was having difficulty in controlling his troops came when a 'source close to the BRA' reported by phone to Rabaul that a group of BRA youths had defied an order by Ona to cease selling fuel to the public in order to conserve diminishing supplies, which at the current rate of consumption would be exhausted in 'two to three weeks'. Ona apparently became 'very angry'. (*Post-Courier*, 7 May 1990)

In the meantime the national government's perennial efforts to solve the Crisis by face-to-face negotiations with the BRA's leaders surfaced in news reports at predictable times: when the government was charged by the Opposition with 'doing nothing'; when nothing else of public interest was taking place. By the beginning of April the only agreed-upon—or not disagreed-upon— aspect of the mirage-like talks was the identity of the govenment's negotiators. Father Momis had been the militants' champion in nearly everything but secession, and for that reason was unacceptable to the government. One negotiator was to be Foreign Minister Michael Somare, the Father—or perhaps, in Bougainvillians' eyes, Mother's Brother—of the Nation. The other was Justice Minister and Attorney-General Bernard Narakobi, who could be expected to have an 'open' mind. As for the agenda: the only item the militant leaders said they would not compromise on was their demand for secession, which of course was the one item the national government would not permit. This led Narakobi to ask, 'What then can we talk about?' (*Post-Courier*, 12 April 1990)

Prime Minister Namaliu announced that a 'fall-back' plan had been constructed for use should the current peace initiative fail. On 24 April he denied that the fall-back plan included the blockading of supplies to Bougainville. Two weeks later, he announced that he had received a message from the North Solomons Premier, Joseph Kabui, the previous night, 'seeking talks to be held at a neutral venue, sometime after May 13th'. To which the Prime Minister added that 'his Government would grant Bougainville greater autonomy'. (Radio Australia, South Pacific News Summary, 1 May 1990)

The announcement was subsequently published in the form of an official statement, *Government Position over the Bougainville Negotiations*. Its principal points were:

- That he, the PM, completely rejected the use of military force as a way of dealing with the situation. He is quoted as saying, 'It would of course be possible for the National Government to use force, not only do we have the Defence Force available but we could also use foreign troops.' (This is an intriguing suggestion, to say the least!)
- That the North Solomons Provincial Government had not exploited the 'full range of powers available to it', and that the national government wished to help it do so.
- In addition, the national government was willing to grant to the provincial government a new and special status of autonomy, which would include: control over 'almost all' internal affairs, especially those affecting the daily lives of the people of the province; greater financial resources, through new funding arrangements under the government's mining policy as well as from other sources; and a comprehensive programme of reconstruction and rehabilitation. 'We have very large sums — perhaps as much as K50 million — available for this purpose, including funds from overseas donors, *provided a peaceful agreement is reached.*'

The national government under Namaliu seemed to be taking a conciliatory position, but events were moving fast. Having made his announcement about greater autonomy for Bougainville, the Prime Minister boarded a flight to the USA. No sooner was he out of the country than an abrupt reversal of policy was announced.

The Executive Council agreed, effective from 2 May, to declare a state of emergency on Bougainville, to outlaw the BRA and to press all-out war against it. This included a blockade of all goods and services except essential medical supplies. These measures were announced by the Minister of State, Ted Diro, who was Acting Prime Minister in Namaliu's absence.

A change of tactics? Or a change of tacticians?

There was more.

Within minutes of PM Namaliu flying out of Port Moresby for a week-long visit to the US, Diro revealed that he had asked PNG's Chief Ombudsman to suspend the North Solomons Premier and charge him with misconduct in office . . . Mr Diro released copies of a letter dated last Thursday [26 April 1990] that he said Premier Kabui had signed along with the leader of the BRA, Mr Francis Ona, seeking Solomon Island's recognition for an independent Bougainville. Mr Diro said there's clear evidence of Mr Kabui's involvement with the secessionist movement and he wanted him suspended and charged with misconduct in office. [Radio Australia, South Pacific News Summary, 1 May 1990]

Various interpretations have been offered for what appears to be a major policy split within the Cabinet concerning the Bougainville Crisis. One construction is that it was a deliberate tactic to influence the BRA leadership through the age-old 'tough cop : soft cop' routine. Another is that it was a true reflection of differences among members of the Cabinet — not just a difference between hardliners and softliners (including proponents of different versions of the Melanesian Way), but a whole range of differences growing out of individual personality attributes, party ideologies, and regional interests.

The blockade as announced by Diro was wide and high. All vessels illegally within 80 kilometres of Bougainville–Buka would be fired upon, as would all aircraft flying over the islands below 20,000 feet. According to Diro, 'only approved vessels and aircraft carrying essential medical supplies and fuel to enable the Arawa General Hospital to operate would be cleared to the island'. (*Australian*, 9 May 1990) In addition, all banking and national government public services were ordered to halt.

As matters turned out, the proposed blockade was discovered to be too comprehensive. It was later explained by Diro that the source of the order had been the Department of Transport, which had erred in overlooking the fact that the distances proposed would have cut international shipping lanes and air routes. Accordingly the blockade was revised to extend only as far as PNG's territorial waters — 12 nautical miles from the shores of Bougainville and Buka. Evidently, someone had done some homework concerning international law, thereby lowering eyebrows in Canberra and elsewhere. For those within the blockaded area, however, the revision was meaningless.

The first publicized reaction came from the Catholic Church of Bougainville. Bishop Gregory Singkai condemned it as 'inhumane and genocidal', and likened it to the 'slaughtering of innocent lives . . . children will die of malnutrition or become affected with mental retardation and physical deformity.' This was a telling commentary on the consequences of the Westernization of a people who had survived and proliferated *without* imported foods during the preceding 28,000 years.

Moses Havini, a prominent Bougainvillian then resident in Australia 'pursuing legal studies', claimed to be the official spokesman of the self-

declared Republic of Bougainville. In a radio interview in Melbourne on 22
July (ten weeks after the imposition of the blockade), Havini declared:

> I don't think malnutrition exists at the moment, because people are still
> living on produce, foodstuffs from the gardens, but it's just that at the
> moment they are missing the usual Western foods, like rice and [tinned]
> fish and sugar, coffee, and that sort of thing.

On the other hand, he said, 'the medical situation there is getting terrible'
because of a cut-off in medical supplies and the departure of many medically-
trained personnel.

As we have seen, in pre-colonial times the principal diseases that afflicted
Bougainvillians were yaws, tropical ulcers, skin diseases, respiratory infections, malaria, leprosy, tuberculosis, and meningitis. In recent decades the
prevalence and severity of most of those afflictions had been reduced by
Western-type medicines and facilities, and by trained medical personnel. By
mid-1990, except for those attached to missions, most of the latter had left
in the general exodus of non-Bougainvillians. As a result of the dwindling
of supplies of fuel to generate electricity, the hospitals and clinics were
becoming incapable of preserving medicines or conducting surgical procedures. Therefore, when the PNG government imposed the blockade, an
exception was made in favour of medicines and fuel, at least for the Arawa
General Hospital. But whether or not such supplies went ashore is not known
to this writer. Nor do I know whether there remain on Bougainville enough
competent individuals to provide adequate medical services.

Like most 'information' about 1990 conditions on Bougainville, that concerning health is scanty and anecdotal. Here, for example, is the assessment
by Havini, the Republic's spokesman in Australia:

> The blockade, as far as we are concerned, is a slow genocide tactic by
> the PNG defence force. Now, in my family, which is a relatively healthy
> family, I have so far lost two people, one of whom, a favourite auntie,
> died from recurrent malaria. Now the reason why she died was the fact
> that there were no longer any malarial drugs available on the island.
> Now, this is a death that is really unforgiving.
>   Another young married [indistinct] in the village recently died also
> from appendicitis, simply because when he had the attack there was just
> nowhere to go, so the appendicitis killed this young man. Now just to
> give you the enormity of what the health situation could be on
> Bougainville, if you could just multiply two deaths in my own family
> times 160,000 people who are on Bougainville, you'll soon find that the
> health situation on Bougainville is, indeed, very critical . . . Other
> diseases have also broken out, diseases which have been contained for the
> last 10 years, but because of no drugs these diseases have surfaced again,
> and these diseases are killers, things like malaria, we have TB which has

come up again, we have yaws which have broken out again, broken out in many many quarters of Bougainville.

Since we do not know the size of Havini's family, his statistics on mortality are not very informative. Nevertheless, it is reasonable to agree with his statement that the health situation on Bougainville–Buka was at the time worse than it was in the 1970s and 1980s, although perhaps still better than in previous decades.

It is difficult to determine the effects of the blockade. It served to strengthen the resolve of the BRA — or weaken it; to unify other Bougainvillians behind the BRA — or turn them against it: and to do all, or none, of the above, depending upon which reports one believes.

Meanwhile during that momentous month of May several other events were occurring that served to complicate the Crisis in one way or another. One was the news, reported on 9 May, that a national government patrol boat had visited Nissan Island on 24 April, had arrested two BRA agents there and had commandeered the province's own cargo-passenger ship, MV *Sankamap*. And further, that a national government patrol aircraft had received and transmitted a letter to the Prime Minister, written by a Nissan official, requesting that troops be sent to Nissan as protection against the BRA. Their agents were said to be 'harassing the people of Nissan, threatening them with guns, chasing their women, stealing goods from trade stores and taking Nissan cargoes, including petrol and other fuel to Buka'. (*Post-Courier*, 9 May 1990) The letter was reported also to contain a request that Nissan be allowed to secede from the North Solomons Province and be included within the New Ireland Province. The national government promptly acceded to the request to station troops on Nissan, but no decision has yet been made public concerning the island's provincial ties.

A second event that may have given the BRA leadership some disquiet was the 'leaking' of a fourteen-page 'secret' report to the national government concerning the situation on Bougainville. Among other things it urged the redeployment of troops to Bougainville in order to capture the BRA leaders 'if present non-military options fail'. The premises underlying that recommendation were that popular support for the BRA was lacking in certain parts of Buka and Bougainville and was diminishing elsewhere, and that the BRA's 'real' leader, Sam Kauona, was 'weak, ineffective and indecisive' and if captured 'could easily break down under pressure and stress'. (*Post-Courier*, 15 May 1990) The national government promptly announced an investigation into the leak, and reiterated its policy against use of military force on Bougainville.

Stirred, perhaps, by these happenings, the leaders of the BRA made a unilateral declaration of independence (UDI). As we mentioned in the previous chapter, on 17 May they held a ceremony at Arawa to announce and celebrate the founding of a new and independent Republic of Bougainville, made up of four 'States': South, Central, North and Islands.

The programme — which got off to a late start on account of the island's usual and politically neutral rain — was scheduled to run as follows:

Venue: Independence Oval
09.00:   Assembly at marketplace
10.00:   March to Independence oval, accompanied by kaur dancers
11.00:   Combined church service
11.30:   Speeches: Supreme Commander's independence declaration
          Speech (Gun Salute) — Bougainville anthem — choir
Others: 10 minutes: S. Kauona, D. Sisito, M. Misimuko, P. Lovitupa,
          Bishop Zale.
13.30:   Celebratory activities opened by choirs Singsing, Kaur, power
          band, Solomon dances, Tsigul, choirs, string bands, marshal
          arts. Donations are in order.
15.00:   Words of thankyou and Farewell.

It is not known whether the programme proceeded according to plan — 'traditionally'-run Bougainville meetings rarely did. It is, however, reported that attending members of the BRA were clad in the resplendent camouflage uniforms recently purchased from Singapore (where, as history would have it, PNG's current Prime Minister, Rabbie Namaliu, was at the time engaged in talks with the World Bank). The *Post-Courier* reported that the ceremony at Arawa was to be led by the (recently semi-sacked) Northern Solomons Premier, Joseph Kabui, 'flanked by local leaders of the Catholic, Uniting and Seventh-Day Adventist Churches', and that the 'self-styled Interim President [of the Republic, Francis Ona] and the BRA's Military Commander, Lt. Sam Kauona, were not expected to attend for security reasons — apparently fearing an attack on their lives by PNG agents or dissident factions within the BRA.'

Other information about this momentous occasion included a list of the republic's principal officials and a copy of the Declaration of Independence, which reads as follows:

Declaration of Independence — Republic of Bougainville. In the name of God the merciful Father, ruler of nations and Lord of the universe,

WHEREAS Bougainville Island was politically separated from the Solomon Group of Islands and seeded [*sic*] to the German colony of New Guinea by an agreement made in 1886 between England and Germany, and was subsequently included in the Australian Trust Territory of Papua and New Guinea in 1918 [not exactly true, but near enough] and WHEREAS Bougainville was then included in the Independent State of Papua New Guinea despite the objections of the people of Bougainville, on the 16th September 1975.

AND WHEREAS Bougainville is geographically apart and its people culturally distinct from Papua New Guinea.

AND WHEREAS it has been a long standing wish and aspiration of the people of Bougainville to become a separate independent nation.

AND WHEREAS in 1989 Papua New Guinea declared and fought a war against the people of Bougainville the result of which has been the withdrawal of all police, army, navy and judicial functions from Bougainville.

AND WHEREAS Papua New Guinea still declares that it has jurisdiction and authority over Bougainville and has begun imposing an economic embargo against Bougainville.

AND WHEREAS the Bougainville Revolutionary Army has since been in total control of Bougainville on behalf of its people.

AND WHEREAS Papua New Guinea has again declared its intention to invade Bougainville and subjugate its people.

AND WHEREAS Papua New Guinea has refused to recognize the democratic rights of the people of Bougainville.

AND WHEREAS it is the inalienable right of a people to be free and independent.

Now therefore, I Francis Ona, Interim President of the Republic of Bougainville, DO HEREBY PROCLAIM AND DECLARE on behalf of the people and the government of the Republic of Bougainville that as from today the seventeenth day of May in the year one thousand nine hundred and ninety Bougainville shall be forever a sovereign democratic and independent nation; founded upon the principals [sic] of liberty and justice and ever seeking the welfare and happiness of her people in a more just and equal society.

Francis Ona
President

The officials of the new interim government were reported to be: President, Francis Ona; Vice-President, James Singko, and the following Ministers:
Defence, Sam Kauona
Justice, Joseph Kabui, former Premier of the Province
State, Joe Pais, a former PNG Defence Force Lieutenant
Foreign Affairs and Trade, Lembias Magasu
Health, (unnamed), a Uniting Church representative
Youth and Recreation, (unnamed), a Seventh-Day Church representative
Primary Industry and Minerals, Michael Laimo
Culture and Tourism, Gerard Sinato
Education, (unnamed) a Catholic Church representative
Communications, Ken Savia
Lands-Atolls, (unnamed)
Doubtless, all had been 'democratically' chosen (despite the preponderance of Nasiois among them), and all were very 'interim', pending the elections to be held at some unspecified date.

By means of facsimile transmission (*whose* machines are not reported) the new nation sent off letters to several Pacific island governments and to foreign embassies located in Port Moresby requesting recognition, before the PNG authorities cut telephone communications. To date no government is known to have replied affirmatively. As the new Interim Minister of Defence, Sam Kauona, is reported to have said a few weeks earlier, 'If Australia does not recognize us [the BRA] we will not stop looking at other world powers, let's say maybe the Communists.' (*Post-Courier*, 27 April 1990).

One motive for the declaration of independence (according to a self-described confidant of Kauona based in Lae) was the desire by some members of the BRA leadership, particularly Kauona, to delegate responsibilities more widely. Reproduced below is the published account of this 'news'; whether it deserves more, or less, credence than other second- or third-hand 'information' emanating from enshrouded Bougainville I cannot judge:

> Bougainville Revolutionary Army leaders Francis Ona and Sam Kauona declared the island's 'independence' last week because they were finding the leadership difficult, a confidante of Kauona said at the weekend. The Lae-based source said he had been in frequent telephone contact with Kauona and other BRA leaders up until the Government imposed a communications blackout on Thursday.
>
> He said one of the main reasons why Ona and Kauona had chosen an early date for their unilateral declaration of independence was that they felt their control of the movement was slipping.
>
> 'Sam Kauona talks to me freely,' the source said. 'On one occasion he personally told me that he was feeling threatened by some young members of the BRA because of the obvious leadership weaknesses. He told me that having to carry all the responsibilities of running the BRA while Francis Ona continued in hiding was having an effect on him. He personally felt he could no longer shoulder everything by himself. That was one reason why he and Ona asked for Bougainville to be declared an independent State, and for an interim authority to be established so that the leadership responsibility could be shaped [shared?].' He said Ona and Kauona, together with North Solomons Premier Joseph Kabui and his cabinet ministers, had chosen to set up an 'unauthorized' interim government quickly because they feared ordinary Bougainvillians would rebel against the BRA and also overthrow the provincial government. [*Post-Courier*, 21 May 1990]

National government reaction to the UDI ranged from the grave to the ribald. The acting Prime Minister, Diro, called it 'illegal'; another Minister called it 'unconstitutional', and Bougainville's principal MP, Father John Momis, described it as 'a pointless exercise'.

For Prime Minister Namaliu the only immediate setback resulting from the UDI was political, and occurred in parliament. On 27 April the leader of the Opposition, Paias Wingti, had written to the Prime Minister offering his co-operation in resolving the Bougainville Crisis — a welcome offer af-

Damage by militants, Kobuan, 10 July 1989. Three civilians were murdered in this attack.

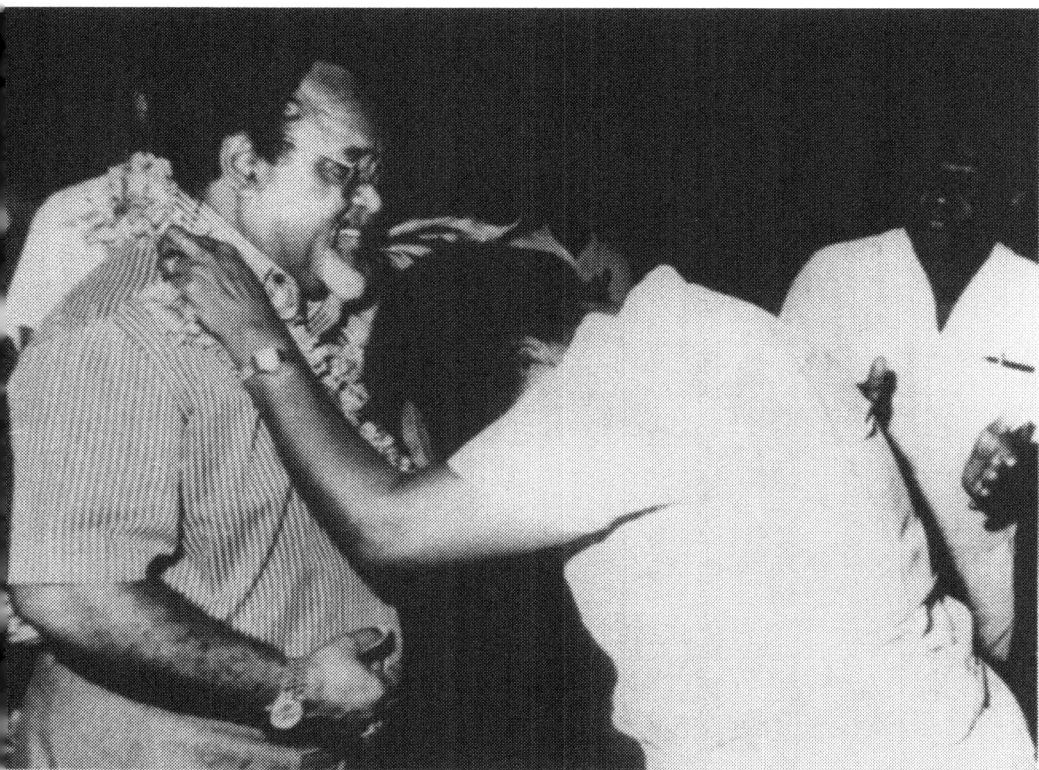

An emotional Mrs Cecilia Kenevi (a Bougainvillian living in Port Moresby) embraces Sir Michael Somare on his return from signing the 'Endeavour Accord', 5 August 1990.
(By courtesy Post-Courier)

Don Carruthers.

Peace ceremony Arawa, 27 October 1989. Prime Minister Namaliu *(left)* joins hands with Joseph Kabui over traditional pig and food assembly. *(By courtesy Post-Courier)*

ter a series of Opposition attacks accusing the government of mishandling and neglect. The Prime Minister set about arranging discussions, but these did not take place. As Wingti wrote in June, Bougainville's UDI was a 'dramatic turn of events' which had added a 'new dimension' to the problem. He expressed a willingness to reconsider this stand, however, 'if I am absolutely certain that your Government is serious in adopting some dramatic measures and approaches to realistically address the situation'. (*Post-Courier*, 6 June 1990) Declaring himself to be discouraged and saddened by Wingti's 'turnabout', and charging him with 'political point-scoring', Namaliu continued his efforts to solve the crisis without the Opposition's help.

Despite the imposition of an official communications blackout, information continued to seep out of Bougainville about conditions there. There was a report about the BRA's angry annoyance at the presence of PNG Navy boats patrolling along the islands' coasts and stopping now and then. This seems to indicate that the BRA was not fully in control of the territory it claimed. In fact, on one occasion a Navy patrol boat came close to the shore and sank a small ferry which the BRA had been using.

Notwithstanding these many complications and (reciprocal) provocations, a tentative agreement was reached between the national government and Bougainville leaders to hold talks—this time at Honiara, at the invitation of the Solomon Islands government. The tentative date was 11 June, and Somare and Narakobi were again designated to lead the PNG delegation. The Interim Government of the Republic of Bougainville also named a delegation, with Joseph Kabui as chairman and the Uniting Church bishop, John Zale, as his deputy. Ona and Kauona were not among them, because, it was reported, they were 'in fear for their lives'. (*Post-Courier*, 12 June 1990) On hearing this, Somare said that he was not willing to participate in the talks (which, he implied, would be pointless without them). His colleague, Narakobi, disagreed and declared himself eager to go on with the talks, which were 'too good a chance to miss' even without the two BRA leaders.

Reaction to the designated composition of the Bougainville delegation served to postpone the meeting. Another episode served to cancel it altogether.

On 12 June two PNG Navy patrol boats arrived at Honiara, without invitation and without prior clearance. They were ordered to leave Solomon Islands waters—which they did only after some suspicious delay. The PNG government, declaring itself 'embarrassed', apologized for the intrusion and explained it as 'an emergency stop-over'. The politically outraged leader of the Solomon Islands Opposition demanded cancellation of the postponed talks. The Solomon Islands government later declared that the matter of the unauthorized visits had been 'resolved', but arrangements for the Honiara peace talks were not resumed.

13

Peace Posturings

D URING THE EXTENDED PERIOD OF SABOTAGE AND CONFLICT ON Bougainville, many attempts had been made to hold peace talks. Like all the others, the talks planned for Honiara on 11 June 1990 were overtaken by events and came to nothing.

In fact the PNG government cannot have had firm expectations about the Honiara meetings, for, according to newspaper reports of 12 June, it had been engaged for 'several weeks' in discussions with the New Zealand government concerning another venue for the meetings that would guarantee safety for both delegations. (*Daily Telegraph*, 29 June 1990) The outcome of those discussions, announced in the same news report, was that New Zealand would host the talks on three of its Navy vessels in waters close to Bougainville at a date to be arranged. New Zealand, it was added, would provide the facilities for the talks but would not be a party to them. The (then) Prime Minister, Geoffrey Palmer, described the plan as being 'compatible with our position in the Pacific', and 'a unique role' for the nation's armed forces.

By 29 June the plan had been accepted by both parties. It involved the use of three vessels: one on which the talks would be held, and one each for housing the two delegations during the talks. Another month was required to finalize arrangements for the meetings, which were to begin on 29 July. During that month there took place several other events that had some bearing on matters to be discussed at the talks, and on the positions of the parties to them.

First, statements by PNG officials indicated either that the government continued to be divided on how to deal with Bougainville, or that a 'soft cop : tough cop' tactic was still in place.

On 3 July Prime Minister Namaliu declared that the government was prepared to deliver goods and services to the North Solomons Province if the BRA would give assurances that no armed confrontations would occur and that BRA 'elements' were prepared to allow normal government services

246

to resume on Nissan and Buka. (*Post-Courier*, 4 July 1990) The identity of those 'elements' was not identified; in any case, the national government had already stationed troops on Nissan and on 17 July the commandeered provincial boat *Sankamap* was sent with supplies to BRA-free Nissan and the province's other outer atolls.

At about the same time, however, the Deputy Prime Minister, Diro, was reported to have 'taken a much tougher line'.

> He told Parliament there were grave lessons to be learned from Bougainville, and serious warnings to be taken heed of about growing lawlessness within our communities . . . [And he] advocated the combined efforts of the police, Defence Forces and prisons services, supported by many other agencies, to deal with increasing crime and violence. [*Post-Courier*, 4 July 1990]

Evidently, in Diro's view, the Bougainville militancy was but a large example of 'lawlessness'.

Information coming out of Bougainville ranged from reports of deaths due to a dearth of foods and medicines, to BRA executions of anti-secessionists; from universal support for Ona and co., to universal lawlessness and anarchy. A widespread reversion to '. . . cargo-cult practices, which include primitive rituals' was reported in the *Age* of 10 July. The Prime Minister confirmed publicly some of the anti-BRA stories; and Australian Catholic Relief conceded that 'food remained sufficient, although protein shortages affected some areas'. (*Post-Courier*, 27 September 1990)

Concern for what was, or was not, occurring on Bougainville was developing amongst other nations in the region. According to items in the *Post-Courier* of 27 July, the sequence of events in one such international episode was as follows.

In early July the Indonesian Defence Minister told the Australian Opposition leader, John Hewson, during the latter's visit to Jakarta, that 'it would not be healthy for the region if Bougainville were allowed to secede', and urged that Australia assist PNG militarily in crushing the rebellion because PNG itself had too few troops to do so. In reaction to the Indonesian request to Australia, the PNG government is reported to have declared its readiness to send its own troops back to Bougainville if the impending peace talks failed. And in any case, Namaliu asserted, the PNG Defence Force was well trained and quite capable of putting down any rebellion in the country. (*Post-Courier*, 27 July 1990)

It was reported in the Radio Australia South Pacific News Summary of 22 July that the governments of PNG and the Solomon Islands had agreed that any of their citizens caught illegally crossing their common border would be prosecuted under the laws of the country in which they were caught. They also agreed to increase surveillance in the border areas, but not to permit troops to engage in 'hot pursuit' into each other's territory.

With these sub-plots and distractions off the stage we are now in position to view the next *main* event of the Crisis.

The three ships provided by New Zealand for the peace talks were the supply ship *Endeavour*, where the meetings would be held; the frigate *Wellington* to house the PNG delegation; and the frigate *Waikato* for the Bougainville delegation. The PNG delegation was headed by Sir Michael Somare, the Foreign Minister, and Bernard Narakobi, the Minister for Justice and Attorney-General, and included a number of other Administration officials. The Bougainville delegation consisted of Joseph Kabui (its head), the Catholic bishop Gregory Singkai, the Uniting Church bishop John Zale, the Adventists' head pastor Jeffry Paul, along with seventeen other prominent Bougainvillians, including Cecilia Gemel, president of the Panguna Land-owners' Association and an aunt of Francis Ona. It did not include Ona nor his BRA Commander, Sam Kauona (for reasons described by many persons, but best known only to themselves).

Joseph Kabui's status at the meetings provides a nice expression of the fundamental differences between the delegations. He attended as Minister for Peace, Justice and Police of the self-declared Interim Government of the Republic of Bougainville, but he was addressed by the PNG delegates throughout the meetings as Mr Premier (i.e. Premier of the North Solomons Province). Diro's earlier efforts to sack him evidently had been postponed.

At the request of both delegations two international observers had been invited to attend; they were Nicholas Etheridge of the Canadian High Commission in Canberra and Tony Browne, the Director of New Zealand's Security Secretariat. Later, in the course of the meetings and upon request of the Bougainville delegation, a third international observer was invited: he turned out to be Fela Molisa, the Finance Minister of Vanuatu, and he arrived at the meetings on 2 August.

Also attending as observers were three members of Moral Rearmament, described by one reporter as 'an international body of God-fearing people who make crisis and problem resolution part of their lives'. (*Post-Courier*, 17 August 1990) Their presence at the meetings was believed to have been at the invitation of Narakobi, who was known to be an admirer of the organization's goals and strategies.

Only three media representatives were allowed to attend. One was Sean Dorney, an Australian and the ABC's long-time representative in PNG; the second was Angwi Hriehwazi, a Niuginian on the staff of the PNG *Post-Courier*. Both of these excellent newsmen had covered the Bougainville Crisis since its beginning, and were personally acquainted with the island and its leading personalities. A third reporter was present from Radio New Zealand.

Another noteworthy set of participants on the stage of the meetings were the Maori sailors aboard the three Navy ships. There were enough of them to perform the well-known *haka*, the customary Maori challenge-and-greeting dance, for both delegations. I do not know whether their numerical prominence among the New Zealand sailors was normal, or had been specially arranged. In any case, whether or not any of the delegations perceived it to be so, their presence provided added credibility to New Zealand's role in the affair as a Pacific nation.

On 28 July the frigate *Wellington* boarded the PNG delegation at Rabaul and ferried them overnight to an assembly place in the waters off Kieta Harbour. There follows a day-by-day résumé of the meetings at this long-awaited event.

July 29.   The two delegations were ferried to the *Endeavour.* The *Post-Courier* correspondent Angwi Hriehwazi describes the initial episode:

> The Bougainville delegates to the peace talks aboard the Royal New Zealand Navy ship *Endeavour* are refusing to accept food, water or other refreshment offered by the Navy. They are also not shaking hands with any member of the PNG Government delegation. They say that while their people are suffering, they will not allow themselves the benefit of access to good food. The delegation leader, North Solomons Premier Joseph Kabui, said the time for eating and drinking would come later. Meanwhile, the Bougainville delegates would return nightly to Kieta, instead of remaining aboard another New Zealand ship, HMNZS *Waikato*, as originally planned by the talks organizers. Panguna Landowners Association president Cecilia Gemel said that as a further symbol of the prevailing troubles they would not shake hands with anybody. [30 July 1990]

In that unpromising ambience the meeting began with a solid half-hour of prayer — which, it seems, was sorely needed! There were plenty of clerics present: two bishops, a head pastor, and a Navy chaplain. Only occasional mention will be made of this aspect of the meetings, but it should be noted that it was omnipresent. According to Sean Dorney, 'we spent almost as much time praying as we did attending the conference.' (interview on ABC radio, 13 August 1990)

After the prayers Sir Michael Somare made a speech, asking both parties to seek out 'seriously' a solution to the Crisis: 'many of the problems have not been addressed over the years, and we [the PNG Government] accept part of the blame.' (*Post-Courier*, 30 July 1990)

The leader of the Bougainville delegation, Premier or Minister Joseph Kabui, also declared his desire for 'an amicable solution', including the finding of 'ways of avoiding past mistakes', but was less inclined than Somare to accept responsibility for those mistakes. Unlike Somare, he added some specific allegations of the Bougainville case, including 'his peoples'' desire for independence, and their desire that government services be restored 'even if the talks failed'. (*Post-Courier*, 30 July 1990)

Indeed, the only hopeful result to emerge from that first day of meetings was the action taken by the New Zealand Navy, at the request of Somare, in transporting a seriously ill woman from Kieta to Nissan, where she was flown to Rabaul for medical treatment.

July 30.   Day two of the meetings was used by the Bougainvillians to vent their anger. In the words of one reporter: 'The Bougainville delegation . . . seized the opportunity, in the presence of the media and international observers, to accuse the PNG Government of human rights violations.' Their litany of wrongs included the following acts allegedly committed by the PNG Defence Forces: torture, mutilation and murder; rapes and disrobings of females; forced evacuation of 25,000 villagers; total destruction of 70 villages and partial destruction of 46 others. The rest of the meeting was used by the Bougainvillians to reiterate their demands: for lifting of the blockade; for restoration of essential services, i.e., those financed by the PNG government; and for freeing of 'political prisoners' (referring, perhaps, to the BRA men captured on Nissan and held in the Rabaul jail). All this was to be granted 'before there could be any negotiations on Francis Ona's original demands of 10 billion kina compensation, the end of mining and a referendum on secession'. (Dorney, interview on ABC radio, 13 August 1990)

The last of these demands may have come as a surprise. Several members of the Bougainville delegation had demanded 'recognition' of secession: acceptance of a fait accompli. In this wording, Kabui appears to have introduced the possibility of *negotiations* on secession, including the holding of a referendum.

The PNG delegation reserved its reply for the following day, although Somare said in an aside that 'it was good that the Bougainvillians were able to "get it off their chests". ' (*Post Courier*, 31 July 1990)

During that second meeting the Bougainvillians targeted the PNG government with their charges, but in an interview afterwards Kabui shifted some blame for the Crisis to Australia. He accused it of ignoring Bougainville's right to independence for selfish financial reasons. (Dorney, interview on ABC radio, 13 August 1990)

July 31.   The third day of the meetings was a surprise to many, who believed that the previous day's angry charges by the Bougainvillians would lead to similar replies. Instead, in his address to the delegates Narakobi asked for 'forgiveness' for the acts of the PNG forces, and stated his readiness to have those acts investigated by an international body for the purpose of compensating the victims. More substantive, and probably more effective in keeping the talks afloat, was Somare's statement. Rather than replying specifically to the previous day's charges, he proposed to put aside the political future of Bougainville and focus first on ways for restoring essential Administration services. To that end he proposed that the two delegations prepare a written memorandum of understanding on how such services should be restored, and requested each delegation to prepare its own draft, overnight, for consideration the following afternoon.

Evidently, the task of preparing such a draft required more than the allotted time, because the Bougainville delegation asked for and received an adjournment until the afternoon of 2 August. The request was understandable in view of the need to obtain agreement from such diverse persons as clerics, politicians and laymen, not to mention Francis Ona and Sam Kauona.

August 2. The meetings were resumed in the afternoon, but were again adjourned upon request of Somare, who said he could not possibly negotiate on the basis of the Bougainvillians' draft. According to Sean Dorney, it was more like a demand for surrender than a memorandum of understanding: 'It obliged PNG to recognize Bougainville's independence, to fund its independent bank, to stock its reserves, and even to pay all the future wages of the Republic's public servants.' In other words, a nation-size cargo!

Before the adjournment, some progress seems to have been made towards postponing the issue of whether secession was to be recognized as a fait accompli. That progress was, however, abruptly halted when James Singko, a member of the Bougainville delegation, refused to continue discussion of the government's offer to restore essential services until the government's representatives recognized Bougainville's independence.

According to Dorney, Singko was the 'strongman' of the Bougainville delegation, the chief of Francis Ona's Supreme Advisory Council, and second only to Ona in the BRA's political structure. Dorney described him as 'a middle-aged brooding man who has a strong traditional following and who's described by PNG intelligence as a cultist leader'.

A two-day adjournment followed the fruitless 2 August meeting (when even the persevering Somare seemed ready to quit). Relations between the delegations were further damaged by receipt of published statements of the Prime Minister, who had told journalists in Vanuatu that a number of prominent anti-secessionist Bougainvillians had recently been tortured and killed by the BRA. One of the men he named was soon proved to be alive. This of course served to strengthen the Bougainvillians' charge of government 'deceit', but it did not nullify the government's charges about some other BRA assassinations, including the earlier one of John Bika.

August 5. On this, the eighth day, the meetings resumed. In the words of Sean Dorney, here is what occurred:

> When the Bougainvillians resumed proceedings by asking again for Papua-New Guinea to lift the blockade without any conditions, the prospects for any sort of agreement seemed hopelessly remote, but right towards the end of the day, with little warning, the breakthrough came. With some advice from the international observers . . . a document . . . was formulated and agreed to by both sides . . . The key sentences were to do with deferring the unilateral declaration of independence and [implementing] the speedy restoration of services.
>
> Reading out the accord, Sir Michael [Somare] said the two delegations agreed that face-to-face dialogue should continue, [adding] 'The long-term political status of Bougainville is to be addressed as part of the continuing dialogue. Meanwhile, all political declaration(s) with respect

to that future status are deferred. The National Government delegation [has] confirmed that at the earliest opportunity it would take all practical steps consistent with the constitution of Papua-New Guinea to bring about return of services to Bougainville.'

Joseph Kabui put the result down to a miracle, and James Singko saw a divine hand at work. In fact it was the very human hands of Michael Somare, and possibly of Bernard Narakobi, which crafted most of the formula that led to this breakthrough. It was promptly named the Endeavour Accord.

Upon the signing of the accord, it is reported, radiant smiles abounded and embracing and hand-shaking took place, even by Cecilia Gemel and the brooding James Singko. And before the government delegation departed for home they sent two tonnes of medical supplies to the Arawa Hospital, as an earnest of more to come and as the beginning of an end to the Bougainville Crisis.

Alas, within weeks the smiles had faded, and by the end of October they were replaced by recriminations and outright animosity. The failure of the Endeavour Accord can be attributed to several circumstances and actions; the following pages will sketch out what I infer, from published reports, to be some of them. It is probable that there are others not yet documented. Not least among those causes was the wording of the accord itself, which reads as follows:

A delegation of the National Government, led by the Right Honourable Sir Michael Somare GCMG MP held broad-ranging talks with a delegation of Bougainville, led by Joseph Kabui, from the 28th July to 5th August 1990 outside of Kieta Harbour, aboard Her Majesty's New Zealand Ship, Endeavour.

International observers from Canada, Vanuatu and New Zealand were present at the Talks.

The two delegations welcomed the opportunity for face-to-face dialogue, and agreed that this process should continue.

They agreed that the long-term political status of Bougainville is to be addressed as part of the continuing dialogue.

Meanwhile all political declarations with respect to that future status are deferred.

The National Government delegation confirmed that at the earliest opportunity it would take all practical steps consistent with the Constitution of Papua New Guinea to bring about the return of services to Bougainville.

Such restoration of services would be done without force. But should personnel responsible for restoring services be interfered with, the Government of Papua New Guinea reserves the right to withdraw services where such interference occurs.

The return of services, particularly health, education and communications are accepted as a matter of urgent priority.

The two delegations agreed that in order for services to be restored on Bougainville, the security of personnel must be fully safeguarded.

The Government of Papua New Guinea will consult regularly with the present Bougainville delegation on the restoration of services.

The two delegations agreed on the desirability of the full participation of Bougainvillians in future security arrangements on Bougainville.

The two delegations thanked the Government of New Zealand and the Royal New Zealand Navy for their generous assistance in making a venue and support facilities available for the talks.

The two delegations also thanked the observers for their cooperation.

Further meetings of the two delegations to review implementation of arrangements for restoration of services will be held. The next such meeting will be held within eight weeks.

In the succeeding months, the Endeavour Accord turned out to be mutually disappointing, both in what it said and in what it did not say. The text did not specify when the promised goods and services would begin to arrive, and for the Bougainvillians the ever-lengthening delay in arrival increased their mistrust and anger towards the PNG government. What proved to be even more divisive was the clause stating that the PNG government would take all practical steps to bring about the return of goods and services 'consistent with the Constitution of Papua New Guinea'. This seemingly innocent, even perfunctory, clause had, or was later on given, diametrically different meanings by the two sides. But before considering that difference, let us take a look at the social and political settings from which the two delegations came — first, that of PNG.

The PNG delegates to the *Endeavour* meeting were representatives of a government whose ascendancy was very precarious and whose leading members, the Cabinet, were outspokenly factionalized regarding Bougainville. All Cabinet members appeared to be opposed to Bougainville's secession, but there were apparent differences among them regarding means. Some members seemed to favour prompt and direct military action to suppress the rebellion; others seemed to favour a more gradual strategy of reducing secessionist sentiment and BRA popularity through restoration of normality, including government largesse. Differences in opinion doubtless exist among the leaders of all democratically elected governments, but in PNG such differences appear to be unconstrained. In the case of the Bougainville problem, the strategy of the Cabinet's doves seemed to prevail while the Prime Minister was in residence, that of the hawks when he was outside the country. Full explanation of this inconsistency would require more space than this book can spare and more information than this writer possesses; among them, however, may be mentioned the ease and frequency of party-switching (encouraged in some cases by financial lures), and unlimited opportunities for parliamentary no-confidence votes. (For a lively and authoritative description of PNG's current governmental institutions the reader is referred to Dorney's 1990 book.)

The instability of PNG's government and the dissonance among its elected leaders were matched by rivalries and lack of co-ordination among some of its administrative bodies. Examples are the premature withdrawal of police after the March 1990 ceasefire on Bougainville, and (as will be described) the September 1990 attempt to resume governmental services on Buka. As for the widely publicized cases of human-rights abuses by police and security forces on Bougainville during their actions against the BRA, that has been attributed to a combination of pay-back mentality and weak discipline.

The lack of cohesion among and within PNG governmental bodies was paralleled by widespread civil disorder throughout the country. In the urban areas gangs of feuding, robbing, bashing, and sometimes raping *raskals* were numerous, and nearly beyond control; in parts of the Highlands tribal feuding intensified. Various factors contributed to those disorders. Most weighty perhaps was an education system which was both inadequate and irrelevant: every year thousands of young people were thrust into an economy that provided no jobs for large numbers of them, or jobs for which many of the lucky ones were ill prepared. Another factor was the persistence of ethnic exclusiveness and chauvinism—not just brownskins versus blacks, or Highlanders versus Coastals and Islanders, but one group of *wantoks* (language-mates) against another, however similar their skin colour or however close their natal communities.

The civil disorder was the result of lack of employment and a shortage of agents to maintain order, and both these factors were exacerbated by the loss of revenue from the Panguna mine, which had provided a large proportion both of the nation's foreign exchange and of the government's normal operating revenues. If all goes well, other mining projects already under construction will eventually make up for the revenue losses from Panguna. But 'satisfying the relevant landowners' in these ventures has proved to be no easy task. For one thing, the problem of deciding which claimants were relevant had become highly complex and fractious. Also, by 1990 the ore-owning claimants, or their opportunistic advisers, had learned the Panguna lesson well enough to demand, and obtain, lion-size shares of mine earnings.

Meanwhile, the nation was suffering another financial loss, not only in government revenues but in the cash incomes of thousands of its citizens. This came about through the fall in world market prices of PNG's principal export crops: copra, cocoa and coffee. The closing of Panguna meant an indirect loss to most PNG citizens in terms of various government services, and a direct loss to BCL's employees (and to many other Bougainvillians whose incomes depended upon the mine). The decline in cash-crop income affected a much larger and more widely distributed segment of the nation's population, amongst them the thousands of youths who were thereby deprived of employment and had little option but to become *raskals*.

It should be added, however, that not all of the nation's citizens suffered financial losses during this time of mine closure and cash-crop recession. In fact, it was widely charged that several prominent citizens, including some in high official positions, sold their 'services'—their influence—for large sums of money, mainly to devious foreign entrepreneurs but also to aspiring or grasping fellow Niuginians. In a word, in the PNG of 1990 corruption was said to be rife.

Such were some of the more salient political and economic situations that prevailed in 'loyalist' PNG during and after the Endeavour Accord. Let us now attempt to discover what comparable situations obtained in rebellious Bougainville during that period.

What any outside observer 'knows' about those situations is very little, and most of that is second- or third-hand. We have only what has come through news broadcasts and articles in the PNG and Australian press. After mid-September, when PNG forces occupied Buka, conditions and events on that island were more fully reported, but communications with Bougainville itself remained tenuous and episodic.

From a few fragments of positive information, and in the absence of contrary evidence, it would appear that the interim government continued in office and that the BRA remained mobilized. The terms of the relationship between the civilian officials and the BRA leadership are not known. Considering pre-revolution arrangements, it is reasonable to suppose that there was some kind of hierarchically organized structure linking the interim government and the island's many village communities, but what functions were performed within that structure, and how it operated, can only be guessed. What is more patent is that members of the BRA were numerous and ubiquitous. They were actively engaged in sustaining and advancing the aims of the revolution, as defined either by the BRA leaders or by their own personal inclinations. Most of them appear to have been young—from mid-teens to mid-twenties. Concerning their numbers, their military commander, now 'General' Kauona, stated that by year's end 20,000 were 'under arms'; these were arms recovered from World War II caches plus, probably, some locally constructed cross-bows. Twenty thousand might strike some readers as a gross exaggeration, but it is not an entirely unlikely figure, considering the overall number of Bougainville's young men and the likelihood that few of them would have wished to, or been permitted to, remain outside the BRA. Inevitably rumours about 'splits' in the BRA were published in the press; while division may not have occurred among the Nasioi-centred leadership, reports of unauthorized rampages (including wanton destruction of BCL property) continued to seep out. Even officials of the interim government complained that 'law and order' remained a major problem, making reference to the far-flung 'troops' taking matters into their own hands.

Meanwhile the interim government named a representative in Australia, who attempted to win sympathy for the rebellion through radio interviews and through links with organizations characteristically dedicated to 'lost causes'. In addition, Amnesty International undertook an investigation of human-rights violations by the PNG forces during their drive against the

BRA. Their widely publicized findings were based on second-hand information, and did not include violations committed by the BRA. Predictably,
the PNG government refused to act on Amnesty's allegations, and told the
organization not to interfere.

Description of Bougainville's governance during this period would be
incomplete without reference to the roles of its Christian churches. The heads
of those churches — Catholic, Uniting and Adventist — were named as holding
'ministerial' offices in the interim goverment (to what effect is not known).
The churches themselves appear to have functioned quite independently as
before. Outside Bougainville questions were raised about the propriety of
the church leaders' roles in the Cabinet; some critics even demanded their
execution as traitors. The most trenchant and telling of the criticisms was
levelled against the Catholic bishop by the historian James Griffin (himself
a Catholic), but even that did not succeed in persuading PNG's other Catholic
leaders to censor the bishop.

The secessionists' relations with Solomon Islanders remained ambivalent.
On the one hand Solomons authorities arrested and jailed several armed BRA
soldiers discovered in the act of forcing supplies from Shortland Island
villagers. On the other, the authorities permitted village leaders from
southern Bougainville to meet with their Shortland Island counterparts 'to
renew traditional ties', and church groups in the Solomons made attempts,
without success, to deliver medical supplies to blockaded Bougainville.

As the months after the Endeavour Accord passed, these remained the only
material items for which the Bougainvillians continued to express dire
need — along with enough fuel to generate electricty to keep medicines viable.
At the beginning of the blockade complaints were widespread about the
shortage, and then the total absence, of such familiar imports as rice, tinned
meat and fish, and powdered milk, and there were predictions of widespread
fatalities from malnutrition. However, after a few weeks, enlarged gardens
began to produce, and those grievances diminished. Later some leaders
reported not only that there was 'enough food', but that the forced reliance
upon locally produced food was proving to be nutritionally beneficial! The
only reported instance of more recent food shortage occurred in southern
Bougainville, where unusually long and heavy rain ruined many crops.

Accompanying the return to traditional food-getting, several communities on Bougainville established neighbourhood markets in which food,
clothing, tools, etc. were bartered two or three times a week — a novel
practice for most Bougainvillians. In pre-European times organized marketing
had been usually limited to regularized bartering between inlanders and coast-
dwellers, who exchanged garden produce for fish. Elsewhere goods were
bartered irregularly, person to person, or were exchanged as reciprocal 'gifts'.

The lack of fuel doubtless proved irksome to the many Bougainvillians
who had become accustomed to travelling long distances in their trucks and
power boats. On the other hand, there were no longer any supermarkets
or hospitals to travel to, and no point in transporting copra or cacao beans
to shipment centres to which no ships came.

However, the blockade caused one deficiency that no amount of human effort or improvisation could overcome, and that was the absence of imported medical supplies. To a population become dependent upon Western medicines, return to native remedies was unacceptable—and patriotic stoicism was unlikely. Precise statistics are unavailable, but credible sources indicate that numerous preventable deaths occurred from lack of medicines, and that a whole cohort of infants had become endangered through lack of accustomed immunization.

Bougainvillians' need for medical supplies was tragically apparent and locally irremediable. Their expressed wants for resumption of children's schooling is understandable, but rather puzzling. Many of the teachers employed in the province's government-financed schools had departed after the withdrawal of the police and security forces. Money for financing the schools, including the teachers' salaries, ceased to flow after the blockade was imposed. Why, though, should the otherwise impoverished Bougainvillians fail to operate their own schools, at least partially? Surely, there were many trained teachers still available, and if the need was perceived to be urgent some means could have been found to induce them to resume teaching. Even more perplexing was the churches' failure to resume what had once been one of their principal, even jealously guarded, functions.

Between the Endeavour Accord and the Honiara Declaration (see below), there was no reliable news on how many Bougainvillians favoured secession, nor how strongly they felt about it. Nor do we know whether support for the BRA was widespread. Those questions were in due course answered with respect to most of the residents of Buka Island, but as to those still living on Bougainville itself the many estimates that have been aired in the press varied as widely as —and in relation to—the political persuasions of the guessers.

In the text of the Endeavour Accord the PNG delegation promised to resume services 'at the earliest opportunity'. While no time frame was specified, the widespread expectation was for 'as soon as possible'. As for priorities, one 'official' sequence was defined the day after the accord signing by Somare:

> Fuel and medicines will be among the first essential supplies to go into Bougainville. Within the next three weeks the Government will allow a large quantity of fuel, food, medical supplies and other goods to be shipped to the island . . . fuel [being] needed to restore electricity, especially for the upkeep of medical drugs and vaccines. The next step would be the reopening of schools and telecommunications. Communication is expected to be fully returned by the end of the month. [*Post-Courier*, 7 September 1990]

Plans for resumption of medical and other goods and services were initiated immediately, but they were delayed by bureaucratic muddles and transport difficulties, and, it was rumoured, obstructed by political opposition within the Cabinet. With respect to the transport difficulties, most of the companies whose ships and planes had served, or could serve, Bougainville were unwilling to return there without insurance cover, and the insurers were unwilling to insure without government-backed guarantees of safe passage and unloading. Doubts about safety also deterred the resumption of government services. Hundreds of Bougainvillian public servants expressed eagerness to return and work in their home province, but were unwilling to do so unless they could be safeguarded from the BRA. (As for non-Bougainvillian public servants, the Bougainvillian delegates to the *Endeavour* meeting made it clear that only Bougainvillians would be allowed to return.)

The specific administrative foul-ups that were rumoured to have added to the delays are difficult to document, but can be attributed both to the universal sluggishness of bureaucracies and, in PNG in particular, to rivalries over 'turf'. As for political obstructions to implementation of the Accord, some backbench MPs continued to speak out for a military solution to the Crisis. And there are grounds for believing that some hawkishness persisted within the Cabinet itself.

Despite all the above, the national government organized three supply ships to carry food, medical supplies, and fifty-three public servants, escorted by two Defence Force patrol boats. They dropped anchor in Buka Passage on 31 August with the expressed intention of commencing the task of returning services to blockaded Buka. They were faced by throngs of armed BRA, who prevented them from approaching the wharf at Buka town. In addition to raising objections to the presence of the patrol boats, the BRA leaders demanded that all PNG vessels headed for Bougainville, including Buka, call first at Kieta for clearance by the BRA there; that all supplies be dropped off at Kieta for distribution by the BRA; and that no former staff of the Northern Solomons Provincial Government or of the Port Moresby Department of the Northern Solomons be involved in the restoration programme. (*Post-Courier*, 13 September 1990)

After three and a half days of stand-off, the PNG flotilla raised anchor and left. Mutual recriminations flared up again. Joseph Kabui charged that the PNG government had violated the Endeavour Accord by sending patrol boats with the supply vessels. The PNG government countered that 'the security forces did not use force in arms and weapons to clear their way before landing' and that 'the Defence Force involvement in the restoration of services [had] been made in accordance with the National Constitution [in order to] assist civilian personnel'. (Joint Ministerial Statement on the Implementation of the Endeavour Accord, *Post-Courier*, 13 September 1990)

The disagreement exemplified in this episode might have resulted from understandable differences in interpretation of the blurry wording of the Endeavour Accord. Such was not the case with another statement attributed to Kabui, namely, that according to the Endeavour Accord all supplies had to be 'landed first at Kieta, where security checks would be made before the vessels [would be] allowed to land supplies in other ports'. (*Post-Courier*,

5 September 1990) Nothing even suggestive of that interpretation is to be
found in the Accord.

In any case, the disagreement was rendered moot when, three weeks later,
the PNG government landed troops on Buka. The announced reason was
to ensure 'resumption of services' but the unmistakeable purpose was to
re-establish national control. The grounds, or justification, for this action
are contained in a statement by Attorney-General Narakobi, which said in
part:

> Between September 5 and 7, 1990, BRA increased its activities of
> harrassment against the innocent leaders and the people of Buka. BRA
> on Buka picked up Sam Tulo and James Togel, to be transported to
> Panguna for questioning by [the BRA] intelligence organisation. But
> with the help of Buka leaders, the movement did not eventuate. Some
> Buka leaders are being held captive at Hanahan Village by BRA. On
> Monday, September 17, 1990, BRA led by Linus Kabutu carried out
> early morning raids on Lemankoa villages. During these raids a number
> of community leaders and villagers were picked up and were to be
> transported to Panguna via the west coast road through Karoola, for
> questioning and detention. The villagers of Hahalis, Lemanumanu and
> Lemankoa put a force against the BRA and the transportation of the
> leaders and villagers was stopped. On the night of Monday,
> September 17, villagers of Hahalis, Lemanumanu and Lemankoa met
> and petitioned the National Government to send in security forces to
> maintain security and safety of the villagers. The petition is signed by
> 125 people. Minister Narakobi said, following these events, leaders of
> Buka, including some BRA, issued a demand to the National
> Government to send in security forces within 48 hours to protect the
> innocent people. We have now acceded to their request. [*Post-Courier*, 24
> September 1990]

The petition had been carried by boat to Nissan Island (whose people had
remained loyal to PNG) and thence by plane to Rabaul.

The number of government personnel landing on Buka was reported to
include fifty Defence Force troops, two police riot squads, and fifty civil
servants — the latter sent presumably to 'restore goods and services'. The
landing was contended by the BRA in a brief gunfire exchange in which
two BRA soldiers were killed. The landing force was also met by members
of a newly formed group which called itself the Buka Liberation Force (BLF).
It consisted of an unspecified number of Buka residents opposed to the
BRA — including some very recent defectors from the BRA. As stated in
the petition drawn up by the Buka leaders: 'The BRA Commander has been
forced to retreat by the popular demand of the Buka People. As a result he
has resigned. He is now commanding a new anti-BRA Buka Liberation

Force.' (*Post-Courier* 24 September 1990) Surely this is one of the swiftest conversions in military history!

During the next four months the following conditons prevailed on Buka. The government forces eventually gained control throughout most of the inhabited parts of the island, plus an outpost at Bonis plantation, on the Bougainville side of Buka Passage. There were occasional clashes, resulting in mutual casualties, with BRA remnants, whose estimated numbers ranged from 'a small handful' to 'a hundred or more'. The clashes included an 'invasion' of 'several hundred' BRA across the passage, which was repulsed amid losses on both sides; according to the security forces, their own losses were 'very small' while those of the BRA were 'very large'. Only late in the period did the government troops move beyond their protective role and commence to root out the elusive and persistently destructive remnants of the BRA. In the beginning the newly formed BLF was greeted as a welcome ally of the government's people, but after a while some of its, largely undisciplined, members proved to be as troublesome as the BRA to ordinary citizens.

Meanwhile, the restoration of goods and services proceeded at a pace characteristic of governmental programmes nearly everywhere. Stops alternated with starts, supplies were delayed in transit or were unevenly distributed, and public servants argued over jurisdiction and complained about lack of support. So many Buka Islanders travelled to Rabaul, to obtain supplies or visit acquaintances or escape discomforts and tedium, that Rabaul authorities attempted to curb their influx.

Predictably, the leaders in Kieta and Panguna erupted in anger at the 'duplicity' of the Port Moresby government. They scornfully rejected the govenment's explanation that it had sent forces to Buka at the request of local 'leaders', and that they were there only to 'protect the restoration of goods and services'. Bougainvillians feared, with good reason, that the PNG government would undertake similar 'restoration' missions on Bougainville itself, beginning in Buin. This fear was reinforced by Narakobi, who stated in late September that 'it was possible that [Bougainville itself] eventually could come under permanent military administration unless the secessionist crisis was resolved'. (*Post-Courier*, 28 September 1990) More importantly, Kabui and his associates said they would refuse to negotiate with the PNG government on a date and venue for the prescribed second meeting until the security forces had been withdrawn from Buka.

Another set of incidents that revealed a hardening of the PNG government's attitude towards Bougainvillian secessionists was the constant presence of its patrol boats just off Bougainville's shores—and not simply for the purpose of enforcing the blockade. There was one report (undoubtedly biased but circumstantially persuasive) that a PNG patrol boat had entered Kieta harbour and fired mortar shells at houses, shops and other buildings, including the Tubiana Catholic mission. (*PNG Times*, 29 November 1990) According to other on-shore witnesses, Kieta mission buildings were again shelled on 4 December, as were shoreside villages in Koromira on 12 December. (*PNG Times* 20 December 1990) Some anti-secessionists dismissed these incidents as fabrications. Even if they actually occurred, they should not be attributed necessarily to official PNG policy; as noted earlier, there were both hawks and doves — along with doveish hawks and hawkish doves — in the Cabinet. Given the nature of PNG's administrative 'flexibility' it is quite possible that the incidents just cited were independently ordered by a Cabinet hawk, or even by a down-the-line official acting on his own.

To many persons viewing Bougainville's Crisis, its most saddening aspect was its effect upon the health of its blockaded civilian residents. Whatever one's sympathy, the plausible reports about scores of individuals suffering or dying from lack of medicines is moving. Many individuals and organizations in the southwest Pacific attempted to intercede, including church and humanitarian aid groups in Australia, the Solomon Islands and Fiji. Officially the PNG government welcomed the offer of contributions from such groups, but insisted that they be delivered under the protection of its own security forces and police. Because the Bougainvillie secessionists refused to admit those troops and police, the profferred aid could not be delivered. Among its many other woeful aspects, the Bougainville Crisis provides yet another example of the martyring of innocent bystanders to unrelenting politics.

What appeared — or was perhaps designed to appear — to be a softening in the government's attitude occurred in early January 1991. The Prime Minister announced, 'We have . . . just lifted the embargo on goods and services going into Bougainville as of today [January 4].' He added that all such activities would have to be co-ordinated 'through the Security Advisory Committee that advises the National Security Council, which in turn advises the National Executive Council'. (*Post-Courier*, 4 January 1991) Such layers of bureaucracy did not promise a speedy resumption of supplies.

Although the positions of the opposing sides seemed to be fixed publicly in cement, it is evident that some unpublicized dialogue was taking place between the PNG government and the Bougainville secessionists. On 29 December Francis Ona stated, in an interview with a correspondent of the Australian Associated Press, that he and his fellow revolutionaries would no longer talk directly with the PNG government, 'because its negotiations can no longer be trusted'. Instead, he and his associates wanted Father John Momis to act as an intermediary. (*Age*, 29 December 1990) This was puzzling, because Momis, a member of the Cabinet, who had repeatedly condemned the rebels' declaration of secession, had been added to their rumoured hit-list! This writer has received no information which might provide clues about the reason for this dramatic *volte-face*, and will not waste words speculating about it (except to surmise that it was initiated by the Bougainvillians). Shortly thereafter Father Momis proceeded to Honiara, where in company with other leaders of Bougainville's interim government (including some members of the BRA) plans were made for a meeting between the latter and the very PNG negotiators whom they 'no longer trusted'! This species of diplomacy was attributed by some observers to 'Melanesian Common Sense'.

The official meeting convened on 22 January and lasted two days. PNG's principal delegate, Michael Somare, was attended by Bernard Narakobi, Father Momis, and Benais Sabumei, the Minister for Defence. Joseph Kabui, the leader of the Bougainville delegation, was attended by the Rev. John Zale (head of the province's Uniting Church), James Singko (Francis Ona's reputed chief counsellor), and by Patrick Etta (whose position I do not know). Unlike the *Endeavour* meetings no journalists were permitted at the official deliberations — reportedly a condition imposed by the Solomon Islands official hosts.

The negotiations yielded an intriguing document labelled the *Honiara Declaration on Peace, Reconciliation and Rehabilitation on Bougainville.* (See Appendix) Although it did not address the question of Bougainville's future political status, and did not even mention the future of the mine, it proposed some radically new arrangements for 'reconciliation and rehabilitation', including supervision of them by a multinational force. We are not informed about the amount of prayer that punctuated the negotiations, but during the signing ceremony there were the usual 'words of gratitude and praise, tears, handshakes and bear hugs', along with references to The Melanesian Way. (*Post-Courier*, 25 January 1991). In addition, the Rev. Zale uttered what must be adjudged an understatement of dizzying implications: having said that he and his fellow delegates were gratified by the achievements represented by the *Declaration*, he stated that 'it was [now] a matter of implementing the proposals to make it a reality'.

Postscript

In present-day high-tech newspaper publishing events covered by a reporter can be printed and circulated within hours of their happening. Not so with books like this one, which require many months to transform the writers' drafts into illustrated, critically evaluated and scrupulously edited printed products. In the case of the present book that interval will have lasted from the end of January (1991) to October, and will have witnessed an accelerating cascade of events that serve to deepen, complicate and reshape the Bougainville Crisis. By the time of the writing of this postscript, there was no way of knowing how or when the Crisis will end, but several events had been reported, and several situations could be discerned, which provide some indications about its course.

While continuing to express commitment, publicly, to the 'peaceful' principles and procedures contained in the Honiara Declaration, the more influential leaders of the PNG Government appeared to be shifting more and more towards a *military* solution of the Crisis — this, despite suspension from Cabinet, on charges of corruption, of its most hawkist member, 'Ted' Diro. Father John Momis, the Cabinet's most dovish member, who had initiated the Honiara meeting, was appointed to the central PNG role in implementing the Declaration but was in fact often thwarted in performing it by other government agencies, including especially the Defence Forces on the ground. And, although Port Moresby's military control over Buka was virtually completed, that Island's return to normalcy continued to be slowed, through bureaucratic incompetence and inter-agency rivalries — plus traditional factional cleavages among its own residents.

Nevertheless, liberation from BRA control had spread from Buka Island to contiguous areas of north Bougainville, and in two ways. Ever-increasing numbers of north Bougainvillians made unhampered visits to Buka, for economic and social reasons, and expressed their wishes for a return of government 'services', including protection from BRA domination. That evidently spontaneous rejection of secession (which, it will be recalled, had long prevailed in this region) was agreeably assisted through the stationing of two contingents of PNG security forces there. But that was not all.

On 13 April, while the PNG Government was still proclaiming its commitment to a non-military implementation of the Honiara Declaration, two detachments from the Buka-based security forces carried out raids along Bougainville's east-central coast. One of them, in a patrol boat, shelled a BRA camp between Kieta and Arawa; the other landed a few miles north of Arawa, exchanged gunfire with some BRA, and destroyed a vital highway bridge, thereby cutting off central from northern Bougainville for all vehicular traffic. These raids took place, characteristically, while the Prime Minister was out of the country, and according to the Defence Minister had not been 'officially' authorized. Father Momis, of course, condemned the raids vehemently. And, although the Prime Minister subsequently 'reprimanded' the field commander who had ordered the actions ('without National Security Council approval'), he added that it had been done with '. . . the support of the local chiefs and community leaders', and that '. . . the Government had been aware for sometime that [the incursions] were going to happen because of the increasing BRA interference in the restoration of services'. [*Post-Courier*, 19 April 1991]

Meanwhile, 'insiders' were predicting that similar incursions, based on similar grounds, and also in the absence of 'official' authorization, would be mounted along the coast of Southern Bougainville.

Another seemingly 'unauthorized' action taken by an individual PNG official to solve (among other problems) the Bougainville Crisis was the effort made by no less an authority than the Foreign Minister, Sir Michael Somare, to promote the sale of CRA's stake in the copper mine, hoping thereby to soften the BRA's demand for secession. CRA's management had consistently expressed its wish and intent to resume mining when and if possible, along with its willingness to consider selling if that were the *only* way to avoid secession. Even so, BCL's Chairman, Carruthers, submitted to Somare's personal 'request' and met with the prospective buyer (an American financier, who turned out to be a friend of a long-time acquaintance of Somare). However, the meeting led to nothing, partly perhaps as a result of the size of the offer, which Carruthers characterized as 'derisory'. In due course the whole episode was officially dismissed by the Prime Minister, who declared that CRA would not be asked to sell unless that were seen to be the only way to avoid effectively *de facto* secession.

In a subsequent academic seminar at Canberra's Australian National University attended by both PNG parliamentary and Bougainvillian speakers, Carruthers emphasized that until political government issues were solved, the company 'can only be a bystander'.

He then emphasized the company's wish to continue, but it remained to be seen whether the necessary money could be raised at all because of investor perceptions. 'Whoever the potential investors are, before they put up the money, they need to be confident not only about the return *they* would get, but also that any arrangements would endure.'

With a low grade resource and huge re-establishment costs, a rejuvenated mine might employ only 2500 rather than the previous 4000 people. 'Much wealth has been destroyed, and a resumed mining operation will not be able to provide for all stakeholders on the same scale as it did before.'

Turning now to the other official signatory of the Honiara Declaration, the Interim Government of the Republic of Bougainville, several signs about the authority and unity of that body were subsequently revealed. The first occurred in Honiara itself.

One of the mutual undertakings of the Declaration was to create a Multi-national Supervisory Team supervise the disarming of both the BRA and the (Buka-based) BLF . . . 'including the surrender and destruction of arms'. Then, within hours of its signing . . . 'Bougainville delegates today denied that they had agreed to the destruction, as well as the surrender, of the BRA's arms, although the final version initialled by the leader of the Bougainville delegation . . . Mr Joseph Kabui, does include this provision. "Either it is [a] mistake or they have tricked us," one delegate said. "But we did not agree to the destruction of the arms".' [*Sydney Morning Herald*, 25 January 1991] Predictably, this position was subsequently reasserted, most firmly, by the BRA leadership (who had not attended the meeting).

A second sign was exposed a few weeks later, when the first PNG-sponsored vessel arrived at Kieta and unloaded food, medicines, and fuel—all of which had been requested, even implored for, by one or another of the Republic's more or less official spokesmen. But not by all of them. For, a representative of the BRA leadership soon appeared at the dock and ordered the food back on board, saying, in effect, 'So long as we can grow our own sweet potatoes we don't need or want charitable hand-outs of food.' (Doubtless, an accurate reflection of the views of BRA's secessionist zealots but not, perhaps, of the average Bougainvillian grown accustomed to supermarket imports.)

For a third kind of sign we can turn to news reports of two more recent events. The first was broadcast on the 6 May South Pacific News Summary over Radio Australia and said:

> The Bougainville Revolutionary Army has ordered a ceasefire in the battle against the PNG government for independence. But the northern commander of the BRA, Peter Barek, says the group has not given up its desire for independence, which it will now pursue through political means. Mr Barek says the BRA is now interested in peace, reconciliation and rehabilitation because the ordinary people have suffered too much in the last two years. He says BRA leader, Francis Ona, ordered the ceasefire so that he could prepare for talks with the PNG government scheduled for July.

Four days later the same news medium broadcast as follows:

> The leader of the secessionist rebel group on Bougainville, Francis
> Ona, has denied ordering a ceasefire. In a statement issued in Honiara,
> Mr Ona said news of the ceasefire . . . was not correct. He said he
> had not spoken recently to the northern commander of the
> Revolution, Peter Barek, who had given news of the ceasefire to the
> media.

Meanwhile, how the rest of Bougainville's people have been and are now
faring is difficult to determine. The blockade imposed by the PNG government has been officially lifted but access to, including supplies for, the Island
is still controlled by Port Moresby—which means very *limited* access and
very *meagre* supplies (and the latter mainly from non-government charitable organizations). Moreover, because of the secessionist Interim Government's requirement that all supplies be channelled through Kieta–Arawa,
it is plausible to conclude that their distribution is uneven, to say the least.

Even slower, and in some cases non-existent, has been implementation
of the other grand programmes envisaged in the Honiara Declaration,
including the establishment of a Multinational Supervisory Team. In fact,
at this (final) writing the only 'hopeful' sign of an end to the impasse is
the reported willingness of both 'President' Ona and 'General' Kauona to
attend the next scheduled meeting with the PNG government . . . 'sometime in July'. The kind of 'hopefulness' expressed by this sign doubtless
differs with the parties. To the PNG government and other anti-secessionists
it may signify progress in their (unannounced) strategy to weaken the BRA
by constricting its territorial domain. And to the BRA and its supporters
it may signify a growing self-confidence in the strength and ultimate success of their cause.

In early June, Buka Island leader Joseph Hapisiria was quoted as calling
for a separate province for his smaller island, where schools had reopened
for more than 4000 children. 'Mr Francis Ona continues to mislead the
people of Bougainville about secession, a cultist dream which has led many
of our people astray and caused so much destruction and suffering.' [*Post
Courier*, 10 May 1991]

A fortnight later, an interim government press release claimed a 'unanimous show of solidarity' by 'more than 400 chiefs representing all regions
including Buka Island and the Atolls' for independence. Which and what
kind of 'chiefs' was unspecified. In the same week, it was reported that
Father John Momis was a shortlisted candidate for the vacant post of
secretary-general of the Fiji-based South Pacific Forum.

Later in June, Australian television reporting accelerated the impasse again.
The previously-reprimanded PNG field commander was replaced after a
meeting of the PNG Executive Council. A fortnight later, he had been
accused by Bernard Simiha, the public servant head of his own govern-

ment's 'Bougainville Task Force' of a 'bashing and kicking' because of the content of a Simiha radio broadcast. Lt Colonel Nuia had appeared, in the edited television program, both to confirm the use of helicopters as firing platforms (p.231) and the dumping at sea of Bougainvillian bodies. Another item was added to the long list of issues of fact which future official enquiries and future historians will attempt to untangle.

30 June 1991

Appendix:
The Honiara Declaration

After listing the delegates and expressing thanks to the Solomon Islands hosts, the *Declaration* reads as follows:

Principles
6. We desire peace and reconciliation with each other and with our Heavenly Father.

We take a joint responsibility to restrain from the use of weapons and arms to help us to create an environment of peace and harmony as well as a precondition to justice and peace.

We agree to defer discussions on the future political status of Bougainville and have further agreed to embark upon a joint programme of Peace, Reconciliation, and Rehabilitation, within the current constitutional framework of the Nation of Papua New Guinea.

We reject violence and seek meaningful consultation as a means of solving the crisis, and deeply mourn the loss of lives and destruction to properties, and trust in the common fatherhood of God and resolve to find lasting justice, peace and security on the Island of Bougainville.

We recognise the importance of establishing legal and representative authority in Bougainville to assist in returning the land to normalcy.

We recognise the constitutional role of the Papua New Guinea Defence Force.

We agree to accept external assistance including a Multinational Supervisory Team (MST) to contribute to the implementation of this programme under the framework determined in this Declaration.

We commit ourselves to the welfare and security of all individuals and organisations who participate in this Programme.

We endorsed maximum Bougainvillean involvement in the implementation of this Programme.

Definition of 'Programme'
7. The 'programme' in this Declaration means the Package phased arrangements for the restoration of services on Bougainville including: —
Phase I
(i) Peace and Restoration.
(ii) Lifting of the Blockade.

(iii) Establishment of Task Force.
(iv) Establishment of Interim Legal Authority.

Phase II
(i) Restoration of Services.
(ii) Rehabilitation Programme including maintenance and reconstruction; and other associated activities determined by the Legal Authority in Bougainville in accordance with this Declaration;
(iii) Future negotiations.

Task Force and Membership
8. In order to facilitate the execution of this Programme, we agree to hereby establish a Task Force which shall consist of representatives appointed by the National Minister for Provincial Affairs, in consultation with an Interim Legal Authority.

Terms of Reference
9. We agree that the terms of reference for the Task Force shall include:
(i) Planning, Co-ordination and implementation of this Programme;
(ii) Monitoring and supervision of this Programme;
(iii) Investigate and determine the scope and components of the project under this Programme;
(iv) Investigate, mobilise and secure all financial avenues at its disposal to finance this Programme;
(v) Develop a detailed timetable to implement this Programme which must be submitted to the Minister for Provincial Affairs for final approval as soon as practicable following their appointments.
(vi) The Task Force shall report to the Legal Authority in Bougainville;
(vii) Furnish monthly reports or otherwise as directed by the Legal Authority in Bougainville; and
(viii) Undertake other responsibilities as directed by the Legal Authority in Bougainville to implement this Programme.

Obligations and Responsibilities
10. Parties agreed to take the following actions:-

National Government
(i) [blank]
(ii) Grand amnesty and immunity from prosecutions to the members of BRA, and BLF in accordance with legal and constitutional requirements of Papua New Guinea.
(iii) Organise a multinational supervisory team to participate in this Programme.
(iv) Commit and disburse funds to the Programme under its normal budgetary allocations.
(v) Allow and facilitate non-government agencies, including Churches and Community groups to contribute towards the successful implementation of this programme.
(vi) Resume all government services including public and statutory administration, law and order and justice.

Bougainville Side
(i) Disarm the BRA, BLF; and its associated militant activities including the surrender and destruction of arms under the supervision of the Multinational Supervisory Team.
(ii) Release of all detainees held as a consequence of the conflict.
(iii) Guarantee the safety and welfare of the members of the Multinational Supervisory Team.

(iv) Assist the Task Force and the Legal Authority in Bougainville to expand funds towards effective implementation of projects to be developed under this programme.

(v) Receive and facilitate non-governmental agencies, including churches and community groups to contribute towards the successful implementation of this programme.

(vi) Provide conditions and environment conducive for the restoration of services under this programme.

Programme Schedule

11. We agree that the following Time Schedule shall be adopted to implement this Programme from the signing of this Declaration:-

(i) One (1) week to one (1) month — establishment of the Task Force and the assembling of the MSD.

(ii) Between one (1) month to 6 months project identification and resource mobilisation.

(iii) Eighteen (18) to thirty-six (36) months completion of Programme, abolition of the Task force, return to normalcy.

Review and Consultations

12. We agree that this Declaration shall be reviewed at least on a six (6) monthly interval until the conclusion of the Programme in accordance with the programme schedule.

13. Notwithstanding the foregoing paragraph, the National Minister for Provincial Affairs may request a review either independently or on advice of the legal authority or the Task Force.

14. The programme under this Declaration does not for the time being include the programme of restoration of services undertaken by the National Government on Buka and adjacent islands.

15. The National Minister for Provincial Affairs shall determine and reconcile the relationships between these two programmes on advice from the legal authority in Bougainville.

Dispute Settlement

16. We agree to resolve any dispute including conflict or misunderstanding arising from the Programme under this Declaration through consultation and dialogue.

17. Where resolution of disputes cannot be reached, each party may recommend arbitrators for approval by the other. When appointed, the arbitrators shall work towards resolving any such conflict and misunderstandings.

Termination

18. This Declaration shall be terminated upon completion of this Programme or by one Party when acts of sabotage or similar action inconsistent with the spirit and letter of this Declaration taken by the other [*PNG Times*, 24 January 1991]

The declaration did not address the question of Bougainville's future political status, and did not even mention the future of the mine. The arrangements for 'reconciliation and rehabilitation', were radically new. The multinational force was subsequently canvassed to consist of contingents from all or some of Australia, New Zealand, Fiji, the Solomon Islands, and Vanuatu.

Then, within hours of its signing, came the following report: Bougainville delegates today denied that they had agreed to the destruction, as well as the surrender, of the BRA's arms, although the final version initialled by the leader of the Bougainville delegation to the Honiara talks, Mr Joseph Kabui, does include this provision. 'Either it is [a] mistake or they have tricked us,' one delegate said. 'But we did not agree to the destruction of the arms'. [*Sydney Morning Herald*, 25 January 1991]

References

CHAPTER ONE

For general summary of human migrations into Melanesia and of their geological and geographic settings see Oliver 1989, Part 1. For specific accounts of the earliest archaeological horizons in the north Solomons see Wickler and Spriggs 1988, Spriggs nd, and Wickler 1989. For works on later horizons on Bougainville and Buka see Spriggs 1990, Irwin 1973 and Terrell and Irwin 1972.

For a general survey of languages in Melanesia see Oliver 1989 (chapter 3). For classification of languages on Bougainville–Buka see Allen and Hurd nd.

For a general survey of physical types ('race') in the Pacific see Oliver 1989 (chapter 2). For specifics on the physical types of indigenes of Bougainville see Friedlaender 1975 and Oliver and Howells 1957.

The data concerning geography are based mainly on Scott et al. 1967.

CHAPTER TWO

This version of Bougainville's voyage is from Bougainville 1967.

For a list of early European voyages to the Solomons see Jack-Hinton 1969.

This version of d'Entrecasteaux' voyage is bsed on de Labillardiere 1800.

For a general summary of labour-recruiting ('blackbirding') in this region see Corris 1973 and Rannie 1912. See also Docker 1970, from which this account of the *Carl* voyage is taken. For accounts of Germany's annexation of Bougainville–Buka see Morrell 1960 and van der Veur 1966.

CHAPTER THREE

The *Encyclopedia of Papua and New Guinea* (P. Ryan, ed., 1972) contains an excellent and comprehensive description of colonial events, institutions and personalities in German New Guinea, including the Bougainville District, written by an Australian historian, Marjorie Jacobs. It also contains a useful bibliography of official and unofficial writings about the colony that were published prior to 1971. Subsequent publications include writings by Moses and Kennedy 1977, and by Sack 1976, 1986. See also Reed 1942, Rowley 1958, Laracy 1976 and Panoff 1979, 1986. For descriptions of Bougainville–Buka indigenes see Parkinson 1907, Ribbe 1903, Frizzi 1914, Wheeler 1914, Burger 1923, and the numerous writings by R. Thurnwald listed in the bibliography.

The most comprehensive official publications on the early post-war era are:

Australia, Administration of the Territory of New Guinea. *Report to the General Assembly of the United Nations*, Canberra, Government Printing Office, 1946.

Australia, *Annual Report of the Territory of Papua*, Canberra, Government Printing Office, 1946.

For Papua New Guinea as a whole in the early post-war years see especially P. Ryan (ed.) 1972. See also Stanner 1953, Clunies-Ross 1969, Epstein, Parker and Reay (eds) 1971, Fisk (ed.) 1966, Hastings 1969, Rowley 1965, Mair 1970, Wolfers 1971, and the International Bank for Reconstruction and Development 1965.

For accounts of the Hahalis Welfare Society see Kiki 1968, J. Ryan 1969 and Laracy 1976.

For an account of the 1964 elections see Bettison, Hughes and van der Veur (eds) 1965 and an article by Ogan in Epstein, Parker and Reay (eds) 1971.

The journal, *New Guinea and Australia, the Pacific and South-East Asia*, published by the Council on New Guinea Affairs, Sydney, contains many relevant articles on this era, as do the *Waigani Seminar Papers*, published by the Research School of Pacific Studies of the Australian National University, Canberra, and by the University of Papua New Guinea, Port Moresby.

The Australian military occupation of German New Guinea is described in Rowley 1958, Lyng 1919, and MacKenzie 1927. See also Laracy 1976.

CHAPTER FOUR

Official information on the Territory of New Guinea during this era is contained in Australia's *Annual Reports to the League of Nations*. For general unofficial publications see Mair 1970, Reed 1942, Rowley 1958 and Stanner 1953.

For information on cargo cults on Bougainville–Buka see Worsley 1957.

For information on missions during this era see Laracy 1976, Luxton 1955, McHardy 1935, and Tippett 1967.

For information on indigenous societies of Bougainville–Buka during this era see Blackwood 1935 and 1936, Chinnery 1924, Oliver 1949, 1955, 1971, H. Thurnwald 1934, R. Thurnwald 1934.

CHAPTER FIVE

The principal sources about military campaigns on Bougainville–Buka are: Long 1963, Miller 1957, and Shaw and Kane 1963.

Accounts of the Coastwatchers on Bougainville–Buka are given in Feldt 1964 and Griffin 1978.

Views of World War II on Bougainville–Buka from the perspective of the missionaries are contained in Decker 1948 (ed.), Laracy 1976, Luxton 1955, and O'Reilly and Sedes 1949. See also Pinney 1990.

CHAPTER SIX

The Paul Hasluck quotation is from Bettison et al. 1965; the Robert Menzies quotation is from the *Sydney Morning Herald*, 21 June 1960; the E. J. Ward quotation and the quotation from the Department of Territories are from Stanner 1953.

Eugene Ogan's account of a local council's beginnings is from Ogan 1972. See also Fingleton 1970.

CHAPTER SEVEN

Some of the publications about these islands' native cultures listed below were based on field research carried out *after* the mine had begun to be developed, and therefore may describe some conditions affected by the mining enterprise. However, since their main focus is on the state of those cultures in pre-mine times they are given here rather than among the references to Chapter 9. Only the more comprehensive studies are listed; some of them contain references to other studies on the cultures in question:

Buka and northwest coastal Bougainville: Blackwood 1935.

Teop (northeast Bougainville): Shoffner 1976.

Konua (northwest Bougainville): Blackwood 1936.

Aita (north-central Bougainville): Rutherford 1977.

Eivo-Simeku (central Bougainville): Hamnett 1977.

Nasioi (east central Bougainville): Ogan 1971B, 1972.

Nagovisi (west central Bougainville): Mitchell 1976, Nash 1974.

Siwai (southwest Bougainville): Oliver 1955.
Buin (southeast Bougainville): H. Thurnwald 1934, R. Thurnwald 1934, Keil 1975.
Buin-Siwai-Nagovisi: Oliver 1971.
For more detailed bibliographies on native cultures of Bougainville–Buka up to 1981 see Oliver 1949, 1973 and 1981.

CHAPTER EIGHT

Publications issued by Bougainville Copper Ltd on its operations include its *Annual Reports*, 1973–89, and the following accounts written by company personnel: King 1978 and Quodling 1990.

Comprehensive accounts of Company operations written by non-Company individuals or organizations include: Applied Geology Associates 1989; Mamak and Bedford 1975, 1977A, 1977B; West 1972; P. Ryan in P. Ryan (ed.) 1972; Faber 1974; Momis and Ogan 1972; and Zorn 1973.

CHAPTER NINE

Articles on 'The Other Bougainville' listed in the Bibliography include the following: Bedford 1974; Bedford and Mamak 1975, 1976A, 1976B, 1979; Connell 1977, 1978A, 1978B, 1985, 1988; Connell (ed.) 1977; Griffin 1970, 1972, 1976, 1982, 1989; Hannett 1969, 1975, 1989; Harris 1975; Makis 1975; Mamak and Bedford 1974A, 1974B, 1977A, 1977B; Moulik 1977; Treadgold 1978; and Ward 1975.

CHAPTERS TEN TO THIRTEEN AND POSTSCRIPT

Most information contained in these chapters is taken from newspapers of Papua New Guinea (*Post-Courier*, *Niugini Nius*, *Times of New Guinea*, and *Arawa Bulletin*) and of Australia (*Sydney Morning Herald*, *Australian Financial Review*, *Age*, *Australian*; and from transcripts of news reports and interviews broadcast by the Australian Broadcasting Corporation and Radio Australia.

See also BCL *Annual Reports* for 1988 and 1989; Carruthers 1990, 1991; Dorney 1990; May and Spriggs (eds) 1990; and Polomka (ed.) 1990.

Bibliography

Allen, Jeremy, *The Languages of the Bougainville District*, Ukurumpa, Papua New Guinea, Summer Institute of Linguistics, nd.

Applied Geology Associates Ltd, *Environmental, Socio-Economic Public Health Review of Bougainville Copper Mine, Panguna*, Wellington, New Zealand, 1989.

Bedford, R. D., 'New Towns on Bougainville', *Australian Geographer*, 12(6) pp. 551–6, 1974.

Bedford, R. D., and A. F. Mamak:

1975A—*Migration to Southeast Bougainville, 1966–1974*, Bougainville Special Publications no. 3, Christchurch, University of Canterbury.

1975B—'A Town Council Election in Bougainville', *South Pacific Commission Bulletin* 25 pp. 45–50, Noumea.

1976A—'Bougainvillians in urban wage employment; some aspects of migrant flows and adaptive strategies', *Oceania*, 46(3) pp. 169–187.

1976B—'Kieta, Arawa and Panguna: Towns in Bougainville', *An Introduction to the Urban Geography of Papua New Guinea*. Ocasional Paper, Department of Geography, University of Papua New Guinea.

1977—*Compensating for Development: The Bougainville Case*, Bougainville Special Publications no. 2, Christchurch, University of Canterbury.

1979—'Bougainville', in A. F. Mamak and A. Ali (eds), *Race, Class and Rebellion in the South Pacific*, Sydney, Allen & Unwin.

Bettison, D. G., C. A. Hughes and P. W. van de Veur (eds), *The Papua New Guinea Elections: 1964*, Canberra, Australian National University Press, 1965.

Blackwood, Beatrice:

1935—*Both Sides of Buka Passage: an ethnographic study of social, sexual, and economic questions in the northern Solomon Islands*, Oxford, Clarendon Press.

1936—'Field work in Bougainville: an interlude', in L. H. Dudley-Buxton (ed.), *Custom is King*, London, Hutchinson's Scientific and Technical Publications, pp. 167–78.

Bougainville, Louis Antoine de, *A Voyage Round the World ... in the Frigate 'La Boudeuse' and the Store Ship 'L'Etoile' ... 1766–1769* (translated by J. R. Forster), London, Nourse 1772. (Reprinted by the Gregg Press, Ridgewood, New Jersey, 1967).

Burger, F., *Unter den Cannibalen der Südsee.* Dresden, Verlag Deutsche Buchwerkstaften, 1923.

Burton, J., 'The land tenure system of the Nasioi people: the bare cupboard of sustained research in mining areas'. Paper presented to the University of Papua New Guinea seminar on the North Solomons Situation: issues and options, September 1989.

Carruthers, D. S.:

1990—'Some implications for Papua New Guinea of the closure of the Bougainville Copper Mine', in May and Spriggs (eds), *The Bougainville Crisis*, pp. 38–44.

1991—'Bougainville—the future of the mining operation', paper delivered at the

Australian National University Conference of 17 May 1991 on the Bougainville Crisis: an update.

Carruthers, D. S., and D. C. Vernon, 'Bougainville Retrospective' in David Anderson (ed.), *The PNG-Australia Relationship: Problems and Prospects*. Canberra, Institute of Public Affairs, 1990.

Chinnery, E. W. P., 'The natives of south Bougainville and Mortlocks (Taku)', *Territory of New Guinea Anthropological Reports*, nos. 4 and 5, Canberra, 1924.

Chowning, Ann, 'The Development of Ethnic Identity and Ethnic Stereotypes on Papua New Guinea Plantations', *Journal de la Société des Océanistes*, 42, pp. 153–64, Paris, 1986.

Clunies-Ross, A. I., 'Economic Nationalism and Supranationalism', Inaugural Lecture, University of Papua New Guinea, 1969.

Connell, John:
    1977 – 'The People of Siwai: Population Change in a Solomon Island Society', *Department of Demography Papers in Demography*, no. 8, Australian National University, Canberra.
    1978A – *Taim Bilong Mani: the evolution of agriculture in a Solomon Islands Society*, Australian Development Studies Centre Monograph no. 12, Canberra.
    1978B – 'The Death of Taro: local response to a change in subsistence crops in the Northern Solomon Islands', *Mankind*, 11(4), pp. 445–52.
    1985 – 'Copper Cocoa and Cash: Terminal, Temporary and Circular Mobility in Siwai, North Solomons, in M. Chapman and R. Prothero (eds), *Circulation in Population Movement: Substance and Concepts from the Melanesian Case*, London, Routledge and Kegan Paul, pp. 119–148.
    1988 – 'Temporary Town folk?' Siwai migrants in urban Papua New Guinea, *Pacific Studies*, 11(3), pp. 77–100.

Connell, John (ed.), 'Local Government Councils in Bougainville', in Bougainville Special Publications, no. 3, Christchurch, University of Canterbury, 1977.

Corris, Peter, *Passage, Port and Plantation: A History of Solomon Islands Labour Migration: 1870–1914*, Melbourne, Melbourne University Press, 1973.

Damon, Albert, 'Human ecology in the Solomon Islands. Biomedical observations among four tribal societies', *Journal of Human Ecology*, 2: 191–225.

Decker, C. F. (ed.), *Saving the Solomons: from the diary account of Rev. Mother Mary Rose S. M.*, Bedford, Massachusetts, Marist Mission, 1948.

Docker, Edward, *The Blackbirders: the recruiting of South Seas labourers for Queensland, 1863–1907*, Sydney, Angus & Robertson, 1970.

Dorney Sean, *Papua New Guinea*, Sydney, Random House, 1990.

Dorsey, George, 'A Visit to the German Solomon Islands', in *Putnam Anniversary Volume*, New York, G. E. Stechert & Company, 1909.

Dove, J., T. Miriung and M. Togolo, 'Mining bitterness', in P. G. Sack (ed.), *Problem of Choice: Land in Papua New Guinea's Future*, Canberra, Australian National University Press, 1974.

Dunmore, John, *French Explorers in the Pacific* (2 volumes), Oxford, Clarendon Press, 1965.

Epstein, A. L., R. S. Parker and Marie Reay (eds), *The Politics of Dependence: Papua New Guinea 1968*. Canberra, Australian National University Press, 1971.

Faber, M. L., 'Bougainville re-negotiated', *Mining Magazine*, December 1974.

Feldt, Eric, *The Coastwatchers*, Sydney, Angus & Robertson, 1964.

Filer, Colin, 'The Bougainville Rebellion, the Mining Industry and the process of Social Disintegration in Papua New Guinea', in May and Spriggs (eds), 1990.

Fingleton, Father W., 'A chronicle of just grievances', *New Guinea and Australia, the Pacific and Southeast Asia*, Vol. 5, no. 2, p. 17, 1970.

Fisk, E. K. (ed.), *New Guinea on the Threshold: Aspects of Social, Political and Economic Development*, Canberra, Australian National University Press, 1966.

Foley, W., *The Papuan Languages of New Guinea*, Cambridge (UK), Cambridge University Press, 1986.

Friedlaender, J. S., *Patterns of Human Variation*, Cambridge (Mass.), Harvard University Press, 1975.

Friedlaender, J. S., and Douglas Oliver, 'Effects of Ageing and the Secular Trend in Bougainville Males', in Giles and Friedlaender (eds), pp. 142–63, 1976.

Frizzi, E., 'Zur Ethnologie von Bougainville und Buka, mit sapeziellen Berücksichtigung der Nasioi', *Baessler Archiv*, Leipzig and Berlin, Druck und Verlag von B. G. Teubner, 1914.

Giles, Eugene and J. S. Friedlaender (eds), *The Measures of Man: Methodologies in Biological Anthropology*, Cambridge (Mass.), Peabody Museum Press, 1976.

Griffin, James:
1970 – 'Bougainville', *Australia's Neighbors*, 68, pp. 7–12.
1972 – 'Bougainville – secession or just sentiment?', *Current Affairs Bulletin*, 48(9), pp. 259–80.
1976 – 'Kieta, Honiara and Port Moresby: hidden but not unknown', *New Guinea*, 10(4), pp. 43–50.
1978 – 'Paul Mason' in J. Griffin (ed.), *Papua New Guinea Portraits*, Canberra, Australian National University Press.
1982 – 'Napidakoe Navitu' in R. J. May (ed.), *Micronationalist Movements in Papua New Guinea*, Department of Social and Political Change, Australian National University, Monograph no. 21.
1989 – 'Bougainville – a people apart', *Times of Papua New Guinea*, 6 April.
1990 – *The PNG-Australia Relationship: Problems and Prospects*, Canberra, IPA.

Guppy, H. B., *The Solomon Islands and their Natives*, London, Swan, Sonnenschein, Lowry and Company, 1887.

Hahl, A., *Gouverneursjahre in Neuginea*, Berlin, 1937 (edited and translated by P. Sack and D. Clark, Canberra, Australian National University Press, 1980).

Hamnett, Michael, 'Households on the Move: settlement patterns among a group of Eivo and Simeku Speakers in Central Bougainville', unpublished Ph.D. dissertation, University of Hawaii, 1977.

Hannett, Leo:
1969 – 'Down Kieta Way, Independence of Bougainville', *New Guinea and Australia, the Pacific and Southeast Asia*, 4, pp. 8–14.
1975 – 'The Case for Bougainville Secession', *Meanjin Quarterly*, Spring.
1989 – 'My stand on the Bougainville Crisis', *Times of Papua New Guinea*, 3 August.

Harris, G. T., 'The 1973/4 Urban Household Survey: the Bougainville Towns', Department of Economics, University of Papua New Guinea, Discussion Paper no. 20.

Hastings, Peter, *New Guinea: Problems and Prospects*, Melbourne, Cheshire, 1969.

Havini, Moses, 'Co-operation among Melanesian Peoples, with special reference to Bougainville', in R. May (ed.), *Priorities in Melanesian Development*, Canberra, Australian National University Press, 1973.

Imbrun, B., ' "In Panguna we are brothers": the social life and consciousness of an urban Engan community', unpublished BA Honors thesis, University of Papua New Guinea, Department of Anthropology and Sociology, 1989.

International Bank for Reconstruction and Development, *The Economic Development of the Territory of Papua and New Guinea*, Baltimore, Johns Hopkins Press, 1965.

Irwin, G. J., 'Man-land relationships in Melanesia: an investigation of prehistoric settlement in the Islands of the Bougainville Strait', *Archaeology and Physical Anthropology in Oceania*, 8(3), pp. 226–52, 1973.

Jack-Hinton, Colin, *The Search for the Islands of Solomon, 1567–1838*, Oxford, Clarendon Press, 1969.

Jacobs, Marjorie, 'German New Guinea', in P. Ryan (ed.), *Encyclopedia of Papua and*

New Guinea, pp. 485–98, Melbourne, Melbourne University Press, 1972.

Keil, J. T., 'Local Group Composition and Leadership in Buin', unpublished PhD. dissertation, Department of Anthropology, Harvard University, 1975.

Kiki, Albert Maori, *Kiki: Ten Thousand Years in a Lifetime*, Melbourne, Cheshire, 1968.

King, Haddon F., *Discovery and Development of the Bougainville Copper Deposit*, Melbourne, CRA, 1978.

Kokere, M., 'The great god technology. The other aspect: from behind brown eyes', *Journal of the Papua New Guinea Society*, 6, pp. 13–31, 1972.

Labillardiere, J. J. H. de, *An Account of a Voyage in Search of La Perouse . . . 1791–1794*, 2 vols, London, Stockdale, 1800.

Laracy, Hugh, *Marists and Melanesians*, Canberra, 1976.

Lewis, David, *We, The Navigators: The Ancient Art of Landfinding in the Pacific*, Honolulu, University Press of Hawaii, and Canberra, Australian National Unversity Press, 1972.

Long, Gavin, *The Final Campaigns*, Canberra, Australian War Memorial, 1963.

Loy, T., M. Spriggs and S. Wickler, 'Evidence for the use of aroids in the northern Solomons from 28000 years ago, Honolulu, nd.

Luxton, C. T., *Isles of Solomon: a tale of missionary adventure*, Auckland, Methodist Foreign Missionary Society of New Zealand, 1955.

Lyng, J., *Our New Possession (Late German New Guinea)*, Melbourne, Melbourne Publishing Company, 1919(?).

MacKenzie, Seaforth, *The Australians at Rabaul*, Sydney, Angus & Robertson, 1927.

McHardy, Emmett, *Blazing the Trail in the Solomons*, Providence, Rhode Island, 1935.

Mair, Lucy, *Australia in New Guinea*, 2nd edn, Melbourne, Melbourne University Press, 1970.

Makis, Ephraim, 'Bougainville Copper and local business on Bougainville', University of Papua New Guinea, Department of Economics, no. 17, 1975.

Mamak, A. F., and R. D. Bedford:

1974A — 'Bougainville's students: some expressed feelings towards non-Bougainvillians, Arawa town, and the copper mining company', *New Guinea and Australia, the Pacific and Southeast Asia*, 9, pp. 4–15.

1974B — *Bougainville Nationalism: Aspects of Unity and Discord*, Bougainville Special Publications no. 1, Christchurch, University of Canterbury.

1975 — *Bougainville Copper and Trade Unionism*, Bougainville Special Publications no. 4, Christchurch, University of Canterbury.

1977 — 'Inequality in the Bougainville Mining Industry: Some Implications', in F. S. Stevens and E. P. Wolfers (eds), *Racism: The Australian Experience, A Study of Race Prejudice in Australia*, 2nd edn, Sydney, Australia and New Zealand Book Company.

May, R. J., 'Micronationalist Movements in Papua New Guinea', Canberra, Australian National University, Department of Social and Political Change, Monograph no. 1.

May, R. J. and Matthew Spriggs (eds), *The Bougainville Crisis*, Bathurst, Crawford House Press, 1990.

Miller, John, Jr, *The Reduction of Rabaul*, Washington, D.C., 1957.

Mitchell, D. D., *Land and Agriculture in Nagovisi, Papua New Guinea*, Boroko Institute of Applied Social and Economic Research, Monograph no. 3, 1976.

Momis, John, 'Bougainville — causes and solutions', *Post-Courier*, 27 September 1989.

Momis, John, and Eugene Ogan, 'A view from Bougainville', in M. W. Ward (ed.), *Change and Development in Rural Melanesia, Papers presented at the Fifth Waigani Seminar, 1972*, Canberra, Australian National University Press, 1972.

Morrell, W. P., *Britain in the Pacific Islands*, Oxford, Clarendon Press, 1960.

Moses, J. A., and Paul Kennedy (eds), *Germany in the Pacific and the Far East*, St Lucia, University of Queensland Press, 1977.

Moulik, T. K., *Bougainville in Transition*, Development Studies Centre, Australian

National University, Monograph no. 7, Canberra, 1977.

Narakobi, B., 'Lo Bilong Yumi Yet: Law and Culture in Melanesia', Suva, Melanesian Institute for Pastoral and Socio-Economic Service and the University of the South Pacific, 1989.

Nash, Jill, *Matriliny and Modernization: The Nagovisi of South Bougainville*, New Guinea Research Bulletin no. 55, Canberra, Australian National University Press, 1974.

Nash, Jill and Eugene Ogan, 'The Red and the Black: Bougainville Perceptions of other Papua New Guineans', *Pacific Studies*, vol. 13, no. 2, pp. 1-18, 1990.

Ogan, Eugene:
1965 – 'An election in Bougainville', *Ethnology*, 4(4), pp. 397-407.
1966 – 'Drinking behavior and race relations', *American Anthropologist*, 68(1).
1970 – 'The Nasioi Vote Again', *Human Organization*, 29, pp. 178-89.
1971A – 'Charisma and Race', A. L. Epstein et al. (eds), *The Politics of Dependence*, Canberra, Australian National University.
1971B – 'Nasioi land tenure: an extended case history', *Oceania*, 42(2), pp. 81-93.
1972 – *Business and Cargo: Socio-Economic Change among the Nasioi of Bougainville*, Canberra, Australian National University Press, New Guinea Research Bulletin no. 44.
1974 – 'Cargoism and Politics in Bougainville 1962-1972', *Journal of Pacific History*, 9, pp. 117-27.
1985 – 'Participant observation and participant history in Bougainville', D. Gerertz and E. L. Schieffline (eds), *History and Ethnohistory in Papua New Guinea*, Sydney, University of Sydney, Oceania Monograph no. 28.

Ogan, Eugene, J. Nash and D. Mitchell, 'Culture change and fertility in two Bougainville populations', in E. Giles and J. Friedlaender (eds), 1976.

Okole, H., 'The Panguna Landowners Organization', Paper presented to the University of Papua New Guinea Seminar on 'The North Solomons Situation: Issues and Options', 1989.

Oliver Douglas:
1949 – *Studies in the Anthropology of Bougainville, Solomon Islands*, Cambridge (Mass.), Papers of the Peabody Museum of American Archaeology and Ethnology, Vol. 29.
1955 – *A Solomon Island Society: Kinship and Leadership among the Siuai of Bougainville*, Cambridge (Mass.), Harvard University Press (also in Beacon Press paperback).
1971 – 'Southern Bougainville', in R. Berndt and P. Lawrence (eds), *Politics in New Guinea*, Nedlands, University of Western Australia Press.
1973 – *Bougainville: A Personal History*, Melbourne, Melbourne University Press (also by University of Hawaii Press).
1981 – *Aspects of Modernization in Bougainville, Papua New Guinea*, Working Papers Series, Pacific Islands Studies, University of Hawaii.
1989 – *Oceania: the Native Cultures of Australia and the Pacific Islands* (2 vols), Honolulu, University of Hawaii Press.
1989 – *The Pacific Islands* (3rd edn), Honolulu, University of Hawaii Press.

Oliver, Douglas, and W. W. Howells, 'Micro-evolution: cultural elements in physical variation', *American Anthropologist*, 59, pp. 965-78, 1957.

O'Reilly, Patrick and Jean-Marie Sedes, *Jaunes, Noirs et Blancs: trois années de guerre aux îles Salomon*, Paris, Edition de Nouveau Monde, 1949.

Panoff, M.:
1979 – 'Travailleurs, recruteurs et planteurs dans l'Archipel Bismarck de 1885 à 1914', *Journal de la Société des Océanistes*, 35, pp. 159-73.
1986 (ed.) – 'Les plantations dans le Pacifique Sud.' *Journal de la Société des Océanistes*, 42.

Parkinson, Richard, *Dreisig Jahre in der Südsee, Land und Leute, Sitten und Gebrauche in Bismarckarchipel und auf den deutschen Salomoinseln*, Stuttgart, Verlag von Strecker & Strecker, 1907.

Pinney, Peter, *The Glass Cannon: a Bougainville Diary, 1944–45*. St Lucia, University of Queensland Press, 1990.

Polomka, Peter (ed.), *Bougainville: perspectives on a crisis*, Canberra, Canberra Papers on Strategy and Defence No. 66, 1990.

Premdas, Ralph, 'Ethnicity and Nation-Building: the Papua New Guinea Case', Paper presented to the U.N. University Symposium, Suva, Fiji, August 1986.

Priests of Bougainville Diocese, 'The Bougainville Crisis: a Church Perspective', *Post-Courier*, 24 and 28 August 1989.

Quodling, Paul, *Bougainville Copper Limited: A History*, Melbourne, CRA, 1990.

Rannie, Douglas, *My Adventures Among South Sea Cannibals ... experiences and adventures of a government official*, London, Steeley Service, 1912.

Reed, Stephen, *The Making of Modern New Guinea*, Philadelphia, American Philosophical Society, 1942.

Ribbe, Carl, *Zwei Jahre unter den Kannibalen der Salomo-Inseln: Reiseerlebnisse und Schilderungen von Land und Leuten*, Dresden-Blasewicx, Beyer, 1903.

Ross, M., 'Proto-Oceanic and the Austronesian languages of Western Melanesia', *Pacific Linguistics*, Series C no. 98, Canberra, 1988.

Rowley, C.D.:
  1958 — *The Australians in German New Guinea 1914–1921*, Melbourne, Melbourne University Press.
  1965 — *The New Guinea Villager: a retrospect from 1964*, Melbourne, Cheshire.

Rutherford, John, ' "Zen" affluence in a subsistence economy of Bougainville, Solomon Islands', in I. R. C. Hirst and D. Reekle (eds), *The Consumer Society*, London, Tavistock, pp. 91–113, 1977.

Ryan, John, *The Hot Land: Focus on New Guinea*, Melbourne, Macmillan, 1969.

Ryan, Peter (ed.), *Encyclopedia of Papua and New Guinea* (3 vols), Melbourne, Melbourne Unversity Press, 1972.

Sack, Peter:
  1977 — 'Law, Politics and Native Crimes in German New Guinea', in J. A. Moses and Paul Kennedy (eds), *Germany in the Pacific and the Far East*, St Lucia, University of Queensland Press.
  1986 — 'German New Guinea: A Reluctant Plantation Colony?', *Journal de la Société des Océanistes*, 42, pp. 109–27.

Sack, Peter, and D. Clark (eds), *German New Guinea: The Annual Reports*, Canberra, Australian National University Press, 1979–80.

Sapper, Karl, 'Eine Durchquerung von Bougainville', *Mitteilungen aus den Deutschen Schutzgebieten*, 23, pp. 206–17, 1910.

Scott, R. M., et al., *Lands of Bougainville and Buka Islands, Territory of Papua and New Guinea*, Commonwealth Scientific and Industrial Research Organization, Land Research Series no. 20, Melbourne, 1969.

Shaw, H. I., and K. T. Kane, *Isolation of Rabaul*, Washington, D.C., 1963.

Shoffner, R. K., 'The Economic and Cultural Ecology of Teop: an Analysis of the Fishing, Gardening, and Cash-cropping System in a Melanesian Society', unpublished Ph.D. dissertation, University of Hawaii, 1976.

Smith, Geoffrey, 'Education, History and Development', in Peter Ryan (ed.), *Encyclopedia of Papua New Guinea*, pp. 315–32, 1972.

Spriggs, Matthew:
  1984 — 'The Lapita Culture Complex: Origins, Distribution, Contemporaries and Successors', *Journal of Pacific History*, 19, pp. 202–23.
  1990 — 'Dating the Lapita Culture: Another View', in M. Spriggs (ed.), *Lapita Design, Form and Composition*, Proceedings of the LaPita Design Workshop, Department of Prehistory, Australian National University, Canberra.
  nd — 'The Solomon Islands as bridge and barrier in the settlement of the Pacific' in W. S. Ayres (ed.), *Penetrations of the South Pacific: Lapita and early Pacific Maritime settlement*. Pullman, Washington State University Press.

Stanner, W. E. H., *The South Seas in Transition*, Sydney, Australasian Publishing Company, 1953.

Terrell, John, *Prehistory in the Pacific Islands*, Cambridge (UK), Cambridge University Press, 1986.

Terrell, John, and G. J. Irwin, 'History and Tradition in the Northern Solomons: an analytical study of the Torau Migrations to southern Bougainville in the 1860s', *Journal of the Polynesian Society*, 81, pp. 317–49, 1972.

Thurnwald, Hilde, 'Women's Status in Buin Society', *Oceania*, 5(2), 1934.

Thurnwald, Richard:

1909 — 'Reisebericht aus Buin und Kieta', *Zeitschrift für Ethnologie*, 41, pp. 512–31.

1910A — 'Im Bismarckarchipel und auf den Salomoinseln, 1906–1909', *Zeitschrift für Ethnologie*, 42, pp. 98–147.

1910B — 'Ermittlungen Über Eingeborenen-rechte der Südsee', *Zeitschrift für Vergleichende Rechtswissenschaft*, 23, pp. 309–64.

1910C — 'Die Engeborenen Arbeitskrafte in Südseeschutzgebiet' *Kolonialen Rundschau*, 10, October.

1912 — *Forschungen auf den Salomoinseln und den Bismarck archipel*, 2 vols, Berlin, Dietrich Reimer (Ernst Vohsen) Verlag.

1913 — 'Ethno-psychologische Studien an Südseevolkern auf den Bismarck-Archipel und den Salomo-Inseln', *Beiheft zur Zeitschrift für Angewandte Psychologie und Psychologische Sammelforschung*, Verlag von Johann Ambrosius Barth, Leipzig.

1934 — 'Pigs and currency in Buin', *Oceania*, 5(2).

Tippett, A. R., *Solomon Islands Christianity*, London, Lutterworth Press, 1967.

Treadgold, M. L., *The Regional Economy of Bougainville: Growth and Structural Change*, Occasional Paper no. 10, Development Studies Centre, Australian National University, Canberra, 1978.

Veur, Paul van der, *Search for New Guinea's Boundaries, from Torres Strait to the Pacific*, Canberra, Australian National University Press, 1966.

Ward, M. W., *Road and Development in Southwest Bougainville*, New Guinea Research Bulletin no. 62, Australian National University, Canberra and Port Moresby, 1975.

West, Richard, *River of Tears: The Rise of the Rio Tinto-Zinc Corporation Ltd*, London, Earth Island Limited, 1972.

Wheeler, G. C., 'Totemismus in Buin', *Zeitschrift für Ethnologie*, 46, pp. 41–4, 1914.

Wickler, Stephen, 'Lapita exchange: recent evidence from the northern Solomons', paper presented at symposium of Society of American Archaeology, Atlanta, Georgia, 1989.

Wickler, Stephen, and Matthew Spriggs, 'Pleistocene human occupation of the Solomon Islands, Melanesia', *Antiquity*, 62, pp. 703–6.

Wolfers, E. P., 'Papua New Guinea and coming self-government', *Current Affairs Bulletin*, 48(5), 1971.

Worsley, Peter, *The Trumpet Shall Sound: a study of cargo cults in Melanesia*, London, MacGibbon and Kee, 1957.

Zorn, S., 'Bougainville: Managing the Copper Industry', *New Guinea and Australia, the Pacific and Southeast Asia*, 8, pp. 23–40, 1973.

Index

[*Note:* Bougainville–Buka place names appearing in the text are indexed in only a few special cases.]

www.ingramcontent.com/pod-product-compliance
Lightning Source LLC
Chambersburg PA
CBHW071638270326
41928CB00010B/1966